Innovations in Neurocognitive Rehabilitation

K. Jayasankara Reddy

Innovations in Neurocognitive Rehabilitation

Harnessing Technology for Effective Therapy

Springer

K. Jayasankara Reddy
Department of Psychology
Christ University
Bangalore, Karnataka, India

ISBN 978-3-031-88116-9 ISBN 978-3-031-88117-6 (eBook)
https://doi.org/10.1007/978-3-031-88117-6

This Springer imprint is published by the registered company Springer Nature Switzerland AG
The registered company address is: Gewerbestrasse 11, 6330 Cham, Switzerland

Preface

> Everything can be taken from a man but one thing: the last of human freedoms—to choose one's attitude in any given set of circumstances, to choose one's own way. (Popova, 2013)

Neurocognitive disorders have been one of the most challenging health issues that patients, caregivers, and clinicians persevere through. Conditions such as Alzheimer's disease, stroke, and traumatic brain injury continue to be adverse experiences with an impending decline of essential functional capacity. Globally, these conditions are strenuous and leave a toll on families and healthcare institutions—of which the vast majority follow traditional rehabilitation strategies to mitigate the challenges. These methods fail to address the diverse range of cognitive impairments that leave patients entirely dependent on caregivers and medications. Amidst these approaches, the rapid evolution of technological innovations is setting the stage for a new era of rehabilitation. Personal therapies, accuracy in assessment and rehabilitation planning, and integration of tech-wear leave us hopeful about the future of patient healing.

This book creates a pathway to introduce a well-balanced interdisciplinary perspective of neuropsychological rehabilitation. Here, we discuss the interplay of traditional approaches with cutting-edge technology, namely, tools such as virtual reality (VR), augmented reality (AR), robotics, artificial intelligence (AI), and wearable technologies that can transform neurorehabilitation. These tools facilitate patient interventions and unbiased, data-driven evaluation that refine therapy and rehabilitation. The constant feedback provided through tech integration will prove to increase the maximization of rehabilitation outcomes and to have clarity in goals. However, we do not overlook the importance of human interventions in patient-centered care and navigating through the nuanced challenges that are incomprehensible to technological development. Neuropsychological principles of neuroplasticity are heavily relied on to understand the role technology and traditional rehabilitation perspectives apply to improve the overall quality of patient life.

> Neuroplasticity is the property of the brain that enables it to change its own structure and functioning in response to activity and mental experience. (Doidge, 2007)

The book is categorized into 5 parts with a total of 16 chapters where we will delve into the foundations of neurocognitive rehabilitation, detailing the various types of cognitive impairments, impact on patient life, and rehabilitation challenges. The second part focuses on the latest technology in neurocognitive rehabilitation (VR, AI, robotics, assistive technology, wearable devices, and sensors), its role in rehabilitation, and avenues to integrate them in traditional rehabilitation settings. The third part deals with the application of technology including case studies and recent research to determine the challenges, motivations, and benefits of technological rehabilitation. The fourth section details the advancement of technology in rehabilitation. This includes neuroplasticity, cognitive training programs, speech and language rehabilitation, motor rehabilitation, and biofeedback. The book concludes with the final part on the challenges in integrating technology in neurological rehabilitation. These include the ethical and privacy violations, the feasibility of integrating the technology, equity in accessibility, and the future scope of rehabilitation technology in neurological conditions.

This book provides clarity in key rehabilitation concepts and the practical application of technology in rehabilitation and substantiates with a holistic approach to rehabilitation. The book encourages its readers to identify the potential of technology in rehabilitation, the possible future areas of study, and the importance of keeping up with the rapidly evolving world of technology. The book sparks the need for collective action to identify that rehabilitation is multidisciplinary and requires sophisticated treatment options to improve the quality of life and neurocognitive disorders. Through ethical innovation and interprofessional collaboration, our field can continue to evolve—delivering targeted, efficacious therapies that help restore function and foster independence for those most affected by these challenging conditions.

Bangalore, India K. Jayasankara Reddy

References

Doidge, N. (2007). *The brain that changes itself: Stories of personal triumph from the frontiers of brain science*. Viking. https://www.penguinrandomhouse.com/books/291041/the-brain-that-changes-itself-by-norman-doidge-md/

Popova, M. (2013, March 26). *Viktor Frankl on the human search for meaning*. The Marginalian. https://www.themarginalian.org/2013/03/26/viktor-frankl-mans-search-for-meaning/

Acknowledgements

This book, Innovations in Neurocognitive Rehabilitation, is the result of collaborative efforts, insightful guidance, and unwavering support from numerous individuals and institutions to whom I owe a deep sense of gratitude.

I sincerely extend my appreciation to the Centre for Research Projects at Christ (Deemed to be University), Bangalore, for their generous financial support and invaluable assistance throughout the writing process. Their steadfast encouragement played a crucial role in the successful completion of this work.

I am immensely grateful to our esteemed Vice-Chancellor for their unwavering support and thoughtful guidance, which have been a source of inspiration throughout this journey.

I also recognize the dedication of the book editors, whose meticulous reviews and constructive feedback significantly enriched this work, ensuring its academic integrity and excellence. Their commitment to scholarly rigor is deeply appreciated.

A special expression of gratitude goes to my research assistant, Priya Prakash, whose exceptional contributions, tireless efforts, and unwavering commitment were instrumental in shaping the depth and scope of this book. Her dedication has been invaluable to this project. To everyone who has supported this endeavor, I extend my heartfelt thanks for your belief in this work and its significance.

Competing Interests The author has no competing interests to declare that are relevant to the content of this manuscript.

Contents

Chapter 1
Introduction to Neurocognitive Rehabilitation

1.1 Overview of Neurocognitive Rehabilitation

The human brain has been a constant source of fascination for its investigators for over millennia. The exploration of the brain has evolved from examining its structure to understanding its functions and advanced to identification and diagnosis. Psychology interweaved with neuroscience through the contributions of Franz Joseph Gall, who recognized psychologists for introducing standardized tests and assessments that facilitated effective treatment. Various studies across the globe during the late 1800s and early 1900s paved the way for treatment and research in the areas of neuropsychology and neurodegenerative disorders. However, it was during the residuals of the World Wars and the Korean War that the need to focus on rehabilitation emerged. Adding to the existing pool of the elderly that experienced neurodegenerative conditions, young soldiers and civilians with mild to severe trauma the brain were in need of long-term support. The younger individuals experienced detrimental effects of these injuries in the everyday functions and behavioral alterations (Gurd, 2010).

The dire need to reintroduce the injured back into the working population initiated a movement to standardized batteries and assessments to further distinguish the brain injuries and the impact to the cognition. Evaluating various aspects of attention, memory, concentration, and motor skills were developed by neuropsychologists during this period. Medical care practitioners observed that treating the brain independently of the body was infeasible, introducing the concept of treating the body as a whole. This generated a strong foothold for neuropsychological rehabilitation (Gurd, 2010). "Neuropsychological rehabilitation (NR) is concerned with the amelioration of deficits caused by insult to the brain. It adopts a goal-planning approach and addresses real-life difficulties. Neuropsychology studies how the brain affects behavior, emotion, and cognition. Rehabilitation is a process whereby people who are disabled work together with professional staff, relatives, and others

K. J. Reddy, *Innovations in Neurocognitive Rehabilitation*,
https://doi.org/10.1007/978-3-031-88117-6_1

to achieve optimum physical, psychological, and vocational well-being" (Wilson, 2020). This definition reiterates the primary goals of rehabilitation, i.e., assessment and diagnosis that will mitigate further risks and create optimal functioning that can ease barriers gradually. Neuropsychological rehabilitation does not substitute treatment nor does it guarantee complete recovery of lost functions. The process may be long-term and time-consuming but approaches the diagnosis with a practical approach that helps the patients accept the challenges while substituting or reducing the cognitive or physical limitation.

Early psychological rehabilitation focused on overt behavioral symptoms, with the introduction of X-rays the trauma to the skull was the only visible evidence. Since X-rays could not explain trauma and all the behavioral abnormalities, neuroimaging techniques started developing at a rapid rate to identify damages to the brain. Neuroimaging techniques such as positron emission tomography (PET), magnetic resonance imaging (MRI), near-infrared spectroscopy (NIRS), magnetoencephalogram (MEG), electroencephalography (EEG), and functional magnetic resonance imaging (fMRI) were gradually introduced and have been crucial in identifying areas of insult to the brain without invasive procedures (Xue et al., 2010). While neuroimaging techniques have progressed to explain cognitive and behavioral abnormalities, these techniques exclusively do not support rehabilitation. Rehabilitation requires a dynamic approach in terms of understanding practical challenges of day-to-day functions. Behavioral and emotional challenges as well as limitations can only be reported by the patient and the caregiver as feedback to supplement rehabilitation. The cost incurred for the imaging is an added burden on the patients. Cumulatively, the degree of lifestyle changes, the cognitive and physical limitations of certain occupations and functional difficulties are incomprehensible through an expensive neuroimaging procedure (Wilson, 2013). This gap is compensated through contemporary advancements in neurocognitive techniques that are detailed in further chapters.

1.2 Theoretical Foundations

A precondition in being an expert in neuropsychological rehabilitation is to assimilate the conceptual and theoretical frameworks for neurocognitive rehabilitation. Grasping these vital concepts enable explaining the etiology and formulate treatment and rehabilitation plans. The essential theories are elaborated below.

1.2.1 Neuroplasticity and Brain Recovery

Early explorers of the brain theorized that the divisions of cells in the brain encourages cognitive development of the individual. "Immaturity" of the neurons existed until birth after which it was believed that existing neurons continue to develop and strengthen the neurons while maintaining the "youth" of the molecular features (Rosa et al., 2020). Until 1890, it was believed that the human brain remained stagnant in regeneration of neurons, that those neurons that were impacted by factors such as brain injury and aging indicated a decline in functions impossible to recover. The first indication of plasticity was identified in rats by William James. This introduced a new school of thought that the human brain may have the capacity to recover. The term neuroplasticity was introduced by Konorski in 1948 but the larger contribution to neuroplasticity was by Tanzi who was credited by Donald Hebb—who contributed to further evidence through his research (Berlucchi & Buchtel, 2009; Puderbaugh & Emmady, 2024).

"Neural plasticity, also known as neuroplasticity or brain plasticity, can be defined as the ability of the nervous system to change its activity in response to intrinsic or extrinsic stimuli by reorganizing its structure, functions, or connections" (Mateos-Aparicio & Rodríguez-Moreno, 2019). Neuroplasticity is the brain's capacity to adapt and reorganize itself, forming new neural connections in response to learning and recovery. Early experimentation with lab rabbits indicated the neuroregeneration of cells in the hippocampal region. Later studies detected similar outcomes in other cell areas. However, much of the research discredited the possibility of neurodegeneration in adults. Adult neurogenesis was introduced by Josef Altman through his experimentation with adult rats. While groundbreaking evidence was discovered in nonmammalian vertebrates that indicated rapid neurogenesis, the human brain had indicated no evidence of regeneration. The biomarkers for adult neurogenesis are identified as the olfactory bulb and hippocampus. Although conclusive evidence is yet to be identified in the human brain, the possibility of compensatory function and partial recovery raises hope in maximizing rehabilitation outcomes (Owji & Shoja, 2020; Rosa et al., 2020).

The key takeaways from neuroplasticity in rehabilitation are neuronal regeneration and reorganization of functions. Neuronal regeneration introduces synaptic plasticity which was theorized through the identification of decline in neuromodulators in the synapses among the aging population and those diagnosed with neurodegenerative disorders. Spike-timing-dependent plasticity (STDP) elaborates that the strengthening of synapses complements timing delay in functions that are utilized less often. Indicating that repetition of certain tasks can increase plasticity of the synaptic neuromodulators. Complimenting this theory is metaplasticity and homeostatic plasticity both of which explain the synaptic plasticity that can impact lager networks through activity-dependent changes and in maintaining homeostasis (Puderbaugh & Emmady, 2024). The introduction of neuroplasticity in rehabilitation can include treatment accompanied by medication to strengthen synaptic moderators. The introduction of selective serotonin reuptake inhibitors (SSRIs), serotonin and noradrenergic reuptake inhibitors (SNRIs), and amantadine has displayed significant recovery rates in patients. Rehabilitation can expand to focus on restoring functionality and treating maladaptive neuroplasticity as in the case of phantom leg syndrome (Maier et al., 2019; Puderbaugh & Emmady, 2024).

1.2.2 Core Principles of Rehabilitation

Elucidated by research in neuronal plasticity, one can identify that rehabilitation strategies must focus on improving neuronal connectivity and reaction time. The core principle of rehabilitation is to establish repetition, training, and pushing the limits of the patients. There are three core principles of neuropsychological rehabilitation—repetition and intensity, task-specific training, and progressive difficulty. Each of which are based on approaches to rehabilitation. The restorative approach emphasizes on retraining of memory, which often requires repetition and increasing levels of intensity. One of the core principles of rehabilitation is to ensure that the patients are able to learn or relearn skills and knowledge required for day to day functioning; this model is the knowledge acquisition approach which relies on task-specific training and ultimately the compensatory approach combines the use of memory systems and task-specific training to tackle situational adversities. This is met through the core principle of progressive difficulty in the interventions (Gopi et al., 2022).

1.2.2.1 Repetition and Intensity

Consistent and repetitive exercises strengthen neural pathways. Various therapeutic models of neuropsychological rehabilitation focus on repetition to strengthen neural networks. The approach to cognitive rehabilitation with repetition of activities emerges from the neuroplasticity theory that perceived the brain capacity to act as a "mental muscle" to strengthen existing and declining neuronal networks. The vast research in cognitive rehabilitation has identified the benefits of task repetition with increase in the intensity of each task. Studies have indicated that patients with TBI, stroke, and other mild injuries have been able to improve their working memory capacity through repetition of tasks. Studies conducted on repetition trails through working memory tasks and attention tasks have observed an increased WM performance, verbal memory, and self-reported changes in daily life (das Nair et al., 2016; Gopi et al., 2022; Weicker et al., 2024). The repetition of tasks increases the familiarity and strengthens knowledge networks in the brain.

1.2.2.2 Task-Specific Training

Training focuses on real-life tasks to promote functional recovery. The training is meant to increase the capacity to retrieve knowledge to actively engage with a situation. Task-specific training relies on the patient to recollect long-term and short-term information to tackle everyday challenges. Strategies applied include errorless learning, vanishing cues, and spaced retrieval of information. Through reduction and elimination of incorrect responses, errorless learning ensures that the patient is able to retain the desired information in the long-term memory. While obtaining an errorless response is time-consuming, this strategy enables long-term memory retention (Baddeley & Wilson, 1994). Spaced learning helps patients with word and name recall. As the name suggests, spaced learning helps with recollecting information through revision and uniform and nonuniform intervals. When an error occurs, the information is re-learnt. Spaced learning applied to adults with dementia and cognitive impairments (Creighton et al., 2013). Vanishing cues are used to encourage recollection gradually. The cues to recollect words or numbers gradually reduce until the patient is able to recollect the information with minimal cues. Vanishing cues are the most time-consuming among the tasks and increase the time and cost while distorting the learning motivation for the patient (Kapur et al., 2004). All the methods of task-specific training are meant to enhance the quality of life of the patients. The tasks are time-consuming and expensive which can be mitigated by involving family members and caregivers into the training—by asking the participants to engage in activities like recollecting phone numbers and names at suggested time intervals (Evans et al., 2020; Gopi et al., 2022).

1.2.2.3 Progressive Difficulty

Starting with simpler tasks and gradually increasing complexity ensures sustained cognitive challenge and engagement. As the name suggests, compensatory approach is utilized in cases where the possibility of recovering lost function (such as memory) is considered impossible. These tasks work to strengthen and encourage restructuring of neural networks to "compensate" for the absence of certain functions. These strategies rely on internal and external aids to optimize neural pathways. The internal aids association strategies for memorization—acronyms, chunking, chaining, rhyming, and mnemonics. Whereas external aids include using sources to store information to reduce the cognitive load. These aids include note-taking, calendars, digital tools to store information and set reminders, signs, and labels for items within the household or work place. Internal aid depends on the ability to recollect information from the long-term storage whereas external aids reduce the burden on the patient and eases navigating through familiar places. Prosthetics and intelligent technological solutions for the house are considered as modern accessories for external aids. Internal aids can play a substantial role in neuroplasticity among patients with mild to moderate cognitive deficits whereas external aids can enhance the quality of life for patients with moderate to severe cognitive deficits (Cicerone et al., 2019; das Nair et al., 2016; Gopi et al., 2022).

1.2.3 Biopsychosocial Model

This holistic approach considers biological, psychological, and social factors in treatment. Biological factors might include brain injury; psychological factors encompass the patient's emotional state and motivation; social factors involve family support and community integration. Unlike previously discussed models that solely focused on the cognitive and physiological aspects of rehabilitation, the biopsychosocial model utilizes a holistic approach to recovery shifting focus from merely quantifiable improvements in biological recovery to improvement in the quality of life. Improving the quality of life is central to observing rehabilitation changes in this model. Various studies have indicated that a shift to this model in rehabilitation has observed a significant improvement in community engagement, metacognition, emotional regulation, and minor improvements in the overall life satisfaction among patients. Applying the biopsychosocial model shifts focus to other areas of impact that may not be identified through testing. Familial and social challenges, acceptance of limited functioning and occupational barriers widens the areas of rehabilitation that was once limited to cognitive improvements (Cicerone et al., 2019; Gopi et al., 2022).

1.3 Conditions and Rehabilitation Needs

The various theoretical models and interventional approaches are used interactively to tailor and prioritize rehabilitation to meet the needs of the patient. Understanding the conditions commonly addressed in neurocognitive rehabilitation helps clinicians tailor interventions to specific impairments and recovery goals. Applying the interventions in neuropsychology requires a need to cognitively train the patients with diverse neurological and psychiatric conditions.

1.3.1 Neurological and Psychiatric Conditions

Traditionally clinical neuropsychological rehabilitation focuses on the treatment and rehabilitation of brain injuries and dementia. The development of various neuropsychological interventions has proved to be beneficial in the rehabilitation of conditions in mental health conditions—schizophrenia and other psychotic disorders along with existing treatment and interventions for stroke, dementia, and traumatic brain injury (TBI). Each of the conditions mentioned here has its own deficits and impairments. The role of rehabilitation in this case is to follow the holistic approach to observe the impact of their biological conditions on other areas of life.

1.3.1.1 Stroke

Stroke is a medical condition that occurs when the blood flow to the brain is interrupted by blood clotting in vessels or damage to the blood vessel occurs. Stroke limits the essential nutrients or the removal of waste from the brain effectively reducing the oxygen and nutrients supplied to the brain. A minor delay in this supply chain of oxygen can lead to severe cognitive deficits. There are two main types of stroke—ischemic and hemorrhagic stroke. Ischemic stroke occurs due a blockage in the blood vessel due to clots or plaque build-up. The patient may observe weakness and numbness of the limbs and face, lack of clarity in speech and/or comprehension, blurry vision, imbalance in movement and coordination, and severe headaches as warning symptoms. Hemorrhagic stroke on the other hand is the outcome of blood vessels rupturing or leaking, leading to high pressure and injury to the brain. Trauma to the brain, aneurysms, congenital conditions such as arteriovenous malformation and high blood pressure are key symptoms. Ischemic strokes are further subclassified on the basis of the location of the clot and hemorrhagic stroke is subclassified on the location of the aneurysm or bleed. Other less prevalent categories of stroke include transient ischemic attack (TIA), cryptogenic stroke, brainstem stroke, and recurrent stroke (Types of Strokes and Treatment, n.d.).

Patients with stroke may experience a range of deficits, from motor skills to memory and language. Recent data suggests a rise in cases of stroke among children and young adults. Previously, rehabilitation for stroke was modeled for the elderly and aging population. Poststroke impact on the overall quality of life is significant. Individuals with stroke have experienced physical and cognitive deficits such as attentional and executive functional deficits, executive function, working and immediate memory. Stroke impacts other areas of life including occupational limitations, difficulty driving, inability to overexert physically, impact on identity, self-esteem, and relationship dynamics (Ibrahim et al., 2024). Rehabilitation for stroke is primarily centered around attention and visual perception using cognitive interventions. Motor training through physiotherapy is highly recommended to focus on compensatory training. Additionally, noninvasive neuromodulation such as transcranial stimulation through currents and magnetism have reported positive results. While rehabilitation in adults focuses on restoring lost functions, in children with stroke, the functions are still under development. The focus in rehabilitation shifts to cognitive behavioral interventions with an emphasis on speech and language training, adaptive and compensatory training must be incorporated into their home and school environment (Krivitzky et al., 2022; Xu et al., 2013).

1.3.1.2 Traumatic Brain Injury (TBI)

A strong external forceful insult or blow to the head causes traumatic brain injury (TBI). Injury can penetrate through the skull (penetrating TBI) or may be caused by blunt-force trauma (non-penetrating TBI). TBI can have either short-term or long-term deficits in cognition and motor control. Physical symptoms of TBI include headaches, seizures, blurry vision, unequal pupil dilation, projectile vomiting, slurring of speech and imbalance. Cognitive behavioral symptoms of TBI are observed through alteration in consciousness, disorientation, memory and attentional difficulties, irritability, decision-making, and concrete thinking deficits. Other sensory symptoms include dizziness, oversensitivity to stimuli, fluctuations in mood, fatigue, and emotional dysregulation. TBI in children are observed to have additional symptoms such as impairments in acquiring skills, sleep disturbance, inattention and imbalance while walking, and seizures. Consciousness is significantly impacted in TBI patients including minimal loss of consciousness to severe cases leading to coma or death.

TBI is classified into various subtypes on the basis of the impact and course of progression. Primary damage is the immediate damage that sets in immediately after the event of injury and secondary damage progresses over the course of hours to weeks post-injury (Traumatic Brain Injury (TBI), 2024).

TBI affects attention, executive functions, and memory, often requiring long-term intervention. TBI draws similar interventional models as stroke. Cognitive training for attention and memory deficits have produced positive outcomes. Motor training is limited in cases with severe TBI and ultimately compensatory training applied to adjust with lost functions. Impaired functions such as verbal, visuospatial organization, working memory, and fine motor coordination are trained through neuropsychological interventions. Since TBI requires patience and long-term rehabilitation, most patients lose motivation and may acquire comorbid mental health conditions such as anxiety and depression. Most patients would require home-based interventions that focus on task completion through checklists and planning to reduce the constant requirement of executive functions. Overall, the quality of life involves reintegrating individuals into the community while accepting complete recovery may be unlikely (Mohanty & Gupta, 2013; Soule, Fish, Winegardner, & Schrieff-Brown, 2024; Traumatic Brain Injury (TBI), 2024).

1.3.1.3 Dementia

Dementia is the decline or loss of cognitive functions that interfere with everyday activities. Dementia is observable through abnormalities in the brain; the gradual decline is prominent in old age. Dementia is a common term applied for a larger group of neurodegenerative diseases whose underlying symptoms include cognitive-motor deficits. Prevalent subtypes of dementia include Alzheimer's disease (AD), frontotemporal dementia (FTD), Lewy body dementia, and vascular dementia. AD is distinguished through deposits of amyloid plaques and tau tangles (a protein molecule) throughout the brain. Typically patients are diagnosed in their mid-60 s with early onset in their 30 s.

AD is classified on the basis of the severity of the symptoms from mild to severe with symptoms ranging from forgetting directions to frequented places to increased difficulty remembering close family and friends and losing the ability to communicate. FTD dementia is marked by significant accumulation of tau protein and TDP-43 proteins in the frontal and temporal regions of the brain. The typical age for diagnosis is between 45 and 64 years with behavioral and emotional symptoms (difficulty in planning, sequencing, organizing, impulsivity, flat affect); motor impairments (tremors and motor imbalance) and language impairments (forming and comprehending speech). Lewy bodies are identified through the abnormal accumulation of alpha-synuclein protein (Lewy bodies) that impact synaptic transmission with a diagnostic age range of 50 years and above. Lewy bodies are characterized by cognitive deficits (low concentration, inattention, and disorganized thought), motor deficits (stiff muscles, poor coordination, and reduced facial expression), sleep disturbances (insomnia), and visual hallucinations. Vascular dementia is identified through blood clots in the brain leading to forgetting past events, misplacing items, hallucinations, and inability to follow through instructions. The age of diagnosis is similar to Alzheimer's disease (Understanding Different Types of Dementia, n.d.).

Alzheimer's and other dementias present progressive cognitive decline, necessitating ongoing cognitive support. Dementia heavily relies on the core principles of neuropsychological rehabilitation. Dementia is untreatable despite advanced research in the area and has additional limitations depending on the levels of literacy and occupational demands of the patient. The primary aim of rehabilitation is to decrease the rate of cognitive decline, improve on existing skills, and to prepare the patient and their caregivers for long-term adjustments to the disease. Dementia requires a combination of rehabilitation with medication. Cognitive rehabilitation to improve on working memory, short- and long-term memory as well as motor balance reduces the accelerating rate of decline. Tele-rehabilitation and home-based interventions would be cost-effective options for rehabilitation. Activities of daily living (ADL) rehabilitation has a significant effect on improving the individual's quality of life (Avila et al., 2004; Elera, 2024; What Is Dementia*?*, 2022).

1.3.1.4 Mental Health Disorders

Conditions such as depression and schizophrenia often involve cognitive impairments that benefit from neurocognitive interventions. Most patients with psychotic disorders such as schizophrenia require psychosocial rehabilitation. However, memory and learning, attention and working memory along with executive functions (planning, self-monitoring, response inhibition, fluency, and cognitive flexibility) have impairments in psychotic disorders. The rise in multifaceted interventions in neuropsychological rehabilitation has gained popularity among patients with psychosis (Rezapur et al., 2017). As previously discussed in the section on stroke, the prevalence of depression as a poststroke mental illness is significantly high. The interventions to improve cognitive symptoms that align with improving attention and memory could improve cognitive symptoms of depression. A holistic neuropsychological approach to mental illness may counter traditional treatment models by tackling gaps in motor-cognitive deficits observed in mental illness and psychosis and supplement psychosocial interventions. Cognitive-behavioral therapies, social skills training, and cognitive rehabilitation programs are additionally applied in cases of post-brain injury depression and schizophrenia (Fleminger et al., 2003; Neuropsychology of Schizophrenia, 2024; Rezapur et al., 2017).

1.3.2 *Cognitive Deficits*

1.3.2.1 Memory

Across all neurological conditions, deficits in different types of memory have been observed to impact the overall quality of life of the individual. Dementia has observed a marked impairment in working, remote, long-term, and short-term memory that lead to confusion, irritability, disorientation, and hallucination. TBI-related impairment has an impact on recollection and working memory leading to occupational and social challenges depending on the severity of the injury. In the case of stroke, patients may observe long-term decline in memory and other cognitive impairments. Addressing short-term, working, and long-term memory deficits to improve daily functioning plays a vital role in estimating recovery of lost functions. Utilizing task-specific training enables the patients to remember fundamental information such as names of people, addresses. Applying internal and external cues such as reminders, alarms, note-taking, labeling can increase the rehabilitation of prospective memory. With progressing difficulty may extend their memory to meet occupational requirements (das Nair et al., 2016; Kapur et al., 2004).

1.3.2.2 Attention

Attentional impairments are prominent among various neurological and psychological illnesses as mentioned in previous sections. Neuropsychological rehabilitation requires understanding specific attentional deficits to tailor interventions. The intensity of the intervention is determined through alertness, sustained attention, and vigilance. Unlike other cognitive impairments and their rehabilitation models, an overexertion of attentional training can lead to further impairment over recovery. Applying simple interventions including digit span test, tests of everyday attention (TEA), cancellation tasks, and paced auditory serial addition tasks (PASAT) could develop in intensity by increasing the string of numbers or the number of words/phrases used. The task could rise in the level of difficulty, frequency, and duration of attention. Training sustained and selective attention is essential for activities requiring focus. Interventional options such as activities of daily living (ADL) and technological interventions could ease the cost while maintaining the motivation of the patients (Soule, Fish, Thomas, & Schrieff-Brown, 2024; Sturm, 2024).

1.3.2.3 Executive Functioning

Executive functions play a vital role in the complex cognitive process such as planning, organization, problem-solving, and error correction. These processes heavily rely on attention and memory therefore, an impairment in any of these areas can cause an impairment in executive functioning. Essentially, these tasks have significant importance in optimal functioning and enable integration into the community. The rehabilitation tasks focus on goal-oriented behavior. Interventions applied focus on planning and sequencing tasks—one handed dressing techniques, testing the grip while holding onto object, creating checklists, and maintaining journals have an effective impact on adapting to functional loss while developing skills related to problem-solving, decision-making, and planning. Occupational therapy and nontraditional interventions have been observed to have positive outcomes on the rehabilitation process (Chung et al., 2013).

1.3.2.4 Processing Speed

Processing speed is essential in timely execution of cognitive responses to stimuli. Individuals with executive dysfunction, attention, and memory often experience slowed processing of information. Post-TBI assessments, stroke, and dementia have indicated that there is a significant delay in processing stimuli. Improving the speed of cognitive processing to enhance response times in daily tasks. Utilizing cognitive training tests such as N-back, digit span, and sorting tests can ameliorate processing speed in working memory and goal orientation in executive functioning (Abelmann et al., 2023; Feizipour et al., 2020).

1.4 Emerging Trends and Technological Innovations

This section explores advancements transforming neurocognitive rehabilitation, highlighting the role of technology in enhancing traditional therapy methods. Earlier rehabilitation models focus on the traditional pen-and-paper methods to conduct interventions. There are various limitations in this method—the time consumed, cost-effectiveness of commuting to the clinics to seek interventions and the decline in the motivational levels of the patients. A process-based approach (PBA) to address interventions require accuracy in the results obtained, ability to track minor changes in response time and errors, reduction of human error and bias, and generate levels of intensity and difficulty in tasks, all while maintaining the motivational levels of the patients. To meet all of these areas efficiently and to increase efficiency of rehabilitation, the use of technology proves to be a valuable asset.

1.4.1 Neuroimaging and Biomarkers

Brain imaging techniques have proved to be a valuable asset in accurately identifying areas of insult to the brain, the regions involved in processing stimuli and the pathways in cognitive process. Imaging tools available to neural exploration are positron emission tomography (PET), near-infrared spectroscopy (NIRS), magnetoencephalogram (MEG), electroencephalography (EEG), and functional magnetic resonance imaging (fMRI). EEG provides insights into the electrical pathways and activities in the brain. The type of reading an EEG provides is spontaneous and event-related. Spontaneous EEG displays the neuronal responses that are unprovoked by stimuli. It is applied in identifying seizure disorders and has been used in interventions in clinical neuropsychological rehabilitation to observe the presence or absence of neural activity and in sleep studies. Event-related potentials are observed neuronal activity in the presence of a stimuli. These activities can be time-based and applied to observe the rate of processing and the rate of response in patients.

Functional magnetic resonance imaging (fMRI) is widely applied to observe activities in the brain without the radiation emitted by PET and NIRS scans. Activities in the brain are observed through the rate of metabolism, blood circulation, and oxygen supply (blood oxygen level imaging). fMRI has a wide variety of applications and methods of use including early detection of brain tumors, inactivity, and neurodegeneration. Imaging techniques such as functional MRI (fMRI) and electroencephalography (EEG) provide insights into neural changes occurring during rehabilitation. Biomarkers help clinicians tailor interventions based on measurable brain responses. Neuroimaging data is considered to provide biomarkers for early treatment and helps plan rehabilitation on the basis of priority. One limitation of fMRI is that the machines used are expensive and have environmental requirements (Xue et al., 2010).

1.4.2 Digital Therapeutics and AI-Based Tools

The most sought after advancement in neuropsychological rehabilitation is digital interventions and artificial intelligence (AI) tools. Applying digital therapeutics, virtual reality (VR) interventions and AI can provide near-accurate data on the cognitive activity of the patients while maintaining the interactive nature of these interventions. It meets all the PBA requirements as mentioned previously. Applying digital interventions in patients with dementia and memory impairments have proved to be effective in cognitive retraining. Using simple tech such as digital reminders, alarms, checklists, and e-journals help reduce cognitive load on the patients. By using cued and uncued reminders, the patient is able to withdraw key information without much worry or anxiety over forgetting the information. For example, setting a reminder to buy groceries can reduce the cognitive load of repetition of information. At the same time and alarm without a cue could push the patient to think about the possible activity or task they should be engaging in at the time. VR has been applied integratively to create simulations of various situations to aid exposure without immediate risk or danger. VR in integrative fMRI has proved to be invaluable in understanding the neural changes occurring while engaging in simulations. VR has proved to be an asset in improving executive functioning among adults with mild cognitive impairments (Jamieson et al., 2017; Kapur et al., 2004;

Yu et al., 2023). Using computer-based tasks and AI in cognitive rehabilitation can provide accurate results including time delays and processing speed. Technologies, including virtual reality (VR), artificial intelligence (AI), and cognitive training apps, enable highly individualized treatment. For example, VR can simulate real-world scenarios for immersive therapy, while AI adapts exercises to the user's progress.

1.4.3 Tele-Rehabilitation

In order to combat limitations by cost and time, tele-rehabilitation has observed a rapid integration in neuropsychology. Tele-rehabilitation promises recovery at home that ensures that the patients can recover at their own pace. Remote rehabilitation through digital platforms has expanded accessibility, especially for patients with mobility limitations or those in remote areas. Tele-rehabilitation also allows for frequent monitoring and feedback, facilitating continuous progress. This form of rehabilitation helps cater to individual needs and strengths with support from the caregivers and community. Additionally, patients with limited resources to access rehabilitation are able to communicate through videoconferences and home telemedicine systems to access high-quality rehabilitation services. Specific challenges in applying telemedicine could be the awareness and ease of using technology among the patients and caregiver, literacy levels, and Internet access and connectivity (Despoti et al., n.d.; Maresca et al., 2020).

1.5 Evidence-Based Practices in Neurocognitive Rehabilitation

Evidence-based practices ensure that interventions are grounded in research and yield measurable outcomes. Any application of technological trends, testing, and assessments requires a cyclic process of research and evaluation. The practice of providing an informed rehabilitation plan is based on the outcomes of the models used. This implies that tailored intervention is backed by research and the practitioner's observation and utilization of tools would further provide insight into the areas that lack focus. The following are key elements of evidence-based rehabilitation:

1. *Standardized Protocols*: Structured protocols, such as the use of cognitive remediation therapies and goal-oriented approaches, are proven to support cognitive recovery. Protocols can be adapted to individual needs while maintaining a foundation in best practices. Structured protocols can cater to various levels of interventions and provide uniformity in the administration of rehabilitation protocols. Assessing the type of protocol and theoretical frameworks used differs across patients yet the knowledge and the method of application is structured to ensure

efficacy and quantifiability of the outcomes (Poulin et al., 2021; Power & D'Amato, 2021).

2. *Assessing Outcomes*: Various cognitive assessment tools are used to measure progress and effectiveness, such as:

   (a) *Neuropsychological Tests*: These measure specific cognitive domains, including attention, memory, and executive functions. Neuropsychological testing mandates that the administration of the tests maintain multiple intrinsic and extrinsic factors to maximize outcomes. The exhaustive number of scales and assessments becomes challenging to select from if the protocol and interpretation lack clarity. Neuropsychological tests can be time-consuming and expensive which indicates the need to prioritize the interventions.
   (b) *Functional Scales*: Assessments like the functional independence measure (FIM) evaluate a patient's daily living abilities. The FIM measures motor and cognitive domains of rehabilitation. The scores obtained through functional scales act as feedback for further clinical practice. FIM and other functional scales are widely used to obtain the prognosis of the intervention and to measure the treatment outcomes in interventional models like ADL (Pedersen et al., 2015).

3. *Challenges in Clinical Implementation*: Multiple challenges precede neuropsychological rehabilitation. In terms of standardization and tools used, there are cultural factors that play a significant difference between testing protocols in the East (with further cultural differences) and West. This implies that all assessments and protocols require customization or a new scale must be constructed to fit the patient's cultural context. As mentioned throughout the interventional challenges, the cost incurred during the long rehabilitation process can become a significant limitation to the individual. Individualized treatment plans are essential, as each patient's cognitive profile and personal circumstances differ. Additionally, sustained patient compliance is a common challenge, requiring a balance between clinical protocols and flexibility.

1.6 Ethical Considerations in Neurocognitive Rehabilitation

Ethical considerations are paramount, especially with vulnerable populations that may experience cognitive impairments.

1.6.1 Informed Consent

Patients with cognitive impairments may face challenges in understanding treatment plans. Ethical guidelines stress the need for clear, simplified explanations and involve family members when necessary.

1. *Data Privacy*: Digital tools in rehabilitation, such as apps or cloud-based platforms, raise concerns about patient confidentiality and data security. Clinicians must follow guidelines to ensure patient information is protected. Data privacy is a rampant issue across various professions. In rehabilitation, the data obtained has various case-specific details including the patient's history, other medical conditions, and qualitative data that may have personal data about their personal lives. The sharing and storage of these details must be handled with cybersecurity awareness and practitioner ethics in mind.
2. *Equity and Access*: Neurocognitive rehabilitation should be accessible to all, regardless of socioeconomic status. Ethical practices include addressing disparities and advocating for funding and resources to expand access. Equity extends to the inclusivity in research and data. Mitigating cultural limitations can play a vital role in improving the standardization of assessments and tools for each patient. Another factor in rehabilitation is the regulation of practice and research. Few organizations and countries have strict regulation of rehabilitation practices.

1.7 Summary and Future Directions

The closing section synthesizes the chapter's main points and introduces emerging areas for further exploration. Various theories and models of neuropsychological rehabilitation are utilized in creating and formulating rehabilitation strategies. The rehabilitation provided is primarily focused on patients with neurodegenerative disorders—Alzheimer's, schizophrenia, brain injury, and stroke. Each of these conditions have observable impairments in attention, concentration, memory, and executive functions. Some functions are recoverable while others may require compensatory rehabilitation to adjust to daily life. The evolved objective rehabilitation is to approach interventions holistically and evaluate the outcomes through the quality of life. The tools used to supplement rehabilitation include neuroimaging, VR and AI and tele-based rehabilitation—all of which improve the efficacy of treatment, with specific limitations of cost, tech-skills, and awareness. Overall, the practitioner must have a strong sense of ethics; maintaining the privacy and positive outcomes of the patient is vital. Ongoing research in the innovative ways to include technology in neuropsychological research clearly dictates the need to include personalized treatment approaches that incorporate AI-driven tools and large-scale data collection promises potential to obtain precise, evidence-driven, and widely accessible interventions. The following chapter discusses neurocognitive disorders—the areas where future research could take on roles in designing accurate diagnosis and designing effective rehabilitation plans.

References

Abelmann, A. C., Kessels, R. P. C., Brazil, I. A., Fasotti, L., & Bertens, D. (2023). Game-supported cognitive strategy training for slowed information processing speed after acquired brain injury: Study protocol for a randomised controlled trial. *BMJ Open, 13*(9), e067108. https://doi.org/10.1136/bmjopen-2022-067108

Avila, R., Bottino, C. M. C., Carvalho, I. A. M., Santos, C. B., Seral, C., & Miotto, E. C. (2004). Neuropsychological rehabilitation of memory deficits and activities of daily living in patients with Alzheimer's disease: A pilot study. *Brazilian Journal of Medical and Biological Research = Revista Brasileira De Pesquisas Medicas E Biologicas, 37*(11), 1721–1729. https://doi.org/10.1590/s0100-879x2004001100018

Baddeley, A., & Wilson, B. A. (1994). When implicit learning fails: Amnesia and the problem of error elimination. *Neuropsychologia, 32*(1), 53–68. https://doi.org/10.1016/0028-3932(94)90068-x

Berlucchi, G., & Buchtel, H. A. (2009). Neuronal plasticity: Historical roots and evolution of meaning. *Experimental Brain Research, 192*(3), 307–319. https://doi.org/10.1007/s00221-008-1611-6

Chung, C. S., Pollock, A., Campbell, T., Durward, B. R., & Hagen, S. (2013). Cognitive rehabilitation for executive dysfunction in adults with stroke or other adult non-progressive acquired brain damage. *The Cochrane Database of Systematic Reviews, 2013*(4), CD008391. https://doi.org/10.1002/14651858.CD008391.pub2

Cicerone, K. D., Goldin, Y., Ganci, K., Rosenbaum, A., Wethe, J. V., Langenbahn, D. M., Malec, J. F., Bergquist, T. F., Kingsley, K., Nagele, D., Trexler, L., Fraas, M., Bogdanova, Y., & Harley, J. P. (2019). Evidence-based cognitive rehabilitation: Systematic review of the literature from 2009 through 2014. *Archives of Physical Medicine and Rehabilitation, 100*(8), 1515–1533. https://doi.org/10.1016/j.apmr.2019.02.011

Creighton, A. S., van der Ploeg, E. S., & O'Connor, D. W. (2013). A literature review of spaced-retrieval interventions: A direct memory intervention for people with dementia. *International Psychogeriatrics, 25*(11), 1743–1763. https://doi.org/10.1017/S1041610213001233

das Nair, R., Cogger, H., Worthington, E., & Lincoln, N. B. (2016). Cognitive rehabilitation for memory deficits after stroke. *The Cochrane Database of Systematic Reviews, 9*(9), CD002293. https://doi.org/10.1002/14651858.CD002293.pub3

Despoti, A., Megari, K., Tsiakiri, A., Toumaian, M., Koutzmpi, V., Liozidou, A., & Tsapanou, A. (n.d.). Effectiveness of remote neuropsychological interventions: A systematic review. *Applied Neuropsychology: Adult, 0*(0), 1–9. https://doi.org/10.1080/23279095.2024.2382814

Elera, C. M. D. (2024, January 29). *Neuropsychological rehabilitation in older adults with dementia: Experience in a day care center with NeuronUP*. NeuronUP. https://neuronup.us/cognitive-stimulation-news/testimonials/neuropsychological-rehabilitation-in-older-adults-with-dementia-experience-in-a-day-care-center-with-neuronup/.

Evans, F. A., Wong, D., & Stolwyk, R. J. (2020). Retrieval practice enhances memory for names in survivors of stroke. *Neuropsychology, 34*(8), 874–880. https://doi.org/10.1037/neu0000701

Feizipour, H., Sepehrianazar, F., Issazadegan, A., & Ashayeri, H. (2020). The effectiveness of cognitive rehabilitation on processing speed, working memory capacity, executive function, and quality of life in multiple sclerosis patients: A quasi-experimental study. *Urmia Medical Journal, 30*, 804.

Fleminger, S., Oliver, D. L., Williams, W. H., & Evans, J. (2003). The neuropsychiatry of depression after brain injury. *Neuropsychological Rehabilitation, 13*(1–2), 65–87. https://doi.org/10.1080/09602010244000354

Gopi, Y., Wilding, E., & Madan, C. R. (2022). Memory rehabilitation: Restorative, specific knowledge acquisition, compensatory, and holistic approaches. *Cognitive Processing, 23*(4), 537. https://doi.org/10.1007/s10339-022-01099-w

Gurd, J. (with Kischka, U., & Marshall, J.). (2010). *The handbook of clinical neuropsychology*. Oxford University Press.

Ibrahim, S., Francis, T., Sheehan, K. A., Kokorelias, K., Stanimirovic, A., Hashmi, S., Kalocsai, C., Ng, S., Berkhout, S. G., Cameron, J. I., Rac, V., & Pikula, A. (2024). Exploring unmet needs and preferences of young adult stroke patients for post-stroke care through PROMs and gender differences. *Frontiers in Stroke, 3*, 1386300. https://doi.org/10.3389/fstro.2024.1386300

Jamieson, M., Cullen, B., McGee-Lennon, M., Brewster, S., & Evans, J. (2017). Technological memory aid use by people with acquired brain injury. *Neuropsychological Rehabilitation, 27*(6), 919–936. https://doi.org/10.1080/09602011.2015.1103760

Kapur, N., Glisky, E. L., & Wilson, B. A. (2004). Technological memory aids for people with memory deficits. *Neuropsychological Rehabilitation, 14*(1–2), 41–60. https://doi.org/10.1080/09602010343000138

Krivitzky, L. S., Westmacott, R., Boada, R., Sepeta, L., Reppert, L., & Mrakotsky, C. (2022). Recent advances in neuropsychological outcomes and intervention in pediatric stroke. *Stroke, 53*(12), 3780–3789. https://doi.org/10.1161/STROKEAHA.122.037294

Maier, M., Ballester, B. R., & Verschure, P. F. M. J. (2019). Principles of neurorehabilitation after stroke based on motor learning and brain plasticity mechanisms. *Frontiers in Systems Neuroscience, 13*, 74. https://doi.org/10.3389/fnsys.2019.00074

Maresca, G., Maggio, M. G., De Luca, R., Manuli, A., Tonin, P., Pignolo, L., & Calabrò, R. S. (2020). Tele-neuro-rehabilitation in Italy: State of the art and future perspectives. *Frontiers in Neurology, 11*, 563375. https://doi.org/10.3389/fneur.2020.563375

Mateos-Aparicio, P., & Rodríguez-Moreno, A. (2019). The impact of studying brain plasticity. *Frontiers in Cellular Neuroscience, 13*, 66. https://doi.org/10.3389/fncel.2019.00066

Mohanty, M., & Gupta, S. K. (2013). Home based neuropsychological rehabilitation in severe traumatic brain injury: A case report. *Annals of Neurosciences, 20*(1), 31–35. https://doi.org/10.5214/ans.0972.7531.200111

Neuropsychology of schizophrenia. (2024, June 10). https://neuronup.us/cognitive-stimulation-news/mental-illness/schizophrenia/neuropsychology-of-schizophrenia/

Owji, S., & Shoja, M. M. (2020). The history of discovery of adult neurogenesis. *Clinical Anatomy, 33*(1), 41–55. https://doi.org/10.1002/ca.23447

Pedersen, A. R., Severinsen, K., & Jørgen, N. F. (2015). The effect of age on rehabilitation outcome after traumatic brain injury assessed by the functional independence measure (FIM). *SAGE Journals, 29*(4), 299–307. https://doi.org/10.1177/1545968314545171

Poulin, V., Jean, A., Lamontagne, M.-È., Pellerin, M.-A., Viau-Guay, A., & Ouellet, M.-C. (2021). Identifying clinicians' priorities for the implementation of best practices in cognitive rehabilitation post-acquired brain injury. *Disability and Rehabilitation, 43*(20), 2952–2962. https://doi.org/10.1080/09638288.2020.1721574

Power, E. M., & D'Amato, R. C. (2021). Using a standardized neuropsychological model to guide a qualitative and quantitative assessment for evidence-based interventions. In R. C. D'Amato, A. S. Davis, E. M. Power, & E. C. Eusebio (Eds.), *Understanding the biological basis of behavior: Developing evidence-based interventions for clinical, counseling and school psychologists* (pp. 213–256). Springer International Publishing. https://doi.org/10.1007/978-3-030-59162-5_9

Puderbaugh, M., & Emmady, P. D. (2024). Neuroplasticity. In *StatPearls*. StatPearls Publishing. http://www.ncbi.nlm.nih.gov/books/NBK557811/

Rezapur, T., Wurfel, B., Simblett, S. K., & Ekhtiari, H. (2017, January). Neuropsychological rehabilitation for psychiatric disorders. In *Neuropsychological rehabilitation: The international handbook* (pp. 136–148). Psychology Press. https://www.researchgate.net/publication/318684516_Neuropsychological_Rehabilitation_for_Psychiatric_Disorders

Rosa, C. L., Parolisi, R., & Bonfanti, L. (2020). Brain structural plasticity: From adult neurogenesis to immature neurons. *Frontiers in Neuroscience, 14*, 75. https://doi.org/10.3389/fnins.2020.00075

Soule, A. C., Fish, T. J., Thomas, K. G. F., & Schrieff-Brown, L. (2024). Attention training after moderate-to-severe traumatic brain injury in adults: A systematic review. *Archives of Physical Medicine and Rehabilitation, 106*, 433. https://doi.org/10.1016/j.apmr.2024.07.010

Soule, A. C., Fish, T., Winegardner, J., & Schrieff-Brown, L. (2024). Implementing neuropsychological rehabilitation following severe traumatic brain injury in a low-to-middle income country: A case report. *Frontiers in Rehabilitation Sciences, 5*, 1393302. https://doi.org/10.3389/fresc.2024.1393302

Sturm, W. (2024). Evidence-based methods in neuropsychological rehabilitation: Retraining of attention deficits. *ResearchGate*. https://www.researchgate.net/publication/289290979_Evidence-based_methods_in_neuropsychological_rehabilitation_Retraining_of_attention_deficits

Traumatic Brain Injury (TBI). (2024). National Institute of Neurological Disorders and Stroke. https://www.ninds.nih.gov/health-information/disorders/traumatic-brain-injury-tbi

Types of Strokes and Treatment. (n.d.). Brigham and women's hospital. Retrieved November 20, 2024, from https://www.brighamandwomens.org/neurosurgery/cerebrovascular/stroke/types

Understanding Different Types of Dementia. (n.d.). National Institute on Aging. Retrieved November 20, 2024, from https://www.nia.nih.gov/health/alzheimers-and-dementia/understanding-different-types-dementia

Weicker, J., Hudl, N., Hildebrandt, H., & Obrig, H. (2024). The effect of high vs. low intensity neuropsychological treatment on working memory in patients with acquired brain injury | request PDF. *ResearchGate*. https://www.researchgate.net/publication/342015877_The_effect_of_high_vs_low_intensity_neuropsychological_treatment_on_working_memory_in_patients_with_acquired_brain_injury

What Is Dementia? Symptoms, Types, and Diagnosis. (2022, December 8). National Institute on Aging. https://www.nia.nih.gov/health/alzheimers-and-dementia/what-dementia-symptoms-types-and-diagnosis

Wilson, B. A. (2013). Neuropsychological rehabilitation: State of the science. *South Africa Journal of Psychology, 43*(3), 267–277. https://doi.org/10.1177/0081246313494156

Wilson, B. A. (2020). Neuropsychological rehabilitation for people with non-progressive brain damage. In *Oxford research encyclopedia of psychology*. Oxford University Press. https://oxfordre.com/psychology/display/10.1093/acrefore/9780190236557.001.0001/acrefore-9780190236557-e-720?d=%2F10.1093%2Facrefore%2F9780190236557.001.0001%2Facrefore-9780190236557-e-720&p=emailAqcJz61Hb3DqI

Xu, X.-D., Ren, H.-Y., Prakash, R., Vijayadas, & Kumar, R. (2013). Outcomes of neuropsychological interventions of stroke. *Annals of Indian Academy of Neurology, 16*(3), 319–328. https://doi.org/10.4103/0972-2327.116909

Xue, G., Chen, C., Lu, Z.-L., & Dong, Q. (2010). Brain imaging techniques and their applications in decision-making research. *Xin Li Xue Bao. Acta Psychologica Sinica, 42*(1), 120. https://doi.org/10.3724/SP.J.1041.2010.00120

Yu, D., Li, X., & Lai, F. H. (2023). The effect of virtual reality on executive function in older adults with mild cognitive impairment: A systematic review and meta-analysis. *Aging & Mental Health, 27*(4), 663–673. https://doi.org/10.1080/13607863.2022.2076202

Chapter 2
Neurocognitive Disorders: Understanding the Landscape

2.1 Introduction to Neurocognitive Disorders

Neurocognitive disorders encompass a range of conditions that impair cognitive functions such as memory, attention, problem-solving, and language due to underlying neurological causes. These disorders, classified as mild or major, significantly impact individuals' daily functioning and quality of life. While mild disorders allow individuals to maintain some independence, major neurocognitive disorders often require intensive care and support.

Globally, the prevalence of neurocognitive disorders is increasing due to aging populations and advancements in medical diagnostics. Their societal impact includes not only the direct cost of healthcare but also the emotional and financial burden on families and caregivers.

2.2 Alzheimer's Disease and Related Dementias

2.2.1 Alzheimer's Disease

Alzheimer's disease (AD) is the most common form of dementia, characterized by progressive cognitive decline that primarily affects memory, language, and executive functions. Typically patients are diagnosed in their mid-60 s with early onset in their 30 s. AD was coined by Dr. Alois Alzheimer who identified that the unusual death of a female patient was marked by symptoms of memory loss, language impairment, and behavioral changes. Upon conducting an autopsy, Dr. Alzheimer identified abnormal clumping of neuronal fibers in her brain (Understanding Different Types of Dementia, n.d.; What Is Alzheimer's Disease*?*, 2021).

K. J. Reddy, *Innovations in Neurocognitive Rehabilitation*,
https://doi.org/10.1007/978-3-031-88117-6_2

AD directly impacts the central nervous system (CNS), especially the brain. Brain scans and assessments have identified a cognitive decline in the hippocampus and prefrontal cortex. Understanding the impact would require a deeper understanding of the biological processes involved in cognitive decline. The underlying pathology includes the accumulation of beta-amyloid plaques, tau protein tangles, and widespread neuronal loss. There is an extracellular accumulation of amyloidβ (Aβ)1–42 peptides and intracellular tangles (more than 40 tangles) caused by Aβ oligomers and phosphorylated tau proteins. While APOEε4 allele is the primary genetic risk factor, early onset AD is linked to gene mutations in the encoding of amyloid precursor protein (APP), presenilin 1 (PSEN1), and presenilin 2 (PSEN2). Tauopathies are large clusters of tau protein plaques that continue to increase in volume leading to cognitive and motor decline along with changes in the levels of CSF in the brain. AD is often diagnosed through a preliminary neuropsychological assessment followed by brain imaging scans of functional magnetic resonance imaging (fMRI), positron imaging tomography (PET), and cerebrospinal fluid (CSF) biomarkers. Consequently, hormonal dysfunction is identified among patients diagnosed with AD (Singh et al., 2024) (Fig. 2.1) (Biorender, n.d.).

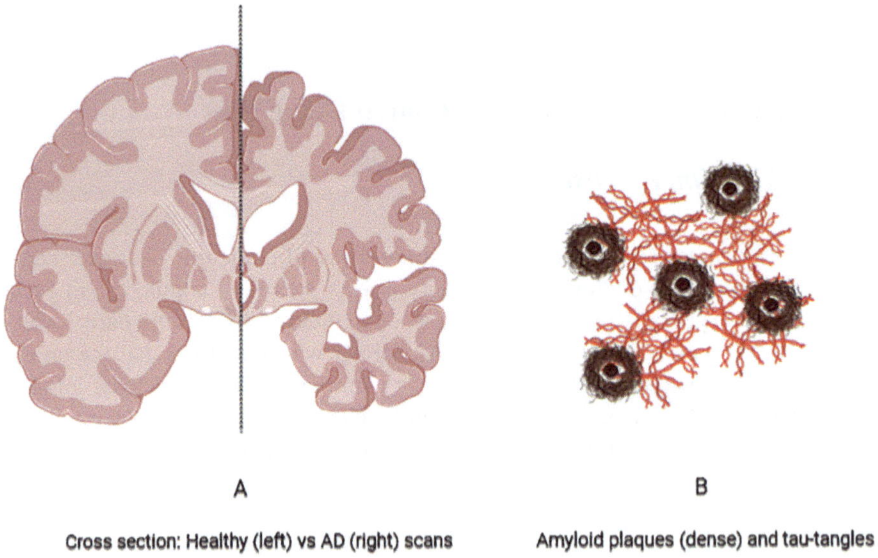

Fig. 2.1 (**a**) Cross section of healthy brain vs AD brain scan. (**b**) Amyloid plaques and tau tangles (Biorender, n.d.)

2.2.2 *Vascular Dementia*

Vascular dementia is the second most common type of dementia after AD. It refers to changes in memory, thinking, and behavior as a result of blood vessels in the brain. The causes are due to an interruption of oxygen flow and blood vessels in the brain. Vascular dementia is characterized by diseased capillaries and altered states of white matter further leading to delays or barriers in neurotransmission. Other medical conditions that impact dementia are cardiovascular diseases and cerebrovascular disease leading to cognitive impairment. There are two predominant ways in which vascular dementia can affect the brain-mixed dementia and multi-fact dementia. Mixed dementia has symptoms that are characteristic to AD and vascular dementia. Multi-infarct dementia are vascular blockages that are unidentifiable and insidious in nature, impairing the blood circulation without significant overt and observable symptoms. A clinical picture of vascular dementia include—difficulty performing simple tasks, inability to follow instructions, learn new information and maintain routines, reading and writing difficulties, memory impairments recollecting past events, losing memory of familiar routes, misplacing items, behavioral impairments changes in personality, mood, poor judgment and decision-making along with a loss of motivation, delusions, and disrupted sleep. An observable and gradual decline is more common than a rapid course decline. Preventative measures for vascular dementia include mitigating risk factors such as high blood pressure, diabetes, high cholesterol and fluctuations in heart rate, and poor lifestyle changes are impactful. A prior history of stroke has a high risk for dementia (Vascular Dementia: Causes, Symptoms, and Treatments | National Institute on Aging, n.d.; Vascular Dementia, 2024).

2.2.3 *Lewy Body Dementia*

An abnormal deposit of the protein alpha-synuclein in the brain leads to Lewy Body Dementia. The chemical alterations in the brain result in the decline in thinking, movement, mood, and behavior; diagnosed around the early 50 s. Lewy bodies were identified by Dr. Friedrich Lewy in 1912 when he identified abnormal protein deposits affecting the normal functions of the brain in patients with Parkinson's disease. The protein alpha-synuclein plays important functions in synaptic transmission and its degeneration causes inhibitions of synaptic neurotransmitters—leading to neuronal death. Lewy bodies form in certain regions of the brain namely the cerebral cortex, limbic cortex, and the midbrain. There are significant impairments in hippocampal and olfactory pathways. The primary neurotransmitters inhibited by Lewy bodies is acetylcholine (ACh) which plays a crucial role in behavior, cognition, motor functions, and sleep. The etiology of Lewy bodies is unclear with some theories predicting that genetic mutation of APOE and GBA increases the likelihood of Lewy bodies dementia (Dementia with Lewy Bodies Symptoms and Causes, n.d.; What Is Lewy Body Dementia?, 2021).

Symptoms of Lewy body dementia (LBD) can be classified as cognitive, motor, and behavioral. The cognitive symptoms of dementia are similar to AD. A significantly declined case of LBD could be impacted with attentional deficits, visuospatial impairments, dysfunctions in executive functions, memory and language impairments, and poor judgment. Motor symptoms have Parkinsonian features along with slow movements, muscle rigidity, shuffling walk and frozen stance, imbalance and stooped posture, poor coordination, difficulty swallowing, weakness of voice, and smaller than normal size in handwriting. Behavioral symptoms include REM sleep behavior disorder, excessive daytime sleepiness, insomnia, restless leg syndrome, depression, apathy, anxiety, agitation and irritability, delusions or misperceptions, paranoia, and hallucinations. Additional symptoms include urinary incontinence, dizzy spells, fainting, fluctuations in blood pressure, heat cold sensitivity, and sexual dysfunction (Lewy Body Dementia, n.d.). Some risk factors of dementia are REM sleep behavior disorder, anosmia, genetic history, and poor lifestyle choices. Dementia is diagnosed through neurological examination followed by neuroimaging, blood tests, and neuropsychological examination. Since LBD is not treatable like all neurodegenerative diseases, rehabilitation and familial support is vital in reducing the possibility of rapid decline (Dementia with Lewy Bodies Symptoms and Causes, n.d.).

2.2.4 Frontotemporal Dementia

The neurodegeneration of proteins TDP43 and tau in the frontal and temporal lobes of the brain causes frontotemporal dementia (FTD). FTD is diagnosed between the ages of 45 and 60 years with perceivable differences in behavior and emotion, language, and comprehension. FTD was previously known as Pick's disease after the physician, Dr. Arnold Pick (1892), who discovered impairments in language distinct from other neurodegenerative diseases. Although the cause is unknown, there is a strong genetic predisposition that increases the likelihood of dementia. FTD is often caused as an extension of brain injuries or other brain diseases. The two most prominent causal factors of FTD are brain disorders caused by degeneration tau protein and brain disorders caused by protein TDP43 (What Are Frontotemporal Disorders*?*, 2021).

FTD subtypes are frontotemporal dementia (bvFTD), primary progressive aphasia (PPA), and movement disorders. The behavioral variant of FTD causes impairments in personality, behavior, and judgment. Difficulty planning and sequencing information and tasks, word repetition and activity repetition, impulsivity and apathy toward family members are common characteristics of the behavioral variant. PPA impacts the patient's ability in speech and language components. Aphasia, slurred speech, mutism, and difficulty forming words are specific PPA characteristics. PPA is characterized as semantic (difficulty with facial and object naming and recognition), agramatic (omission of articles in speech that connect nouns and verbs; gradual loss of speech), and logopenic (difficulty choosing the right words to form sentences but can comprehend speech). Movement disorders are classified as corticobasal syndrome and progressive supranuclear palsy (PSP). Corticobasal syndrome is caused by the atrophy of brain cells in the corticobasal region. This causes a progressive loss of movement control, apraxia, muscular rigidity, and loss of strength. While motor functions are impaired it does not guarantee the loss of cognitive functions. PSP is associated with decline in balance during movement and eye movement control. Slow movements, unpredictable and unexplained falls, loss of facial expression, stiffness of the muscles (neck and upper body) are shared symptoms with Parkinson's disease. A hallmark characteristic is trouble with eye movement control leading to difficulty looking down and fixed staring. FTD is associated with other diseases including multiple sclerosis and Parkinsonism (What Are Frontotemporal Disorders?, 2021). Studies have indicated that aging, family history, genetic conditions such as Down syndrome, Creutzfeldt-Jakob disease, Gerstmann-Sträussler-Scheinker syndrome, and insomnia are significant risk factors of developing FTD. High levels of cholesterol and amino acid in the bloodstream, diabetes, and mild cognitive impairments are significant contributors to FTD. Smoking and alcohol abuse are other risk factors of FTD. The diagnostic procedure is common for Lewy body dementia and FTD (Frontotemporal Dementia, n.d.).

2.2.5 Challenges in Rehabilitation

Neuropsychological rehabilitation in cases of neurodegenerative disorders require special attention in the multiple aspects. Rehabilitation requires a team of healthcare professionals, caregivers, and mental health professions. Special facilities that make daily tasks simple and easy to navigate without a declining sense of autonomy. Integrated rehabilitation elevates the overall quality of life and prevents further deterioration. The model below adopted by Richter states that neurodegenerative disorders require an array of interventions from occupational therapists, doctors, nurses, day care and assisted living facilities, local support groups, and caregiver presence (Fig. 2.2).

In neuropsychological rehabilitation, the primary focus is in applying the compensatory interventional approaches along with restorative strategies. Certain lost cognitive functions are restored through cognitive retraining activities such as maintaining language skills and using new tools and other methods of communication that address specific deficits in cases with aphasia. Using communication notebooks, photo albums to recollect names (cued retrieval), speaking slowly and clearly to form words and sentences encourages patients with language impairments to gradually recover language skills. In cases where compensatory strategies are required, utilizing notebooks to communicate, hand gestures and sign language have proven to be effective. In cases of memory impairment, long-term memory and attention interventions are required to recover lost functions. This includes a focus on retention and retrieval strategies along with reducing the cognitive load on the

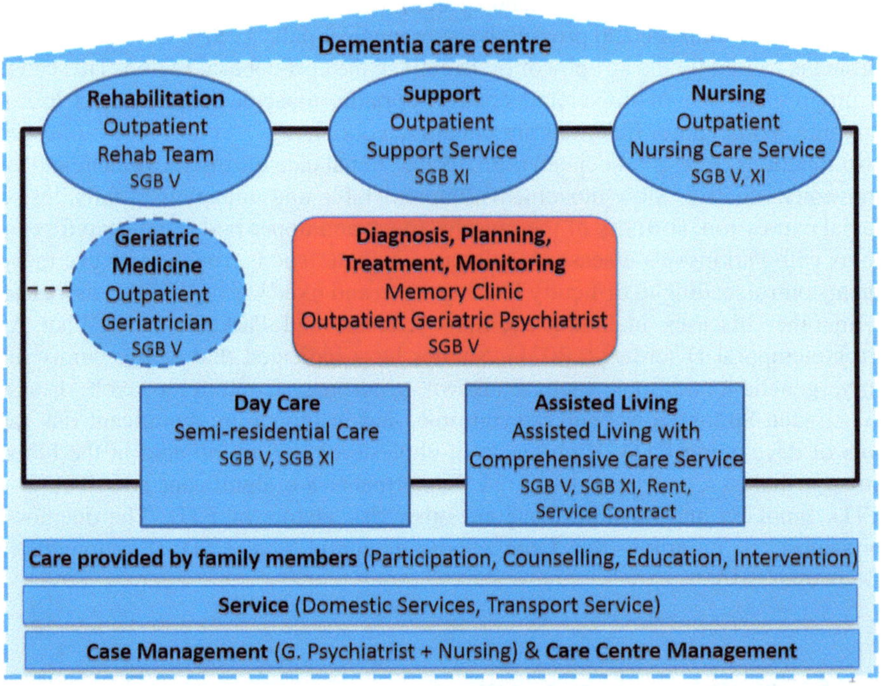

Fig. 2.2 The integrative model of care for dementia care (Richter, 2017)

individual. The strategies include task-specific training and repetition with increasing intensity. Along with these strategies, the use of compensatory technology plays a vital role in reducing the cognitive load on patients: using simple reminders, checklists, important phone numbers, and photos of family members would be crucial in improving the interventional strategies. The progressive nature of dementia complicates rehabilitation efforts. Therapies must balance cognitive stimulation with strategies to manage behavioral and psychological symptoms, often requiring caregiver involvement. While technological advancements have made drastic changes in every aspect of rehabilitation, several challenges limit the effectiveness of rehabilitation. Geographical limitations in accessing rehabilitation facilities and professionals, poor healthcare facilities, inefficient management of rehabilitation teams and inexperienced professionals act as barriers to the maintenance of the disorder. Another limitation is the financial challenges experienced by caregivers; long-term rehabilitation is essential in maintaining the neuropsychological prognosis. Expensive neuro assessments, consultation fees, medication charges, occupational therapy charges, travel, and living expenses are significant stressors in the rehabilitation process. Additionally, patients face multiple comorbid illnesses both mental and physiological. Depression, anxiety, diabetes, cholesterol, and cases of visual impairment are challenges to implementing care. Medication, if provided for these conditions, could have side effects on the rehabilitation processes. The need for community support such as ergonomically designed living spaces, easy access to facilities and local members who might aid the patient (should the experience difficulty remembering their way) to familiar places or in case of confusion improves the overall quality of care for the individual. However, this is challenging to implement in urban settings and for patients living independently and/or in isolation. Challenges in rehabilitation can be supplemented by creating a multidisciplinary approach (Das et al., 2022; Hussain et al., 2018; Richter, 2017).

2.3 Cognitive Effects of Stroke

On the basis of the area and severity of the impairment, stroke is observed to have an effect on memory, attention, and language. When blood flow to the brain is interrupted by a blockage in the capillaries in the brain (ischemic stroke) or by a ruptured blood vessel (hemorrhagic stroke), the brain is deprived of essential oxygen and nutrients. Ischemic stroke is caused by clots or plaque that builds in the blood vessels leading to symptoms such as weakness, numbness, speech difficulties, blurry vision, imbalance, and headaches. Hemorrhagic stroke is caused by rupturing blood vessels, often due to trauma, aneurysms, high blood pressure, or congenital conditions, leading to brain injury. Strokes are classified based on the area of the clotting or bleeding. Other subtypes include transient ischemic attack (TIA), cryptogenic stroke, brainstem stroke, and recurrent stroke. The recovery process after a stroke is highly variable, influenced by factors such as the individual's age, overall health, and the availability of early intervention (Types of Strokes and Treatment, n.d.).

2.3.1 Memory

Memory deficits are challenging for many stroke patients, but research in this area is surprisingly limited compared to other cognitive impairments. The effectiveness of cognitive therapies remains uncertain due to methodological weaknesses in existing studies, including small sample sizes, inadequate randomization, and the inclusion of mixed populations (e.g., stroke and traumatic brain injury patients). Some studies suggest that group-oriented memory training and the use of mnemonic strategies can improve specific skills like verbal memory or face-name recall. However, these improvements often fail to extend to general memory function or subjective experiences of memory improvement. Other interventions, such as the use of external aids like paging systems, have demonstrated short-term benefits in helping stroke patients manage memory and planning deficits. However, these gains frequently diminish after the aids are removed, especially for individuals with compromised executive functioning. Experimental studies, including single-patient designs, have explored novel approaches like verbal working memory training, yielding promising but limited and non-generalizable results. Despite some evidence supporting strategies like errorless learning and combined approaches using both internal and external memory aids, the overall body of research suffers from significant limitations. Issues such as insufficient randomization, lack of blinding, small and biased samples, and absence of long-term follow-up data make it difficult to draw firm conclusions. Many studies also fail to distinguish between spontaneous recovery and treatment effects, further complicating interpretations. The need for more rigorous and stroke-specific research is clear. Future studies should prioritize larger sample sizes, stricter randomization and blinding protocols, and long-term evaluations to better understand the potential of cognitive therapies in addressing poststroke memory deficits (Cramer et al., 2017; Xu et al., 2013).

2.3.2 Attention

Stroke-related cognitive impairments, particularly in the realm of attention, can severely impact daily life. Patients often face difficulties with focusing, shifting attention, and managing multiple tasks, which can hinder their independence and quality of life. These challenges are typically linked to damage in brain regions like the prefrontal cortex, which is crucial for regulating attention and cognitive control. As a result, people who have had a stroke may experience trouble concentrating on one task for an extended period, and may find it challenging to divide their attention effectively between different activities. Interventions targeting attentional deficits are crucial in stroke rehabilitation. Cognitive rehabilitation, including personalized cognitive training programs, aims to strengthen attention skills. Such programs often focus on improving both sustained and selective attention. Mindfulness-based techniques have also emerged as an effective strategy, helping individuals stay present and improve their attention by reducing distractions. In addition to these individual-focused interventions, environmental adjustments, such as using external memory aids or creating structured routines, can help support those affected by attention deficits. Group-based interventions that encourage social interaction have proven beneficial by providing cognitive stimulation in a supportive setting. These rehabilitative approaches not only help stroke patients manage attention impairments but also enhance their overall cognitive flexibility, enabling them to better adapt to the challenges that arise from the cognitive changes poststroke. However, ongoing research is needed to fine-tune these interventions and evaluate their long-term effects, ensuring they are both effective and sustainable for those in need (Loetscher et al., 2019; Navarro et al., 2020).

2.3.3 *Language*

Stroke can have significant cognitive effects, particularly on language, leading to a variety of deficits that complicate communication and daily functioning. When a stroke affects areas of the brain involved in language processing, such as the left hemisphere, it can result in aphasia. Aphasia manifests in several forms, depending on the specific area and severity of the brain injury. For instance, individuals with Broca's aphasia often struggle with speech production, finding it difficult to form grammatically correct sentences, while Wernicke's aphasia involves impaired comprehension, leading to fluent but nonsensical speech. These language deficits can be frustrating and isolating, as they hinder the ability to express thoughts clearly or understand spoken or written words. However, there is promising evidence that rehabilitation can significantly improve outcomes. Structured speech-language therapy remains a cornerstone of recovery, focusing on restoring language abilities through repetitive exercises and interactive activities. This therapy is personalized to individual needs, as some stroke patients may benefit from using alternative communication aids, such as picture boards or speech-generating devices, to support their recovery. Recent advancements have also introduced the use of technology in rehabilitation, which can supplement traditional therapies. Computer-assisted programs and virtual reality environments allow patients to engage in language exercises in a more dynamic and personalized way, offering a sense of progression and adaptability to their specific challenges. This integration of technology has been shown to enhance motivation and adherence to therapy, which is crucial for long-term recovery. Efforts are made to restore language functions but also to reduce the emotional and cognitive burden of language impairment, helping individuals reintegrate into social and professional life (Rehabilitation to Improve Language and Communication, n.d.; Gerstenecker & Lazar, 2019).

2.3.4 *Barriers to Recovery*

Stroke rehabilitation is uniquely personalized to each patient, taking into account factors such as age, preexisting health conditions, and personal goals. A critical aspect of recovery is the patient's ability to regain independence in daily activities, such as bathing, walking, and grooming, which not only enhances their quality of life but also reduces the strain on caregivers. Rehabilitation faces obstacles, such as physical impairments, emotional challenges, and limited access to specialized care, particularly in rural or underserved areas. The integration of both physical and cognitive therapies is key to addressing the broad range of recovery needs. In addition, emotional support for both patients and caregivers is vital, as caregivers often experience stress and burnout. Addressing mental health issues, such as depression, is essential for maintaining motivation and improving adherence to therapy. Socioeconomic factors also play a significant role in the accessibility of quality rehabilitation services, with solutions like tele-rehabilitation and community-based programs proving essential in providing equitable care (Mroczkowska et al., 2019; Xu et al., 2013).

2.4 Traumatic Brain Injury (TBI)

Traumatic brain injury (TBI) is caused by an external force, such as a blow or jolt to the head or body, or an object piercing the skull. The severity of the injury determines the potential for long-lasting damage to cognitive, physical, and behavioral functions, with some injuries being temporary and others more permanent or even fatal. Common causes of TBI include falls, motor vehicle accidents, sports injuries, assaults, and military blasts. Symptoms range from headaches, dizziness, and memory issues to more severe cognitive and emotional disturbances, including mood swings and confusion. TBIs are divided into two categories: penetrating (open) and non-penetrating (closed). Penetrating TBIs occur when an object enters the brain, while non-penetrating TBIs result from blunt forces like falls or car accidents. Within non-penetrating TBIs, there are various types such as concussions, which are typically mild but can have long-term effects. More severe forms include diffuse axonal injury (DAI), which disrupts the brain's communication pathways, and brain bruises (contusions), which can cause swelling and bleeding. Hematomas, or internal bleeding in the brain, also present different risks depending on the location, like epidural, subdural, or intracerebral. Recognizing the different types and symptoms of TBIs is essential for early diagnosis and treatment. The consequences of TBIs can vary, so timely intervention plays a critical role in recovery. Preventative measures such as wearing helmets, avoiding risky behaviors, and driving safely are important steps in reducing the risk of TBI (Traumatic Brain Injury Symptoms & Causes | AllSpine, 2024; Traumatic Brain Injury (TBI), 2024).

The cognitive effects vary with the severity of the injury and may include the following.

2.4.1 *Attention and Concentration*

Traumatic brain injury (TBI) frequently leads to significant attention difficulties, which can be some of the most disruptive and common cognitive challenges for those affected. This occurs because TBI can damage the brain's attention networks, particularly in areas like the frontal lobes that are essential for managing complex cognitive tasks. As a result, individuals may struggle with maintaining focus, filtering out distractions, or shifting attention between tasks, all of which can severely impact daily life. Rehabilitation efforts, such as cognitive training and therapy, aim to enhance attention, though the results can differ based on the severity of the injury and the person's unique circumstances. Continued research is needed to refine these treatments and better support people with TBI. While various rehabilitation approaches, including computerized and traditional methods, show potential, it is still unclear which are most effective, especially for individuals with moderate to severe injuries. That said, there is hope for recovery, with some individuals making progress even years after their injury, suggesting that age and the extent of the damage may not be as limiting as once thought (Shah et al., 2017; Soule, Fish, Thomas, & Schrieff-Brown, 2024).

2.4.2 Processing Speed

Processing speed is naturally impaired due to impairments in attention and concentration. The working memory functions in patients with TBI are limited to slower responses to stimuli or information. Speed of information processing is significantly impacted in those with TBI in tasks of performance. Brain injuries can cause a significant decline in processing speed, leading to slower reactions, difficulty multitasking, and delays in completing tasks. The severity and location of the injury play a key role in these impairments. However, recovery is possible through cognitive rehabilitation, coping strategies, medications, and lifestyle changes. With consistent support, many individuals can see improvements in processing speed, even long after the injury has occurred (Abelmann et al., 2023; Madigan et al., 2000).

2.4.3 Executive Functions

Individuals with traumatic brain injury (TBI) often face cognitive challenges, such as altered risk-taking behavior and difficulty adapting to new situations. Individuals with executive function deficits have difficulty making decisions, planning, and prioritizing information. These issues are linked to broader disruptions in brain systems, including dopamine and norepinephrine. While treatments using dopamine-related medications show promise, more research is needed, particularly regarding the timing and effectiveness of these therapies. Home-based interventions that focus on task completion through checklists and planning, reduce the load on executive functions (Gonçalves et al., 2014; Ozga et al., 2018).

2.4.4 Rehabilitation Strategies

Interventions for TBI often involve a multidisciplinary approach, including occupational therapy, speech therapy, and psychological support. Long-term follow-up is critical for addressing persistent cognitive and emotional challenges. Traumatic brain injury (TBI) occurs when an external force strikes the head with significant impact, potentially causing severe damage. This can either be through a penetrating injury, where an object breaks through the skull, or a non-penetrating injury resulting from blunt force trauma. The consequences of TBI can range from temporary effects to lasting impairments in cognitive abilities and motor control. Physically, individuals with TBI may experience symptoms like persistent headaches, seizures, blurry vision, uneven pupil dilation, vomiting, slurred speech, and difficulties with balance. On a cognitive level, there might be disruptions in memory, attention, confusion, irritability, decision-making struggles, and issues with abstract thinking. TBI can also affect sensory functions, leading to dizziness, increased sensitivity to light or sound, emotional instability, fatigue, and mood swings. In children, TBI might also cause difficulties with skill development, sleep disturbances, trouble focusing, and issues with walking or balance. The severity of TBI can vary, impacting consciousness. Some individuals may only experience brief confusion, while others could slip into a coma or even face life-threatening consequences. There are two distinct stages of brain damage: primary, which occurs immediately after the injury, and secondary, which evolves in the hours or days following. TBI often disrupts functions such as memory, attention, and decision-making, requiring long-term therapeutic interventions. Cognitive rehabilitation, especially for memory and focus, can yield positive results. However, severe cases may present challenges with motor skills, and patients may need alternative strategies to cope with lost abilities. Treatments frequently focus on enhancing verbal communication, spatial reasoning, working memory, and fine motor coordination through targeted neuropsychological therapy. Since recovery is often slow and requires considerable effort, many individuals with TBI experience loss of motivation and are at higher risk of developing mental health issues like anxiety or depression. Home-based strategies, such as using checklists and organizing tasks, can assist in managing daily activities and reduce cognitive strain. Overall, the goal is to help individuals reintegrate into their communities, despite the fact that complete recovery may not always be attainable. The emphasis is on improving the individual's quality of life and adapting to new ways of functioning (Khan, 2003; Min & Shin, 2022; Soule, Fish, Winegardner, & Schrieff-Brown, 2024).

2.5 Additional Neurocognitive Disorders

2.5.1 Parkinson's Disease

Parkinson's disease (PD) is a progressive and chronic disorder that impacts movement and balance. It's one of the most widespread movement disorders, affecting people aged 60 and older. A hallmark of Parkinson's is the gradual loss of muscle control, which becomes observable as disease progresses. Bradykinesia, or slowed movements impact speed and regularity. They may also struggle with balance, shuffling, dragging feet, and an increased risk of falling. As muscles tighten, they can become rigid, which can either be a continuous form of stiffness (known as lead-pipe rigidity) or intermittent (cogwheel rigidity), leading to slow and painful movements. People with Parkinson's tend to have a hunched posture and diminished facial expressions, making it harder for others to read emotions. Drooling or dry eyes due to reduced blinking, cramped and smaller handwriting (micrographia), and resting tremor, where rhythmic shaking occurs, often in the hands or fingers, especially when the body is at rest. Parkinson's can also bring on a variety of non-motor symptoms that may develop even before the classic movement difficulties. These can include comorbid depression, memory deficits, and difficulty concentrating, anosmia, restless legs syndrome, rapid eye movement (REM), sleep behavior disorder, and other disturbances are also common. The autonomic nervous system, which controls involuntary bodily functions, can also be affected, leading to problems such as bowel and bladder control issues, sexual dysfunction, and a sudden drop in blood pressure when standing (orthostatic hypotension). The exact cause of Parkinson's remains unclear, but the disease develops when nerve cells in the brain that produce dopamine—an essential neurotransmitter for controlling smooth movement—become damaged and eventually die off. As dopamine levels decline, the symptoms related to movement become more prominent. Most cases are considered idiopathic, meaning there is no known cause, though research suggests that abnormal protein accumulation plays a role. Specifically, a protein called alpha-synuclein misfolds and forms clumps in nerve cells, which are known as Lewy bodies, and these clumps contribute to cell dysfunction. Environmental factors, such as exposure to certain toxins, are thought to increase the risk of developing Parkinson's, as are genetic factors (Parkinson's Disease—Symptoms and Causes, n.d.; Deal et al., 2019).

2.5.2 *Huntington's Disease*

Huntington's disease is a genetic mutation that progressively affects cognition, motor control, and behavior. Huntington's disease (HD) causes the neurodegeneration in the brain, leading to a variety of physical, cognitive, and emotional deficits. HD is caused by a genetic mutation in the HTT gene, passed down from the affected parent. Rare cases have appeared in individuals without a family history. Diagnosing Huntington's involves a history taking and neurological evaluation supplemented by genetic testing to confirm the diagnosis. This progressive condition is characterized by involuntary, jerky movements known as chorea, which typically start in the hands and face before spreading to other parts of the body. The movements interfere with daily functions including speaking, walking, and self-care—increasingly difficult to maintain independence. Other symptoms include changes in mood and personality along with irritability. Cognitive decline in memory, concentration, decision-making, and multitasking is prevalent. Muscle stiffness, balance problems, and issues with posture become more pronounced, further impacting physical functioning. There is no cure for Huntington's disease, but treatments are available to manage symptoms and improve quality of life. Alongside medications, rehabilitative therapies can assist individuals in coping with the disease. Physical therapy exercises to improve strength, balance, and flexibility can reduce falls and maintain mobility. Occupational therapy helps individuals use assistive devices to manage daily tasks, and speech therapy provides exercises to improve speaking and swallowing. Psychotherapy offers emotional support to both individuals with Huntington's and their families, helping them navigate the psychological challenges of the disease (Huntington's Disease—Symptoms and Causes, n.d.).

2.5.3 Psychiatric Disorders with Cognitive Impairments

Conditions such as schizophrenia and depression often present cognitive deficits, including memory problems and reduced attention spans. Cognitive remediation therapy is increasingly used in such cases to improve cognitive functioning.

2.6 Barriers and Challenges in Rehabilitation

Rehabilitation for neurocognitive disorders faces several challenges.

2.6.1 Patient Variability

Patient variability acts as a major challenge in neuropsychological rehabilitation. Various sub-factors influence patient variability including diversity in the recovery trajectories, complexity of subjective and objective factors. In cases of patients with traumatic brain injury (TBI), patients indicate marked variability in the recovery and outcomes when injuries appear to be similar in severity. Patients can differ in terms of cognitive, emotional, and functional recovery which creates obstacles for interventions that accommodate individual differences and recovery patterns. Subjectivity plays a significant role in patient variability. Variable such as self-efficacy, motivation, and emotional well-being play a decisive role in determining functional outcomes. Research indicates that subjective factors act as prominent barriers in the implementation of rehabilitation skills in real-world settings. Occupational challenges, familial roles, and social reintegration are often limited by personal differences. Patient's ability to confidently apply their cognitive abilities and their perceptions of performance plays a significant role in influencing engagements and progress in therapy. Addressing patient variability fosters long-term well-being of the patients encouraging positive attitudes and emotional resilience. Irrespective of subjective factors, assessment tools and rehabilitation may fail at meeting complex everyday functions. Standardization of cognitive tests in the traditional approach does not translate to its full capacity in real-world settings. For example, patients may excel in structured clinical tests but fail in performing unstructured daily activities such as financial management and social interaction. In order to bridge this gap, innovative assessment methods are required such as technologically aided tools that simulate naturalistic settings and enable remote testing. Socioeconomic disparities and geographical barriers further limit access to comprehensive care, compounding the challenges faced by patients with complex needs. Tele-rehabilitation and virtual assessments provide potential solutions by expanding access to high-quality care while minimizing logistical burdens. PC-based and tele-rehabilitation assessments have been particularly valuable during pandemics and in reducing patients and care-giver burnout (Rath et al., 2011; Torregrossa et al., 2023).

2.6.2 Comorbidities

Rehabilitation for traumatic brain injury (TBI) faces numerous systemic challenges that hinder patient recovery. Many patients, especially those with mild TBI, are discharged from emergency care without structured follow-ups. As a result, persistent symptoms such as cognitive and emotional dysfunction often remain untreated. Even for those with moderate-to-severe injuries, access to rehabilitation services is limited, with only a fraction receiving the inpatient or outpatient care they need. The conditions detailed in this chapter have a strong influence on pharmacotherapy. In conditions like Huntington's and Parkinson's, medication is used to control motor activity and address psychiatric comorbidities such as depression and irritability. Medications reduce jerky movements, while antidepressants and antipsychotics help with mood regulation. Mood stabilizers are used to manage the highs and lows of bipolar disorder. The issue with medication is the constant side-effects it can have on therapy. The rehabilitation may not be successful in cases where psychiatric conditions impair cognition and motor control (Huntington's Disease—Symptoms and Causes | Penn Medicine, n.d.; Understanding Different Types of Dementia, n.d.; Maas et al., 2022).

2.6.3 Access to Care

Geographic disparities exacerbate the issue; low- and middle-income countries lack the infrastructure and trained professionals necessary for comprehensive rehabilitation. Socioeconomic limitations impact access to care, poor communication and unmet needs between acute and post-acute care providers. Another significant barrier is the lack of early rehabilitation. Delays in starting therapy result in worse outcomes and increased disability. Rehabilitation is frequently dismissed as unnecessary, despite lingering symptoms. There is a need for early interventions and better discharge planning to address this gap. Tackling these challenges pushes for integrated care systems, trained healthcare providers, and prioritizing structured rehabilitation strategies for long-term recovery.

2.6.4 Caregiver Burden

Caregivers are pathfinders in supporting patients with chronic or serious impairments. Caregiving can be physically and emotionally draining with several studies detailing overwhelming stress, anxiety, and depression, and health issues as a result of the process. A compromised immune system, risk of heart disease, and reduced shortened lifespan contribute to the burden. The struggle can affect the care they provide. Leading to delays in hospital discharge, breakdowns in home care and readmissions. Different conditions, like dementia, stroke, and brain injuries, come with their own unique challenges, demanding the inclusion of caregivers in treatment and rehabilitation. Addressing caregiver adversities is a pressing priority in healthcare to ensure caregiving is more manageable and sustainable for everyone involved (Harding et al., 2015).

2.7 Advances in Understanding and Therapy

Modern advancements are transforming how neurocognitive disorders are diagnosed and treated:

1. *Neuroimaging*: Imaging tests like CT and MRI scans are used to check for physical changes in the brain, while neuropsychological assessments help evaluate the cognitive and emotional effects of the disease. Tools like MRI and PET scans enable early detection and tracking of disease progression, facilitating timely intervention. Integrating biomarkers into rehabilitation could help doctors tailor treatment plans more precisely, monitor how well patients are recovering, and possibly enhance their recovery outcomes. By tracking specific biomarkers like NfL and ac-tau, clinicians could adjust therapies to better suit individual needs. However, further research is needed to confirm the effectiveness of these biomarkers in real-world clinical settings and to fine-tune their application in rehabilitation strategies.
2. *Digital Therapeutics*: Usage of apps, AI tools are wearable devices that allow for personalized monitoring and rehabilitation plans is a rising trend in limiting motor and visual impairments in patients with neurodegeneration. Digital therapies and AI-based tools are making a significant impact in neuropsychological rehabilitation, offering new ways to support patients with cognitive challenges. These technologies, including virtual reality (VR) and artificial intelligence (AI), provide personalized and interactive treatments. Simple digital tools like reminders, checklists, and alarms help ease the cognitive load for individuals,

reducing stress and anxiety related to memory tasks. For example, setting a reminder to buy groceries can take the mental burden off remembering repeatedly, while an alarm without a cue encourages patients to think about the task they should be working on. VR technology allows patients to safely practice real-life scenarios in a virtual environment, improving cognitive function and helping researchers understand brain activity through brain imaging. AI-powered cognitive exercises can provide personalized support, adjusting to the user's progress and delivering accurate data on cognitive performance, such as processing speed.

3. *Emerging Therapies*: Techniques such as virtual reality training and transcranial magnetic stimulation are used to address cognitive decline. Cognitive rehabilitation often focuses on improving deficits while managing fatigue and motor issues. Emerging therapies that incorporate cutting-edge technologies like digital therapeutics, artificial intelligence (AI), and virtual reality (VR) are transforming rehabilitation practices. AI can monitor and analyze patient progress, adjusting therapy to focus on areas such as memory, attention, or motor skills, making therapy personalization simple. This real-time adaptability enhances the effectiveness of treatment and ensures patients receive the most relevant support. These technologies also address issues of accessibility, especially for patients with mobility challenges or those living in areas without easy access to specialized care. Home-based therapy permits patients to continue their rehabilitation without the logistical barriers of traditional rehabilitation. VR leads the patients in virtual environments where they can safely practice physical or cognitive training. This approach helps maintain motivation and adherence to the rehabilitation process. Consistent interventions can help patients and therapists can track and adjust the course of therapy as needed. Integrating digital tools reduced caregiver burden. Applying digital reminders, VR exercises, and other forms of tech-based therapies enable patients to become more independent in performing daily tasks. This alleviates some of the physical caregiving responsibilities, allowing caregivers to focus more on providing emotional support, which is vital for the patient's overall well-being (National Academies of Sciences, 2022).

2.8 Summary and Future Directions

This chapter provided an overview of common neurocognitive disorders, their cognitive and behavioral manifestations, and the challenges they pose to rehabilitation efforts. It also highlighted advancements in diagnostic and therapeutic approaches, emphasizing the need for personalized and multidisciplinary interventions.

Future research and innovation should focus on integrating emerging technologies and improving accessibility to ensure equitable rehabilitation services. The next chapter will delve into the tools and methodologies used to diagnose neurocognitive disorders, laying the groundwork for effective intervention planning.

References

4: Rehabilitation after traumatic brain injury—Khan—2003—Medical Journal of Australia—Wiley Online Library. (n.d.). Retrieved November 13, 2024, from https://onlinelibrary.wiley.com/doi/full/10.5694/j.1326-5377.2003.tb05199.x

10. Rehabilitation to Improve Language and Communication. (n.d.). Canadian stroke best practices. Retrieved December 1, 2024, from https://www.strokebestpractices.ca/en/recommendations/stroke-rehabilitation/rehabilitation-to-improve-language-and-communication/

Abelmann, A. C., Kessels, R. P. C., Brazil, I. A., Fasotti, L., & Bertens, D. (2023). Game-supported cognitive strategy training for slowed information processing speed after acquired brain injury: Study protocol for a randomised controlled trial. *BMJ Open, 13*(9), e067108. https://doi.org/10.1136/bmjopen-2022-067108

BioRender. (n.d.). Retrieved November 26, 2024, from https://app.biorender.com/illustrations/674590d381b517b4934d270a

Cramer, S. C., Wolf, S. L., Adams, H. P., Chen, D., Dromerick, A. W., Dunning, K., Ellerbe, C., Grande, A., Janis, S., Lansberg, M. G., Lazar, R. M., Palesch, Y. Y., Richards, L., Roth, E., Savitz, S. I., Wechsler, L. R., Wintermark, M., & Broderick, J. P. (2017). Stroke recovery and rehabilitation research. *Stroke, 48*(3), 813–819. https://doi.org/10.1161/STROKEAHA.116.015501

Das, R., Paul, S., Mourya, G. K., Kumar, N., & Hussain, M. (2022). Recent trends and practices toward assessment and rehabilitation of neurodegenerative disorders: Insights from human gait. *Frontiers in Neuroscience, 16*, 859298. https://doi.org/10.3389/fnins.2022.859298

Deal, L. S., Flood, E., Myers, D. E., Devine, J., & Gray, D. L. (2019). The Parkinson's disease activities of daily living, interference, and dependence instrument. *Movement Disorders Clinical Practice, 6*(8), 678. https://doi.org/10.1002/mdc3.12833

Dementia with Lewy Bodies—Symptoms and Causes. (n.d.). Penn Medicine. Retrieved November 29, 2024, from https://www.pennmedicine.org/for-patients-and-visitors/patient-information/conditions-treated-a-to-z/dementia-with-lewy-bodies

Frontotemporal dementia. (n.d.). Retrieved November 29, 2024, from https://stanfordhealthcare.org/medical-conditions/brain-and-nerves/dementia/types/frontotemporal-dementia.html

Gerstenecker, A., & Lazar, R. M. (2019). Language recovery following stroke. *The Clinical Neuropsychologist, 33*(5), 928–947. https://doi.org/10.1080/13854046.2018.1562093

Gonçalves, P. D., Ometto, M., Sendoya, G., Lacet, C., Monteiro, L., & Cunha, P. J. (2014). Neuropsychological rehabilitation of executive functions: Challenges and perspectives. *Journal of Behavioral and Brain Science, 2014*, 27–32. https://doi.org/10.4236/jbbs.2014.41004

Harding, R., Gao, W., Jackson, D., Pearson, C., Murray, J., & Higginson, I. J. (2015). Comparative analysis of informal caregiver burden in advanced cancer, dementia, and acquired brain injury. *Journal of Pain and Symptom Management, 50*(4), 445–452. https://doi.org/10.1016/j.jpainsymman.2015.04.005

Huntington's Disease—Symptoms and Causes | Penn Medicine. (n.d.). Retrieved December 1, 2024, from https://www.pennmedicine.org/for-patients-and-visitors/patient-information/conditions-treated-a-to-z/huntingtons-disease-hd

Hussain, R., Zubair, H., Pursell, S., & Shahab, M. (2018). Neurodegenerative diseases: Regenerative mechanisms and novel therapeutic approaches. *Brain Sciences, 8*(9), 177. https://doi.org/10.3390/brainsci8090177

Lewy Body Dementia. (n.d.). National Institute of Neurological Disorders and Stroke. Retrieved November 29, 2024, from https://www.ninds.nih.gov/health-information/disorders/lewy-body-dementia

Loetscher, T., Potter, K., Wong, D., & das Nair, R. (2019). Cognitive rehabilitation for attention deficits following stroke. *The Cochrane Database of Systematic Reviews, 2019*(11), CD002842. https://doi.org/10.1002/14651858.CD002842.pub3

Maas, A. I. R., Menon, D. K., Manley, G. T., Abrams, M., Åkerlund, C., Andelic, N., Aries, M., Bashford, T., Bell, M. J., Bodien, Y. G., Brett, B. L., Büki, A., Chesnut, R. M., Citerio, G., Clark, D., Clasby, B., Cooper, D. J., Czeiter, E., Czosnyka, M., et al. (2022). Traumatic brain injury: Progress and challenges in prevention, clinical care, and research. *The Lancet. Neurology, 21*(11), 1004–1060. https://doi.org/10.1016/S1474-4422(22)00309-X

Madigan, N. K., DeLuca, J., Diamond, B. J., Tramontano, G., & Averill, A. (2000). Speed of information processing in traumatic brain injury: Modality-specific factors. *The Journal of Head Trauma Rehabilitation, 15*(3), 943.

Min, J. H., & Shin, Y.-I. (2022). Treatment and rehabilitation for traumatic brain injury: Current update. *Brain & NeuroRehabilitation, 15*(2), e14. https://doi.org/10.12786/bn.2022.15.e14

Mroczkowska, D., Bałdyga, E., Żechowicz, T., & Gromadziński, L. (2019). Neuropsychological rehabilitation of patients with symptoms of depression after ischemic stroke. *Polish Annals of Medicine, 26*(2), 130–133. https://doi.org/10.29089/22019.19.00082

National Academies of Sciences, E, Division, H. and M, Services, B. on H. C, Policy, B. on H. S, Care, C. on A. P. in T. B. I. R, Matney, C., Bowman, K., & Berwick, D. (2022). Rehabilitation and long-term care needs after traumatic brain injury. In *Traumatic brain injury: A roadmap for accelerating progress*. National Academies Press. https://www.ncbi.nlm.nih.gov/books/NBK580075/

Navarro, M. D., Llorens, R., Borrego, A., Alcañiz, M., Noé, E., & Ferri, J. (2020). Competition enhances the effectiveness and motivation of attention rehabilitation after stroke. A randomized controlled trial. *Frontiers in Human Neuroscience, 14*, 575403. https://doi.org/10.3389/fnhum.2020.575403

Ozga, J. E., Povroznik, J. M., Engler-Chiurazzi, E. B., & Haar, C. V. (2018). Executive (dys) function after traumatic brain injury: Special considerations for behavioral pharmacology. *Behavioural Pharmacology, 29*(7), 617–637. https://doi.org/10.1097/FBP.0000000000000430

Parkinson's Disease—Symptoms and Causes | Penn Medicine. (n.d.). Retrieved December 1, 2024, from https://www.pennmedicine.org/for-patients-and-visitors/patient-information/conditions-treated-a-to-z/parkinsons-disease

Rath, J. F., Hradil, A. L., Litke, D. R., & Diller, L. (2011). Clinical applications of problem-solving research in neuropsychological rehabilitation: Addressing the subjective experience of cognitive deficits in outpatients with acquired brain injury. *Rehabilitation Psychology, 56*(4), 320–328. https://doi.org/10.1037/a0025817

Richter, S. (2017). Integrated care for people with dementia—Results of a social-scientific evaluation of an established dementia care model. *Geriatrics, 2*(1), 1. https://doi.org/10.3390/geriatrics2010001

Shah, S. A., Goldin, Y., Conte, M. M., Goldfine, A. M., Mohamadpour, M., Fidali, B. C., Cicerone, K., & Schiff, N. D. (2017). Executive attention deficits after traumatic brain injury reflect impaired recruitment of resources. *NeuroImage: Clinical, 14*, 233–241. https://doi.org/10.1016/j.nicl.2017.01.010

Singh, M. K., Shin, Y., Ju, S., Han, S., Kim, S. S., & Kang, I. (2024). Comprehensive overview of Alzheimer's disease: Etiological insights and degradation strategies. *International Journal of Molecular Sciences, 25*(13), 13. https://doi.org/10.3390/ijms25136901

Soule, A. C., Fish, T. J., Thomas, K. G. F., & Schrieff-Brown, L. (2024). Attention training after moderate-to-severe traumatic brain injury in adults: A systematic review. *Archives of Physical Medicine and Rehabilitation, 16*, 433. https://doi.org/10.1016/j.apmr.2024.07.010

Soule, A. C., Fish, T., Winegardner, J., & Schrieff-Brown, L. (2024). Implementing neuropsychological rehabilitation following severe traumatic brain injury in a low-to-middle income country: A case report. *Frontiers in Rehabilitation Sciences, 5*, 1393302. https://doi.org/10.3389/fresc.2024.1393302

Torregrossa, W., Torrisi, M., De Luca, R., Casella, C., Rifici, C., Bonanno, M., & Calabrò, R. S. (2023). Neuropsychological assessment in patients with traumatic brain injury: A comprehensive review with clinical recommendations. *Biomedicine, 11*(7), 1991. https://doi.org/10.3390/biomedicines11071991

Traumatic Brain Injury (TBI). (2024). National Institute of Neurological Disorders and Stroke. https://www.ninds.nih.gov/health-information/disorders/traumatic-brain-injury-tbi

Traumatic Brain Injury Symptoms & Causes | AllSpine. (2024, February 1). https://allspinesurgerycenter.com/traumatic-brain-injury-symptoms-causes/

Types of Strokes and Treatment. (n.d.). Brigham and Women's Hospital. Retrieved November 20, 2024, from https://www.brighamandwomens.org/neurosurgery/cerebrovascular/stroke/types

Understanding Different Types of Dementia. (n.d.). National Institute on Aging. Retrieved November 20, 2024, from https://www.nia.nih.gov/health/alzheimers-and-dementia/understanding-different-types-dementia

Vascular Dementia. (2024, May 20). https://www.hopkinsmedicine.org/health/conditions-and-diseases/dementia/vascular-dementia

Vascular Dementia: Causes, Symptoms, and Treatments | National Institute on Aging. (n.d.). Retrieved November 29, 2024, from https://www.nia.nih.gov/health/vascular-dementia/vascular-dementia-causes-symptoms-and-treatments

What Are Frontotemporal Disorders? Causes, Symptoms, and Treatment. (2021, July 30). National Institute on Aging. https://www.nia.nih.gov/health/frontotemporal-disorders/what-are-frontotemporal-disorders-causes-symptoms-and-treatment

What Is Alzheimer's Disease? (2021, July 8). National Institute on Aging. https://www.nia.nih.gov/health/alzheimers-and-dementia/what-alzheimers-disease

What Is Lewy Body Dementia? Causes, Symptoms, and Treatments. (2021, July 29). National Institute on Aging. https://www.nia.nih.gov/health/lewy-body-dementia/what-lewy-body-dementia-causes-symptoms-and-treatments

Xu, X.-D., Ren, H.-Y., Prakash, R., Vijayadas, & Kumar, R. (2013). Outcomes of neuropsychological interventions of stroke. *Annals of Indian Academy of Neurology, 16*(3), 319–328. https://doi.org/10.4103/0972-2327.116909

Chapter 3
Traditional Approaches and Limitations

3.1 Neurocognitive Rehabilitation

3.1.1 Introduction

Neuropsychology and its philosophical underpinnings have significantly contributed to the evolution and origins of neurocognitive rehabilitation. Early discoveries of mental functions associated with the brain were found to have dated back to papyrus descriptions in Egyptian civilization. The Hippocratic corpus laid the foundations to understand the brain and its central functions in mental processes. Franz Gall's localization theory was introduced through the specialization of brain regions for independent cognitive functions. These insights were further refined during the "golden age" of neuropsychology by Broca (Broca's aphasia), Wernicke (Wernicke's aphasia), and Jackson all of whom studied the brains of their patients posthumously through autopsy findings that were correlated with their clinical observations. The discoveries sparked discussions about the relationship between brain injuries and behavioral and functional disabilities. These frameworks further contributed to the designing of cognitive evaluations and exercises (Gurd, 2010).

The rapid development of neuropsychology during World Wars I and II had the significant influence of moving the approach to brain damage and defects from that which the elderly population experienced to the young survivors of war. A large population that was observed to be healthy, young adults were then impacted by brain injuries and other comorbid deficits. This created an urgency in introducing standardized testing and assessments to understand and localize the area of cognitive impairment. The process of testing evolved to rehabilitation; clinics by physicians such as Hermann Gutzmann in Berlin introduced the necessity to collaborate with speech and language therapists and psychologists—introducing the new subfield of clinical neuropsychology (Gurd, 2010).

K. J. Reddy, *Innovations in Neurocognitive Rehabilitation*,
https://doi.org/10.1007/978-3-031-88117-6_3

Pioneers in rehabilitation shifted their focus from assessment and impairment-related life adjustments to actively engaging with patients through holistic paradigms. The concept of "whole person" was introduced to complement cognitive rehabilitation with behavioral aspects. Luria was a pathfinder in neuropsychological rehabilitation focusing on introducing concepts of functional systems that integrated characteristics of the human brain with psychological functions. The systematic method to understand the interrelationship between behavioral symptoms and cognitive processes reinforced the importance of individualized rehabilitation strategies. Continued contributions of Leontiev's activity theory and models of cognitive plasticity have led to the development of cognitive plasticity and neurocognitive rehabilitation. Cultural and social factors were combined to customize rehabilitation for different individuals within the contemporary context (Gurd, 2010; Puderbaugh & Emmady, 2024).

> Rehabilitation is the active participation of a disabled person and others to reduce the impact of disease and disability on daily life. (Greenwood, 2003)

Understanding neurocognitive rehabilitation is essential for the innovation and application of technology. Rehabilitation in the contemporary context has strong associations with the physiological and psychological interventions. These are two approaches to rehabilitation, the biological approach focuses on the imbalance and abnormality in body functions that causes an impairment in the individual's capacity to function. The biological approach focuses on the impairment through external factors including trauma to the brain. The psychological approach relies on the impact of psychological factors that increase the rate of impairment. Psychological factors include stress, worry, and motivation that may be hindered by external factors. Psychological factors significantly contribute to the degree of awareness of the illness. Traditional approaches to neurocognitive rehabilitation discuss the importance of neurocognitive training in mitigating further psychological and biological abnormalities through interventions and medication. The attributes of rehabilitation focuses on the differences and similarities in patients; the active participation of the patient and the caregiver and in observing the outcomes of rehabilitation through time-limited interventions. The process of rehabilitation is multidisciplinary and requires a varied team of experts—nurses, therapists, clinical neuropsychologists, social workers, and caregivers—to name a few. Traditional approaches to rehabilitation focus on the impact of rehabilitation on the changes in behavior (motor behavior) and cognition (attention, concentration, memory, executive functions, thinking, and reasoning) (Greenwood, 2003).

3.2 Traditional Approaches

3.2.1 Behavioral Therapy

Behavioral therapy focuses on the externally observable motor modifications. Despite the abstract nature of cognitive processes, the theoretical constructs of behavior therapy has been to have successful outcomes in prioritizing skills in an observable and quantifiable manner. Observable behavior includes attention span, memory recall, and ability to execute instructions. This provides clarity and the ability to manage the course and outcomes of the treatment process. Some theoretical frameworks within the behavioral context are reinforcement, extinction, and behavior shaping. Reinforcement training is founded on the works of Skinner's operant conditioning. The introduction of positive reinforcement for expected outcomes and negative reinforcement for negative outcomes are enablers of recovery in rehabilitation. Tasks are reinforced by verbal praise or appreciation, and rewards to maintain the motivation of the individual. Time-based reinforcement helps increase the intensity of rehabilitation and quantifiability of the outcomes. Extinction works through preventing the reinforcement of undesirable behaviors. The effectiveness of maladaptive behaviors are combated through challenging the undesirable behaviors and negative reinforcement. Behavior shaping reduces complex tasks to simple goals that are achieved sectionally. This approach helps identify problems that limit training within each component and builds the motivation of the individual (Buccilli et al., 2024; Jennings & Jacoby, 2024). In case of conditions such as stroke, traumatic brain injury (TBI), and other cognitive impairments, rehabilitation through errorless learning strategies, metacognitive training, problem-solving training activities, and self-regulation are approaches that reduce the overt behavioral symptoms (Buccilli et al., 2024; Gonçalves et al., 2014; Wood & Alderman, 2024).

3.2.2 Cognitive Retraining Techniques

Cognitive retraining programs have evolved to focus on enhancing the quality of life and the well-being of patients. The programs are aimed at targeting neurological conditions, aging-related cognitive decline, or psychiatric impairments and in improving memory, attention, and executive functioning. Some of the principles of cognitive retraining include—personalization, structure, and goal orientation. Cognitive retraining programs are designed to strengthen functional skills—attention and encoding. The primary focus of cognitive retraining is strengthening neural pathways. The scaffolding approach introduces basic skills and gradually progresses in intensity and complexity. Sohlberg and Mateer's process-oriented approach model is the foundation of cognitive retraining. One of the most important process aspects of process-oriented approach is attention process training (APT) program focuses on attentional training: sustained, selective, alternating, and divided attention with structured repetition. Studies indicate that APT can lead to observable improvements in previously untrained tasks. Recent modifications of this process-oriented approach includes strategies to monitor attention and improve self-control and cognitive challenges. Memory retraining indicates to be less successful—until a paradigm shift introduced by Jennings and Jacoby showed promise. This model's approach to recollection training improves effortful memory for contextual details by leveraging intact familiarity memory. This involves tasks such as—word repetition, recognition, word recollection—through cues. Results indicated significant improvements in memory tasks like n-back tests. Early studies suggest that this type of training could also benefit people with brain injuries, although more research is needed. These findings highlight the potential of cognitive retraining to support individuals in overcoming attention and memory challenges (Cicerone et al., 2019; Greenwood, 2003; Kallio et al., 2017).

3.2.3 Physical and Occupational Therapy

Neurological rehabilitation has evolved from a focus on compensatory techniques and passive approaches to treatment to incorporating recovery, multidisciplinarity, and functional improvement. Since early schools of thought believed that an insult to the brain areas indicated permanent deficits and complete loss of functions, the approaches to therapy focused on medication, passive exercises, massages, and electrotherapy. However, these methods only led to long-term disabilities. The mid-twentieth century introduced alterations in early approaches. Bobath focused on facilitating "normal movement patterns" such as sitting, standing, posture, and walking. These techniques were practiced with a goal of improving neuronal reflexes and muscular contractions while enabling balance and development of childhood neurological deficits in cases of cerebral palsy and stroke to encourage simple actions such as rolling over and crawling. The methods were further developed to work in the contexts of adults who suffered neurodegeneration and deficits (Avila et al., 2004; Elera, 2024; Greenwood, 2003).

Physical and occupational therapy are vital in the overall improvement of the patient. These therapies are incorporated to combine physical activities and mental tasks in a manner that would improve cognitive and motor coordination. Physical therapy and neurorehabilitation for athletes have seen significant growth in recent years, with a notable increase in publications since 2017 highlighting the trend towards utilizing cutting-edge technologies like robotics and virtual reality. This shift reflects the evolving understanding of the complex interplay between physical and cognitive functions in recovery processes. Early neuropsychological rehabilitation in the area of TBI, stroke, and dementia-treated physical challenges as independent of the deficit. The course of evolution has identified ways in which incorporating motor activities can improve on the physical functions of the patient. For example, tasks related to attention, concentration, and memory retrieval can be applied through recollect motor movements individuals observe, retain, and further execute.

Mirroring activities also encourage speedy recovery of lost physical and occupational skills—by the simple act of observing the therapist/caregiver performing the task, the individual can perform the task through repetitive observations that the mirror neuron stores. Asking the patient to recollect dance sequences would require the patient to engage in complex cognitive-motor tasks such as planning and sequencing (Day et al., 2016; Greenwood, 2003).

Consistency is crucial in physical and occupation therapy. Intensive rehabilitation between 2 and 4 h everyday would be more effective in improving the lost functions or in the compensatory approach. Early rehabilitation in physical and occupational therapy displays stronger recovery of lost functions over later rehabilitation. The long-term goals of therapy would be to observe maintenance of trained functions and reintegration of the patient into occupational and social settings (Oberholzer & Müri, 2019). Recent advancements in rehabilitation have shifted gaze to task-specific and goal-oriented approaches such as motor learning to improve activities of daily living (ADL) therapy. Incorporating biomechanics to provide feedback on cognitive and behavioral factors reinforces the importance of consistent practice for neuroplasticity and recovery (Greenwood, 2003).

3.2.4 Neuropsychological Assessments

Neuropsychological assessments are foundational in diagnosing, treating, and supporting individuals with epilepsy and brain injuries, even in cases that might initially appear mild. The assessments are important in all cases of neurodegeneration and other neurological conditions. The assessments provide clarity in the areas of cognitive deficits—attention (selective, sustained), concentration, memory (immediate, working, long-term), reasoning, thinking, language, and other executive functions. The process of testing requires standardization and integrating patient history in order to effectively choose the assessment batteries. The information obtained from testing is crucial in designing rehabilitation and in providing assisted interventions in other disciplines (Lassonde et al., 2006; Lezak et al., 2012).

Pertaining to TBIs, early neuropsychological evaluation is essential to identify the prognosis of the injury. The assessment helps assess the degree of damage to the brain and the possible limitations in personal and professional life. Additionally, the order of interventions is prioritized on the basis of the significance and urgency of the cognitive impairment. Since recovery is focused on preventing long-term challenges at work and daily life, there is a dire need to reduce the risk of choosing incorrect rehabilitation strategies (Oberholzer & Müri, 2019). Assessments used within this context include WAIS-IV, MMSE, and Rey Complex Figure Test (RCFT). Table 3.1 provides a brief detail of the assessments used and its applications.

While all the assessments provide an overview of the cognitive functions, it is important to choose the appropriate measures to understand the areas and subsections of cognitive deficits. Assessments enable early detection, monitors progress, informs the differential diagnosis such as subtypes of dementia, stroke, and TBI,

Table 3.1 Assessments and measures used in neuropsychological rehabilitation

Assessment and measures	Functions	Age range	Applications
Case history and mental status examination	To obtain a complete clinical picture of the patient	It is not dependent on age as it is on the levels of consciousness and insight. Children who can articulate to older adults are supplemented with a case history. In case of patient limitations, the caregiver provides key information.	To understanding the pre-deficit/injury functioning—cognitive, affective, and motor; to estimate the prognosis and social support that is accessible to the individual. Developmental and case history questionnaire is administered.
MMSE (Mini-Mental State Examination) (Bennett, 2001)	Brief screening tool used to detect areas of cognitive impairment	Age 18 yrs. and above	Assesses orientation, attention, memory, language, and visuospatial skills. Administered in clinical settings due to its preciseness and ease of administration.
WAIS-IV (Wechsler Adult Intelligence Scale—Fourth Edition) (Torregrossa et al., 2023)	Assess cognitive functions—attention, executive functions, memory, language, and visuospatial abilities	Age 16 yrs. and above	Evaluates the severity of impairment and the levels of cognitive difficulty—working memory, language, and visuospatial skills.
Rey Complex Figure Test (ROCF) (X. Zhang et al., 2021)	Assess visuo-constructional ability and visual memory. Also assesses attention, concentration, fine motor coordination, visuospatial perception; nonverbal memory—Planning, spatial orientation.	Ages 5–14 yrs	Applied to understand cognitive dysfunction and other neurological conditions such as poststroke dementia.

helps design personalized rehabilitation plans and aids in research. Through assessments, the therapist is able to provide accurate details about the patient's condition and the possible challenges and limitations in later life cognitive functions (Lassonde et al., 2006; Oberholzer & Müri, 2019; Zhang et al., 2021).

3.3 Strengths of Traditional Approaches

1. Established Frameworks with Evidence-Based Efficacy:

Evidence-based interventions require scientific knowledge that grounds the research in empirical findings. Designing a rehabilitation plan is essential in understanding the evidence-based interventions that appraises the efficacy of the intervention. Tailoring rehabilitation in lieu of cognitive deficits and individual requirements and utilizing a personalized approach relies on the empirical evidence obtained from research. Personalization reduces the time required to engage in cognitive training of unaffected areas. Evidence-based approach improves patient outcomes and targets independent functional capacity of the individual. Standardization of measures and interventions improve on culture-specific requirements. Evidence-based practices reduce the risk of interventional failure and side-effects further holding the rehabilitation team accountable for collaboration and interdisciplinary planning of interventional training (Cicerone et al., 2019; De Luca et al., 2018).

2. Universal Application Across Diverse Patient Populations:

Other comorbidities include psychological illness such as depression, anxiety, and psychosis. Interventions must provide holistic programs to improve cognitive capacity.

Interventions are only effective if cognitive rehabilitation is customizable. Interventions are tailored to the degree of severity, the area of injury and the consequential cognitive impairment. In cases of TBI, the location, area(s) of impact, the severity of the insult and the hierarchy of interventional urgency differs from one patient to another. Comorbidities with this psychological illness such as depression, anxiety, and psychosis are components that need to be addressed during the process of rehabilitation to reduce its effects on rehabilitation outcomes. Addressing multiple areas of improvement and shifting gaze to multidisciplinary approaches and integrating different theoretical frameworks to compensate different needs are essential in planning and implementing rehabilitation (Lassonde et al., 2006).

3. Cost-Effective and Accessible for Initial Rehabilitation:

Established traditional interventions call for changes in cost-effectiveness of the interventions further focusing on speedy recovery and reducing long-term care. A major limitation of rehabilitation is the inconvenience of obtaining care that is accessible and affordable for patients. Patients often lose motivation to continue treatment due to the cost of commute, appointments, and other related expenses. Focusing on brief screening tools (as mentioned in Table 3.1) would be vital in aiding early intervention, limit the costs incurred and applying tele-rehabilitation can increase the cost-effective care to improve accessibility. A lack of accessibility and well-trained professionals are additional barriers that could be mitigated for cost-effectiveness (Langenbahn et al., 2012).

4. Well-Suited for Acute and Chronic Conditions in Controlled Settings:

The high adaptability of neurocognitive rehabilitation has proven to be effective in addressing patients from diverse backgrounds in terms of the severity of their impairment. An essential aspect of ensuring positive outcomes are obtained after rehabilitation, the course, intensity, and duration of care plays a significant role. Interventional programs are designed to address the unique needs and allow for personalized interventions that evolve with the patient's progress. The aim is to improve everyday functioning by equipping patients with practical strategies that improve on everyday activities. The evidence-based approach ensures meaningful, accessible, and impactful recovery and promotes long-term independence.

3.4 Limitations of Traditional Approaches

Multiple factors affect the process of rehabilitation. The lack of personalization, gradual recovery, the time intensity, and accessibility are few key issues with the traditional approaches to rehabilitation.

1. Lack of Personalization:

Rehabilitation planning heavily relies on cognitive determinants of impairment. The area of impairment, the size of impact, the type of lesion can vary for different patients. The levels of consciousness—coma, partial loss of consciousness, altered states of consciousness; the level of memory—anterograde, retrograde amnesia, inability to retain information; and language impairment—aphasia, inability to form words and sentences, visuospatial impairment, motor skill deficits area crucial in rehabilitation. One of the key factors in rehabilitation is to identify the areas of impact and further prioritize the rehabilitation. Personalization has additional areas such as: assessing the cognitive and behavioral functioning of the individual such as attention, memory, language functioning, visual-spatial skills, motor skills, problem-solving/mental flexibility, information processing speed (Gurd, 2010; Langenbahn et al., 2012; Wilson, 2020).

2. Slow Progress and Time-Intensive:

The process of rehabilitation is gradual and the time taken to complete rehabilitation generally fits between 6 months and 1 year—depending on the severity of illness. A significant factor that limits traditional approaches to therapy is the loss of motivation and curiosity in task training. One method to mitigate this issue was through increasing the progressive load of the task along with the intensity. Despite much of these efforts it becomes increasingly difficult to maintain the consistency and enthusiasm that the patient experiences (Gurd, 2010; Langenbahn et al., 2012; Wilson, 2020).

3. Inadequate Address of Complex Needs:

Formulating cognitive training and rehabilitation heavily depends on the social and cultural factors. The cultural factors involved in this aspect include the beliefs and practices of the individual, the diets that they can consume, the lifestyle changes that they would have to incorporate within the purview of their culture. Another factor is the demographic background of the individual—the age, personality, prior levels of functioning, the family system that they live in, and the income status of the individual. Other factors such as families' awareness of the disease, the levels of education and ability to cope with these behaviors play a vital role in deciding the course of rehabilitation. Aside from the patient's personal limitations, the motivation and the efforts taken to ensure the rehabilitation is consistent comes from the caregivers. Should the caregivers lack time, experience burnout or financial constraints, the process of rehabilitation becomes limited (Gurd, 2010; Langenbahn et al., 2012; Wilson, 2020).

4. Accessibility Issues:

In some cases where the patient seeks rehabilitation by commuting long distances to access support, the neuropsychologist must identify strategies that might be applicable during their time away from the rehabilitation facility. The lack of resources acts as a significant barrier in the process of rehabilitation. Insufficient access to facilities, the lack of proper transportation, and other personal challenges can act as barriers. Another requirement is the accessibility of resource-intensive rehabilitation that requires highly specialized professionals. Furthermore, the controlled environments of the rehabilitation spaces can limit comprehending challenges in real-world settings which can limit the safety and repetition of rehabilitation. Certain functions such as driving, cooking, navigation can be limited due to the closed environment of rehabilitation facilities (Gurd, 2010; Langenbahn et al., 2012; Wilson, 2020).

3.5 Challenges in Implementation

As detailed in accessibility issues, a primary factor in the poor rehabilitation is the increased dependence of the expertise of the therapist and the availability as per their schedule. In cases of expertise—there is a high likelihood that most well-qualified therapists are easily accessible within urban space in comparison to rural areas. This would require that the patients be the expert in their urban spaces or move to an urban space during the course of rehabilitation. The financial burden incurred during the rehabilitation—medication, consultation fees, training fees, accommodation, and transportation are few aspects that require large sums of money. Often the rehabilitation process dwindles in consistency as the affordability of care weighs a significant burden on the caregiver and the family. Another barrier is the lack of interdisciplinary collaboration during the process of rehabilitation which often limits the training and recovery to certain cognitive and physiological functions. Several patients report significant changes in affect either by environmental factors or social reasons. Various studies report on the significant rise in the levels of depression, stroke, anxiety, and psychosis during the course of treatment.

Patients often develop comorbidities with depression and psychosis. This would indicate the need for counselling and therapy. Integration of AI, VR, wearable technology, and neurofeedback—biofeedback methods could improve the motivation levels and provide real-time feedback to the patients. There is a need to improve the inclusivity of personalized rehabilitation strategies to address the wide variety of patients and conditions. As discussed in the limitations of traditional approaches, there is a need to increase the accessibility to different resources and planning the interventions on the basis of need and priority (Sohlberg & Mateer, 2024; J. Zhang & He, 2023).

3.6 Emergent Needs in Neurocognitive Rehabilitation

While traditional approaches are strongly grounded in empirical evidence, its many limitations call for holistic and adaptive approaches. The traditional approach to neurocognitive rehabilitation has strict frameworks and approaches to ensure that the outcomes are measurable and long-term. However, each theoretical approach presents barriers: the behavioral approach heavily relies on the motor behavioral outcomes and the interventions focus on the same. Whereas the cognitive approach largely prioritizes the cognitive functions such as attention, memory, and perception. Physical and occupational therapy addresses practical challenges in reintegration to work spaces and social activities. These approaches independently would be ineffective but a combination of these approaches would focus on the holistic approach that would encourage rehabilitation planning that influences multiple areas of the patient's life. Consecutively, there is a need to include other areas and experts in the field such as psychologists (to address mental health comorbidities), social workers (to adjust into community life and to provide representation in case of financial issues and psychoeducation), and computer science and neuroscience researchers—in order to integrate technology to improve the effectiveness of rehabilitation and to increase the accessibility (Gonçalves et al., 2014).

Recent advancements in technology have been observed to have revolutionized behavioral therapy and cognitive retraining. The utilization of innovative technological methods and applying safe-practice scenarios through simulations bridges the gap between controlled rehabilitation environments and real-world implications. Virtual reality creates immersive environments where individuals can navigate social situations and cognitive-behavioral training with active feedback and immersive experiences. Integrating biofeedback sensors with wearable technology is a groundbreaking step in monitoring physiological and behavioral responses continuously. The degree of feedback enabling fine-tuning interventions ensures that the responses are effective. AI and machine learning is applied in patient progress and personalization. The ability to track improvements over time has provided large data that promotes research and further modifications of the technology. Application of neurofeedback technology in personalizing brain activity and enabling target-specific tasks for cognitive and emotional deficits. Application of transcranial magnetic stimulation (TMS), a noninvasive technique to stimulate neural activity and recovery, further broadens the scope of therapeutic options. The advancements have increased the effectiveness of rehabilitation and improved accessibility and engagement. Modern advancements in technology have innovations that represent a shift in traditional approaches to therapy that combine cutting-edge technology with customized care to enhance recovery and subsequently the quality of life of the patients (Ho et al., 2022; Jamieson et al., 2015; Jennings & Jacoby, 1993; Kallio et al., 2017).

3.7 Conclusion

Neurocognitive rehabilitation is developed through foundational knowledge from traditional methods. Traditional approaches apply structured testing environments to identify specific weaknesses and further direct rehabilitation and therapy. The effectiveness in controlled settings inadequately identifies real-world functionality. For example, individuals who perform well on tests may still experience significant difficulties in dynamic, unstructured situations. This further highlights the importance of aligning assessments and training with real-world settings to ensure that interventions meaningfully influence everyday functions. A crucial aspect of neuropsychological rehabilitation is the development of compensatory strategies to aid patients with cognitive challenges. Strategies can improve test scores but do not indicate that true and complete recovery of cognitive functions is possible. The long-term viability of these tasks do not indicate significant improvements. Measures and sustainable cognitive improvements are essential in advancing neuroimaging and ecological assessments further helping researchers evaluate effective interventions more comprehensively. The tools aim to bridge the gap between clinical progress and real-world outcomes ensuring that the outcomes of therapy are long endured. Future research on the long-term efficacy of rehabilitation strategies emphasizes the importance of integrating learned skills in daily routines further providing continued support to help individuals regain independence and functionality. While traditional approaches provide a solid foundation, modern technologies and real-world validation are essential in ensuring that the research is compatible with the evolution of a world where all aspects of care are re-worked to maximize outcomes. Integrating technology and improving on traditional approaches are vital in ensuring that rehabilitation remains compatible and aligns with the present time.

References

Avila, R., Bottino, C. M. C., Carvalho, I. A. M., Santos, C. B., Seral, C., & Miotto, E. C. (2004). Neuropsychological rehabilitation of memory deficits and activities of daily living in patients with Alzheimer's disease: A pilot study. *Brazilian Journal of Medical and Biological Research = Revista Brasileira De Pesquisas Medicas E Biologicas, 37*(11), 1721–1729. https://doi.org/10.1590/s0100-879x2004001100018

Bennett, T. L. (2001). Neuropsychological evaluation in rehabilitation planning and evaluation of functional skills. *Archives of Clinical Neuropsychology, 16*(3), 237–253. https://doi.org/10.1016/S0887-6177(00)00082-2

Buccilli, B., Alan, A., Baha', A., Shahzad, A., Almealawy, Y. F., Chisvo, N. S., Ennabe, M., & Weinand, M. (2024). The importance of behavioral interventions in traumatic brain injury. *Surgical Neurology International, 15*, 22. https://doi.org/10.25259/SNI_776_2023

Cicerone, K. D., Goldin, Y., Ganci, K., Rosenbaum, A., Wethe, J. V., Langenbahn, D. M., Malec, J. F., Bergquist, T. F., Kingsley, K., Nagele, D., Trexler, L., Fraas, M., Bogdanova, Y., & Harley, J. P. (2019). Evidence-based cognitive rehabilitation: Systematic review of the literature from 2009 through 2014. *Archives of Physical Medicine and Rehabilitation, 100*(8), 1515–1533. https://doi.org/10.1016/j.apmr.2019.02.011

Day, J., Gillespie, D., Rooney, A., Bulbeck, H., Zienius, K., Boele, F., & Grant, R. (2016). Neurocognitive deficits and neurocognitive rehabilitation in adult brain tumors. *Current Treatment Options in Neurology, 18*, 1–16. https://doi.org/10.1007/s11940-016-0406-5

De Luca, R., Calabrò, R. S., & Bramanti, P. (2018). Cognitive rehabilitation after severe acquired brain injury: Current evidence and future directions. *Neuropsychological Rehabilitation, 28*(6), 879–898. https://doi.org/10.1080/09602011.2016.1211937

Elera, C. M. D. (2024, January 29). *Neuropsychological rehabilitation in older adults with dementia: Experience in a day care center with NeuronUP*. NeuronUP. https://neuronup.us/cognitive-stimulation-news/testimonials/neuropsychological-rehabilitation-in-older-adults-with-dementia-experience-in-a-day-care-center-with-neuronup/

Gonçalves, P. D., Ometto, M., Sendoya, G., Lacet, C., Monteiro, L., & Cunha, P. J. (2014). Neuropsychological rehabilitation of executive functions: Challenges and perspectives. *Journal of Behavioral and Brain Science, 2014*, 27–32. https://doi.org/10.4236/jbbs.2014.41004

Greenwood, R. (Ed.). (2003). *Handbook of neurological rehabilitation* (2nd ed.). Psychology Press.

Gurd, J. (with Kischka, U., & Marshall, J.). (2010). The handbook of clinical neuropsychology. Oxford University Press, Incorporated.

Ho, H.-Y., Chen, M.-D., Tsai, C.-C., & Chen, H.-M. (2022). Effects of computerized cognitive training on cognitive function, activity, and participation in individuals with stroke: A randomized controlled trial. *NeuroRehabilitation, 51*(1), 79–89. https://doi.org/10.3233/NRE-210271

Jamieson, M., Cullen, B., McGee-Lennon, M., Brewster, S., & Evans, J. (2015). Technological memory aid use by people with acquired brain injury. *Neuropsychological Rehabilitation, 27*(6), 919–936. https://doi.org/10.1080/09602011.2015.1103760

Jennings, J. M., & Jacoby, L. L. (1993). Automatic versus intentional uses of memory: Aging, attention, and control. *Psychology and Aging, 8*(2), 283–293. https://doi.org/10.1037/0882-7974.8.2.283

Jennings, J. M., & Jacoby, L. L. (2024). Improving memory in older adults: Training recollection. *Neuropsychological Rehabilitation, 13*(4), 417–440. https://doi.org/10.1080/09602010244000390

Kallio, E.-L., Öhman, H., Kautiainen, H., Hietanen, M., & Pitkälä, K. (2017). Cognitive training interventions for patients with Alzheimer's disease: A systematic review. *Journal of Alzheimer's Disease, 56*(4), 1349–1372. https://doi.org/10.3233/JAD-160810

Langenbahn, D., Ashman, T., Cantor, J., & Trott, C. (2012). An evidence-based review of cognitive rehabilitation in medical conditions affecting cognitive function. *Archives of Physical Medicine and Rehabilitation, 94*, 271. https://doi.org/10.1016/j.apmr.2012.09.011

Lassonde, M., Sauerwein, H. C., Gallagher, A., Thériault, M., & Lepore, F. (2006). Neuropsychology: Traditional and new methods of investigation. *Epilepsia, 47*(2), 9–13. https://doi.org/10.1111/j.1528-1167.2006.00680.x

Lezak, M. D., Howieson, D. B., Bigler, E. D., Tranel, D., Lezak, M. D., Howieson, D. B., Bigler, E. D., & Tranel, D. (2012). *Neuropsychological assessment*. Oxford University Press.

Oberholzer, M., & Müri, R. M. (2019). Neurorehabilitation of traumatic brain injury (TBI): A clinical review. *Medical Science, 7*(3), 47. https://doi.org/10.3390/medsci7030047

Puderbaugh, M., & Emmady, P. D. (2024). Neuroplasticity. In *StatPearls*. StatPearls Publishing. http://www.ncbi.nlm.nih.gov/books/NBK557811/

Sohlberg, M. M., & Mateer, C. A. (2024). *Cognitive rehabilitation, an integrative neuropsychological approach* (p. 492). Guilford Press. https://www.researchgate.net/publication/247695476_Cognitive_rehabilitation_an_integrative_neuropsychological_approach_Edited_by_M_M_Sohlberg_and_C_A_Mateer_Pp_492_4195_Guilford_Press_New_York_2001_ISBN_1_57230_613_0

Torregrossa, W., Torrisi, M., De Luca, R., Casella, C., Rifici, C., Bonanno, M., & Calabrò, R. S. (2023). Neuropsychological assessment in patients with traumatic brain injury: A comprehensive review with clinical recommendations—PMC. *Biomedicine, 11*(7), 1991. https://doi.org/10.3390/biomedicines11071991

Wilson, B. A. (2020). Neuropsychological rehabilitation for people with non-progressive brain damage. *Oxford Research Encyclopedia of Psychology*. https://doi.org/10.1093/acrefore/9780190236557.013.720

Wood, R., & Alderman, N. (2024). Applications of operant learning theory to the management of challenging behavior after traumatic brain injury | request PDF. *Journal of Head Trauma Rehabilitation, 26*(3), 202–211. https://doi.org/10.1097/HTR.0b013e318217b46d

Zhang, J., & He, C. (2023). Evidence-based rehabilitation medicine: Definition, foundation, practice and development. *Medical Review, 4*(1), 42–54. https://doi.org/10.1515/mr-2023-0027

Zhang, X., Lv, L., Min, G., Wang, Q., Zhao, Y., & Li, Y. (2021). Overview of the complex figure test and its clinical application in neuropsychiatric disorders, including copying and recall. *Frontiers in Neurology, 12*, 680474. https://doi.org/10.3389/fneur.2021.680474

Chapter 4
Technological Innovations in Rehabilitation: Virtual Reality

4.1 Introduction

Virtual reality (VR) refers to a computer-generated simulation of an environment that designs and creates immersive experiences of real-life scenarios (simulations). In neurorehabilitation, VR is rising in its utility to promote and support recovery and improvement of cognitive, motor and sensory functions pertaining to brain injury, neurodegeneration, and other neurological conditions. VR has revolutionized rehabilitation by leveraging immersive and interactive digital environments to aid patients that seek neurorehabilitation. An advantage of these immersive experiences is the engaging, safe spaces that are created to practice essential skills that are tailored for unique needs. Unlike traditional therapy settings, VR provides flexibility to generate real-world settings in a controlled and safe manner that minimizes risks while enhancing patient confidence. The safety and adaptability of VR makes it an invaluable addition for rehabilitation (Nieto-Escamez et al., 2023; Qian et al., 2020).

The prime goal of VR is to create an optimal environment for skill enhancement, motivation, and retraining. VR addresses engagement in personalized therapy by creating multiple simulations and complexity in tasks. The uniqueness platform where patients can practice with speed and intensity. The gamified nature of VR indulges the patients with enjoyment and challenge unlike conventional practices. Various rehabilitation studies have focused on recovery from memory loss and virtual exercises such as shopping have proven to be highly effective. This approach not only stimulates cognitive functions but also ensures patients remain invested in their therapy sessions, promoting adherence to rehabilitation programs (Rojo et al., 2022).

An equally significant advantage of VR lies in its capacity for objective assessment and feedback. VR systems are designed to monitor cognitive and motor performance in real time, offering precise measurements of skills such as attention,

K. J. Reddy, *Innovations in Neurocognitive Rehabilitation*,
https://doi.org/10.1007/978-3-031-88117-6_4

memory, and problem-solving. These assessments, conducted within the therapy itself, enable clinicians to track progress and adjust interventions based on measurable outcomes. Unlike traditional methods, which often rely on subjective observations, VR provides reproducible data, enhancing the accuracy and efficacy of rehabilitation programs. Research has shown that these detailed analytics can improve recovery trajectories by identifying subtle changes in patient performance that might otherwise go unnoticed (Brown et al., 2008).

Another critical feature of VR in neurocognitive rehabilitation is its ability to promote neural plasticity, the brain's inherent capacity to reorganize and form new neural pathways. Immersive VR environments stimulate targeted areas of the brain through specific tasks, encouraging the recovery of damaged neural tissue or compensatory mechanisms for lost functions. This is particularly relevant for conditions such as traumatic brain injury, stroke, or neurodegenerative disorders, where promoting neural recovery is a central goal of rehabilitation. For instance, patients with hemiparesis—a common poststroke condition—have shown significant improvements in motor function after engaging in VR-based exercises that mimic real-world movements in an immersive setting (Hao et al., 2022).

The history of VR in neurorehabilitation reveals a gradual evolution from basic technologies to advanced, fully immersive systems. Early applications of VR were primarily limited to motor rehabilitation, where simple, non-immersive simulations on desktop computers were used to guide patients through basic exercises. Over time, as VR technology became more sophisticated, its scope expanded to include cognitive rehabilitation, targeting skills such as memory, attention, and executive function. These advancements allowed therapists to design increasingly complex and interactive interventions, fostering greater engagement and efficacy. For example, early VR systems might have simulated simple arm movements, while modern systems can immerse patients in detailed environments, such as virtual kitchens or cityscapes, where they can practice daily activities in a lifelike manner (Qian et al., 2020).

The versatility of VR has also extended its applications beyond motor and cognitive training to include emotional regulation and social skills development. For individuals with conditions such as social anxiety, autism spectrum disorder, or post-traumatic stress disorder (PTSD), VR offers a controlled environment to practice challenging scenarios. A patient with PTSD, for instance, might gradually confront feared situations in a virtual setting, allowing for desensitization and emotional processing without the risks associated with real-world exposure. Similarly, children with autism can engage in social interactions within virtual environments designed to improve communication and interpersonal skills. These applications highlight the broader potential of VR as a tool not only for rehabilitation but also for enhancing overall quality of life (Brown et al., 2008).

Despite its numerous advantages, implementing VR in neurocognitive rehabilitation comes with challenges. The cost of high-quality VR systems can be prohibitive, limiting access for patients in underserved communities. Additionally, the technical expertise required to set up and maintain VR equipment poses barriers for smaller rehabilitation centers. Patients may also face difficulties adapting to the technology, particularly older adults or those with severe impairments. Ethical considerations, such as ensuring patient privacy and the secure handling of data collected through VR systems, must also be addressed. Nonetheless, ongoing advancements in technology and growing evidence of VR's effectiveness continue to drive its adoption in clinical settings (Catania et al., 2024).

The benefits of VR in neurocognitive rehabilitation are well-supported by research, which consistently demonstrates its effectiveness in improving functional outcomes. Studies have shown that patients who engage in VR-based interventions often achieve better results than those receiving standard therapy alone. For instance, a meta-analysis of VR applications in stroke rehabilitation found significant improvements in motor function, balance, and independence among patients who used VR systems compared to control groups (Prats-Bisbe et al., 2024). These findings underscore the importance of integrating VR into rehabilitation programs to maximize patient recovery and overall well-being.

In addition to its therapeutic applications, VR offers exciting possibilities for future innovations in neurorehabilitation. Integrating artificial intelligence and machine learning into VR platforms could enable even greater personalization and adaptability, tailoring interventions to individual needs in real time. Wearable devices, such as motion sensors and haptic feedback systems, are also being explored to enhance the realism and effectiveness of VR experiences. As these technologies continue to evolve, they are likely to expand the scope and impact of VR, making it an indispensable tool in neurocognitive rehabilitation (Hao et al., 2022; Khokale et al., 2023).

4.2 Theoretical Foundations of Virtual Reality in Rehabilitation

4.2.1 Immersive Experiences

Virtual reality provides a highly immersive environment where the immersive experience captivated the attention of the patient and profoundly enhances real-world engagement of rehabilitated skills. Traditional methods of neuropsychological rehabilitation heavily relies on manual methods and pen to paper techniques to engage the patients. A large portion of these methods are heavily influenced by lab-controlled setups and rely heavily on repetition and monotony of the activities. Traditional methods have proven to reduce the effectiveness of rehabilitation through loss of motivation and curiosity towards tasks by patients. This significant gap is compensated through VR. The power of designing multisensory experiences increases patient engagement and is instrumental in generating realistic scenarios and engaging activities for the patient. Patients in VR therapy feel invested in achieving therapeutic goals for longer periods than patients obtaining interventions through traditional methods. Studies have utilized VR in cases with stroke to retrain skills associated with navigation through scenarios such as grocery shopping, selecting items from a grocery checklist and simulating motor movements associated with grocery shopping including motor coordination, balance, and cognitive processing. These experiences provide realistic applications in the interventional conditions (Aderinto et al., 2023; Crowe et al., 2024; Guzmán et al., 2024).

The enhanced and immersive experience promotes the training of multiple cognitive and behavioral areas. VR technology combines visual, auditory, and tactile stimuli to design realistic scenarios. For example, in the case of the motor balance rehabilitation program, the VR simulates walking by the beach through the visual stimuli of beaches and the small nuances such as the sand, the waves, and other elements that would simulate the same. Auditory stimuli such as sounds of the ocean waves and people in the background are other added features to enable the immersive training experience (Buele et al., 2023; Lv & Guo, 2022). The immersive quality of the simulation produces responses that are similar to the responses that they may produce in the real world. By engaging in multiple sensory experiences, the patients are able to dedicate their focus and maintain consistency. The experiences make the rehabilitation significantly effective and permits the therapist to focus on specific skills and movements that require isolated focus-walking, standing, posture, etc. Other qualities include the flexibility of the systems for adjustments in real-time, which increases the tailored approach to interventions and therapy. Low-intensity mobility-based exercises and gradual transition to complex scenarios are key to rehabilitate individuals with severe deficits in motor behavior. The tailored approach has multiple benefits such as increasing the trails, changing the intensity of the tasks and increasing the number of interactive simulations (Crowe et al., 2024; Drigas & Sideraki, 2024; Guzmán et al., 2024).

4.2.2 Ecological Validity

Ecological validity of VR-based interventions are key to addressing cultural and social inclusivity of neuropsychological rehabilitation. The tasks designed can be customized to suit relevant skills that would encourage the patient to recover lost functions in social and occupational areas. Traditional occupational and physical therapy exercises focus on applying weight and resistance training to improve muscle functions, the application of these strengths do not indicate successful outcomes when translated to work-related tasks and realistic scenarios. Functional activities through VR bridges this gap by incorporating tasks that are simulations of their work-life such as lifting objects from table or the floor, applying fine motor skills to navigate through tasks such as writing, turning pages or even switching the lights on/off. The replicability of these tasks in the patients' life increases the validity of these challenges (Aderinto et al., 2023; Crowe et al., 2024). Other mobility tasks include crossing busy streets, boarding public transports or preparing a meal in the kitchen—each of which requires safe spaces to train prior to confidently applying it to real-world scenarios. The safe and controlled environment permits the practice of regular exposure with minimal risk, encouraging multiple trials and increasing intensity. A patient rehabilitating from poststroke effects would rather engage in activities that have minimal risk of injury and failure, rather than causing further harm (Aderinto et al., 2023; Lv & Guo, 2022).

The ecological validity expands to social and cognitive domains of rehabilitation. Patients that are gradually recovering from TBI and other cognitive disorders would be able to be involved with tasks that challenge memory, attention, and problem-solving with practical aids. Studies conducted in VR-based grocery store simulations require the patients to apply various cognitive skills such as planning, scheduling, decision-making and reasoning in an effective and realistic setting. These scenarios require cognitive burden to shift from passive and hypothetical scenarios to mirrored experiences of real-world situations. By doing so VR training will ensure that the skills required by patients to acquire during therapy are directly applicable to everyday routine. The ecological validity of these tasks translate their significance from clinical interventions to independent living and in improving the overall quality of life (Aderinto et al., 2023; Crowe et al., 2024).

4.2.3 Feedback Mechanisms

As feedback plays a critical role in addressing rehabilitation plans, effective learning and VR excels in providing immediate, real-time feedback that enhances self-correction and skill acquisition. Traditional therapeutic methods are limited by self-reported feedback, observations, and assessments during the training sessions. With the application of VR to address the lack of precision in rehabilitation, there is a need for optimal motor and cognitive learning. VR addresses this gap by offering continuous, objective data that contributes to feedback and in advancing analysis and behavior tracking (Riva et al., 2020). Studies focusing on neurorehabilitation of behavior have indicated that a VR system might track the trajectory of a patient's arm movement during a reaching exercise, visually highlighting deviations from the desired path and suggesting corrections. Instant feedback from these systems allow patients to adjust their movements in real time, encouraging speedy and accurate motor learning. In cases of patients working on balance and coordination, VR systems can provide visual or auditory cues when their posture deviates from the optimal range, helping them maintain proper alignment and avoid potential falls. Feedback mechanisms in VR-based interventions contribute to cognitive and emotional domains, through measuring response times, decision accuracy, and task completion rates, providing patients with detailed insights into their progress. Patients practicing memory recall in a VR simulation receive feedback on their recall speed and the accuracy of providing them with insights on areas that are affected and require attention (Aderinto et al., 2023; Lv & Guo, 2022).

Therapeutic techniques that target anxiety and emotional regulation in cases of dementia and TBI apply VR to monitor physiological changes and biomarkers in providing feedback on stress levels and relaxation techniques. The role of feedback is to be instrumental in interventions and encourage feedback to create impact. VR technology in this aspect heart rate or skin conductance, providing real-time feedback on stress levels and relaxation techniques. Various VR systems integrate gamified aspects such as scoring, rewarding and achievement trackers to motivate the patients and in improving performance-related feedback. Patients could seek a sense of confidence and independence through reaching new levels in the VR game and in reinforcing neuronal pathways. This integration of feedback into a gamified framework ensures that therapy remains engaging while fostering a deeper understanding of progress and areas for improvement (Buele et al., 2023; Guzmán et al., 2024).

4.2.4 Repetitive Practice

Repetition is key to rehabilitation, especially with neurorehabilitation (the brain's potential to reorganize and reconnect neural networks). Early schools of neuropsychology believed that the brain is incapable of regenerating or regaining lost functions from insult or degeneration. However, contemporary theoretical frameworks on neural plasticity focus on the possibility that all functions are not fixed to specific areas of the brain leading to the possibility of regaining lost functions through rewiring and restructuring of neuronal pathways. This introduces the possibility of shifting the focus of rehabilitation from a deficit approach to a restorative-compensatory approach. VR is ideal for facilitating the practice of engaging various exercises and in performing tasks repeatedly without becoming fatigued or losing motivation. Unlike traditional approaches, the usage of VR specifically caters to highly immersive experiences and meaningful activities that eases the repetition and retraining of practices—increasing the sustainability of long-term rehabilitation (Aderinto et al., 2023; Buele et al., 2023; Guzmán et al., 2024). Gamification of tasks increases engagement in rehabilitation. Patients recovering from hand injuries may find it easier to practice repetitive tasks through gamified activities that apply reward tools and mechanisms (solving puzzles, collecting objects, etc.). The reward mechanism increases a sense of achievement and progression can be quantifiable. The ability to tailor exercises to specific levels and increasing the complexity of the tasks transitions rehabilitation from static environments. For example, virtual reality tasks such as "FarmDay" increase the use of cognitive functions and motor activities such as mobility and muscle coordination (Catania et al., 2023; Rojo et al., 2022). The scalability of VE ensures that continuous improvement is possible without overwhelming the patients. Neuropsychological rehabilitation requires a holistic approach and VR can enable the same through emotional rehabilitation. The simulations working on attention training tasks create characters that provide safe and controlled environments to regulate social and task-specific anxiety. Engaging in these tasks increases the wide range of skills and confidence of the patients further amplifying the potential for repetitive practice. Home-based interventions permit the convenience of time and convenience of practicing and training of skills without the traditional barriers of time, space, and commute. Patients are able to achieve high-quality therapy (Buele et al., 2023; Catania et al., 2023; Crowe et al., 2024; Drigas & Sideraki, 2024; Hadjipanayi et al., 2024).

4.3 Applications of VR in Neurocognitive Rehabilitation

Virtual reality (VR) has emerged as a transformative tool in neurocognitive rehabilitation, offering dynamic and immersive interventions for a variety of conditions. This paper outlines the applications of VR in stroke rehabilitation, traumatic brain injury (TBI), Parkinson's disease, autism spectrum disorder (ASD), and post-traumatic stress disorder (PTSD), with detailed exploration of motor, cognitive, and behavioral benefits.

4.3.1 Stroke Rehabilitation

1. *Motor Rehabilitation*: VR interventions are widely used for stroke survivors to improve motor and cognitive functions. Interactive VR games and tasks significantly enhance coordination, balance, and fine motor skills. For example, systems like RehabVR use immersive activities, such as reaching for virtual objects or simulating physical exercises, to promote motor recovery. Research indicates that VR-based motor rehabilitation is more effective than traditional physiotherapy alone, leading to quicker and more sustained improvements in motor function (Quan et al., 2024; Wang et al., 2023).

   Case Example: A stroke survivor with limited hand dexterity engaged in a VR task simulating virtual painting. Over 12 weeks, the patient demonstrated improved grip strength, increased range of motion, and regained the ability to hold and manipulate real-world objects, such as utensils and writing tools (Quan et al., 2024; Scott et al., 2022).
2. *Cognitive Training*: Stroke rehabilitation often includes cognitive exercises to address deficits in memory, attention, and problem-solving. VR platforms are equipped with games that challenge these cognitive domains in an engaging and interactive manner. For instance, patients may solve virtual puzzles or navigate mazes, which require focus, recall, and strategic thinking. Studies have shown that such activities improve neuroplasticity, fostering cognitive recovery (Lv & Guo, 2022; Wang et al., 2023).

   Case Example: A VR program designed to simulate daily activities, like grocery shopping, helped a 65-year-old stroke survivor improve memory recall and decision-making skills. By repeatedly practicing these scenarios in VR, the patient regained confidence and independence in performing similar real-world tasks (Scott et al., 2022).

4.3.2 Traumatic Brain Injury (TBI)

1. *Personalized Cognitive Training*: Patients with TBI benefit from VR's ability to create customized cognitive training modules. These modules target specific deficits, such as impaired attention, executive dysfunction, or memory problems. VR tasks often incorporate elements of real-world complexity, making them both challenging and relevant to everyday life.

   Case Example: A VR program designed for a 28-year-old TBI survivor involved a series of timed tasks, such as categorizing virtual objects or solving spatial puzzles. Over 6 months, the patient demonstrated enhanced problem-solving ability and improved short-term memory recall (Larson et al., 2014; Wang et al., 2023).
2. *Simulating Complex Environments:* VR excels in stimulating environments that are difficult or impossible to replicate in traditional therapy. Patients with TBI can practice navigating busy streets, managing workplace stress, or even driving in a virtual environment. These tasks improve decision-making and stress management, essential for reintegration into everyday activities (Wang et al., 2023).

   Case Example: A 40-year-old patient recovering from a car accident-related TBI participated in VR simulations of a busy urban intersection. Over time, they developed greater situational awareness and confidence, enabling them to safely resume driving (Larson et al., 2014; Scott et al., 2022).

4.3.3 Parkinson's Disease

1. *Motor Rehabilitation*: VR applications have shown promise in managing gait disorders and motor symptoms in Parkinson's disease. Tasks such as stepping over virtual obstacles or practicing balance in simulated terrains have led to significant improvements in mobility and posture. VR's gamified approach makes therapy more engaging, encouraging greater participation (Catania et al., 2023; Wang et al., 2023).

   Case Example: A 72-year-old Parkinson's patient used a VR treadmill system that provided feedback on gait symmetry and step length. After 10 weeks of training, the patient reported fewer instances of freezing episodes and an overall increase in walking speed and stability (Quan et al., 2024).
2. *Cognitive Training*: Cognitive decline in Parkinson's disease, including impaired memory and executive functioning, can also be addressed through VR. Programs that involve memory games, visuospatial challenges, and logical puzzles help maintain cognitive sharpness. These activities are particularly effective in early- to mid-stages of the disease (Lv & Guo, 2022; Nieto-Escamez et al., 2023).

   Case Example: A VR program incorporating memory-based treasure hunts in a virtual park helped a 65-year-old patient with early-stage Parkinson's improve recall and problem-solving abilities. The patient reported increased confidence in handling day-to-day planning tasks (Larson et al., 2014; Nieto-Escamez et al., 2023).

4.3.4 Autism Spectrum Disorder (ASD)

1. *Social Cognition Training:* For individuals with ASD, VR provides a safe, controlled platform to develop and practice social skills. Simulated scenarios such as group discussions, job interviews, or classroom interactions help individuals learn to interpret social cues, engage in dialogue, and manage anxiety (Quan et al., 2024).

   Case Example: A 15-year-old with ASD participated in a VR program simulating a school lunchroom. Over the course of 8 weeks, the individual learned to initiate conversations, interpret nonverbal cues, and respond appropriately to peer interactions (Catania et al., 2023; Quan et al., 2024).
2. *Sensory Integration:* Many individuals with ASD experience sensory sensitivities that affect their ability to engage with the world. VR environments allow for gradual exposure to sensory stimuli, such as lights, sounds, or textures, in a controlled manner. This helps individuals build tolerance and reduces anxiety (Nieto-Escamez et al., 2023; Wang et al., 2023).

   Case Example: A VR therapy program for a 10-year-old with hypersensitivity to noise involved gradual exposure to ambient sounds in a virtual park. Over time, the child was able to visit real parks with significantly reduced distress (Teo et al., 2016).
3. *Cognitive Skill Development:* Interactive VR tasks, such as navigating a virtual maze or assembling puzzles, help strengthen cognitive functions like planning, attention, and spatial awareness. These activities are tailored to the developmental level and unique needs of each individual (Quan et al., 2024; Teo et al., 2016).

4.3.5 Post-Traumatic Stress Disorder (PTSD)

1. *Exposure Therapy:* Exposure therapy using VR has been a groundbreaking approach for treating PTSD. By immersing patients in realistic simulations of their traumatic experiences, VR enables gradual confrontation of fears in a controlled therapeutic setting. This process reduces avoidance behaviors and helps patients reprocess traumatic memories (Scott et al., 2022).

   Case Example: A veteran with combat-related PTSD participated in a VR program simulating battlefield scenarios. Over 12 weeks, the patient experienced a marked reduction in flashbacks and hyperarousal symptoms (Larson et al., 2014).
2. *Gradual Exposure:* VR allows for incremental exposure to trauma-related triggers, enabling patients to confront their fears at a manageable pace. For instance,

a civilian PTSD patient traumatized by a car accident was exposed to increasingly realistic driving simulations. This approach reduced their avoidance of driving and associated anxiety (Teo et al., 2016).

3. *Coping and Relaxation Techniques:* Beyond exposure therapy, VR also provides environments for practicing coping strategies. Simulated tranquil settings, such as beaches or forests, help patients practice relaxation techniques like mindfulness and controlled breathing (Scott et al., 2022).

   Case Example: A VR mindfulness program designed for a 35-year-old PTSD patient involved guided meditation in a serene virtual forest. After 8 weeks, the patient reported improved emotional regulation and a reduction in anxiety attacks (Nieto-Escamez et al., 2023).

4.4 Key Benefits of VR in Rehabilitation

4.4.1 Personalization

Virtual reality (VR) technology enables unparalleled customization, a crucial factor in its efficacy for rehabilitation. Unlike traditional approaches that often rely on generalized therapy techniques, VR systems are designed to adapt to the unique needs of each patient, offering tailored interventions. For instance, the NeuroR system exemplifies this capability by incorporating visual tracking approaches that can be adjusted based on the specific conditions of stroke patients (Aderinto et al., 2023; Massetti et al., 2018). Such adaptability ensures that patients receive therapy that directly targets their individual challenges. Customization in VR extends to replicating real-world scenarios tailored to a patient's impairments or goals. This replication allows patients to practice functional activities in safe, simulated environments that mirror their daily lives. For example, a patient recovering from a spinal cord injury might navigate virtual versions of their home or workplace, helping them regain independence in familiar settings (Catania et al., 2023; Marougkas et al., 2024). Additionally, many VR systems include baseline assessments that determine a patient's current abilities. These assessments help to calibrate the difficulty levels of tasks and exercises, ensuring patients are neither under- nor over-challenged during therapy. The immersive nature of VR also plays a vital role in personalizing rehabilitation. Unlike traditional physical therapy, VR allows patients to engage in repeated practice of specific skills or activities outside formal therapy sessions (Radianti et al., 2020). This flexibility is particularly beneficial for patients who may not have frequent access to clinical facilities. By enabling remote engagement, VR not only democratizes access to personalized rehabilitation but also empowers patients to take an active role in their recovery process.

4.4.2 *Motivation and Engagement*

Motivation is a critical factor in the success of rehabilitation, and VR's gamified, interactive features excel in fostering engagement. Traditional therapy can sometimes feel repetitive and monotonous, leading to decreased patient compliance. VR addresses this challenge by transforming therapy into an enjoyable and stimulating experience. Many VR programs incorporate game-like elements, also known as exergames, which merge physical exercise with video game mechanics. These exergames captivate patients' attention, making therapy sessions more enjoyable while promoting physical and cognitive activity. The immersive nature of VR further enhances motivation by creating a strong sense of presence and agency. In a virtual environment, patients feel directly involved in their rehabilitation journey, which boosts intrinsic motivation. For example, stroke patients engaging in VR therapy might see themselves manipulating objects or navigating through spaces, reinforcing their confidence and belief in their recovery (Insidea, 2024; Radianti et al., 2020; Stefan et al., 2023; Zanatta et al., 2022).

Interactive VR environments also provide continuous feedback and rewards, key elements in maintaining long-term engagement. Feedback systems, such as visual progress indicators or auditory cues, inform patients about their performance in real-time. Rewards, whether in the form of virtual achievements or tangible milestones, foster a sense of accomplishment that motivates patients to persist. The novelty of VR technology itself can also increase willingness to participate, particularly among younger patients or tech-savvy populations. This increased engagement has tangible benefits for rehabilitation outcomes. Studies show that patients who are actively engaged in their therapy demonstrate better adherence and achieve faster s compared to those undergoing conventional methods. As a result, VR not only enhances the therapy experience but also contributes to its overall effectiveness (Aderinto et al., 2023; Insidea, 2024; Zanatta et al., 2022).

4.4.3 *Safety and Accessibility*

Safety is a cornerstone of effective rehabilitation, and VR excels in creating controlled, risk-free environments for therapy. Patients recovering from severe injuries or disabilities often face limitations in practicing real-world activities due to the associated risks. VR offers a solution by simulating such activities in a safe and supervised setting. For example, a patient with balance impairments can practice walking across uneven surfaces in a virtual environment without the danger of falling. Similarly, VR can simulate challenging or hazardous scenarios that would otherwise be unsafe to encounter in real life. A patient with post-traumatic stress disorder (PTSD), for instance, might gradually confront their fears in a virtual setting, such as walking through a crowded marketplace, under the guidance of a therapist. This controlled exposure therapy enables patients to develop coping mechanisms without the risks associated with real-world exposure (Catania et al., 2023; Massetti et al., 2018; Scott et al., 2022; Stefan et al., 2023).

Beyond safety, VR also enhances accessibility to rehabilitation services. Traditional therapy often requires patients to visit specialized facilities, which can be a significant barrier for those with mobility issues or those living in remote areas. Home-based VR therapy eliminates this obstacle by allowing patients to engage in therapy from the comfort of their homes. This approach reduces transportation costs and provides equal access to care, particularly for underserved populations. Additionally, VR systems are designed to accommodate a wide range of physical and cognitive abilities, making them accessible to diverse patient groups (Zanatta et al., 2022).

The accessibility of VR has broader implications for health equity. By reducing barriers to rehabilitation, VR ensures that more patients can benefit from advanced therapeutic interventions, regardless of their socioeconomic status or geographic location. This inclusivity underscores the potential of VR to transform rehabilitation into a universally accessible healthcare service.

4.4.4 Quantifiable Progress

One of VR's most significant advantages in rehabilitation is its ability to provide objective, quantifiable data on patient progress. Traditional rehabilitation often relies on subjective assessments, which can vary between clinicians. In contrast, VR systems leverage advanced technology to track and analyze a wide range of performance metrics. For example, data on movement kinematics, force dynamics, and reaction times can be collected during therapy sessions, providing detailed insights into a patient's recovery journey (Massetti et al., 2018; Radianti et al., 2020).

These objective metrics enable clinicians to monitor improvements over time and identify areas that require additional focus. For instance, a stroke patient using a VR system for hand rehabilitation might receive data on grip strength and dexterity, which can inform adjustments to their therapy plan. This precision not only enhances the effectiveness of interventions but also ensures that therapy remains aligned with the patient's evolving needs. In addition to individual assessments, VR also facilitates standardized evaluation across patients and studies. By using uniform tools and metrics, clinicians can compare outcomes and identify best practices for different conditions. This standardization is particularly valuable for research purposes, as it allows for the aggregation and analysis of data from large patient populations. The ability to track progress in real-time also has motivational benefits for patients. Seeing tangible evidence of their improvement can boost confidence and encourage continued participation in therapy. Moreover, the data collected by VR systems can be shared with patients and their families, fostering a collaborative approach to rehabilitation that involves all stakeholders (Lorenz et al., 2024; Massetti et al., 2018; Radianti et al., 2020; Zanatta et al., 2022).

4.5 Challenges and Limitations

Despite its advantages, VR in neurocognitive rehabilitation faces challenges.

4.5.1 *Cost*

The incorporation of VR in neurocognitive rehabilitation programs presents significant financial challenges due to the high costs in acquiring the hardware and software. The need for advanced equipment such as head-mounted displays (HDMs), motion capture devices, high-performance computers and specialized software tailored to therapeutic applications. These costs are compounded in hybrid systems and robot-assisted rehabilitation—which, combined with VR are highly sophisticated and expensive components (Bang et al., 2024; Zanatta et al., 2022). The financial burden limits the accessibility of VR rehabilitation programs, especially in resource-constrained settings. Remote areas, rural clinical, small-scale healthcare facilities and institutions within low-middle income countries face a significant challenge in prioritizing VR assistive technologies due to its cost and lack of trained professionals. The cost incurred through VR assistive technological rehabilitation may be scrutinized by insurance providers. Insurance agencies continue to perceive contemporary developments in technology as experimental or elective that leaves patients with the burden of covering the financial costs of treatment. Patients from underserved communities can perceive the large dependence on VR-based rehabilitation as a significant barrier to health inequity (Catania et al., 2023; Prats-Bisbe et al., 2024).

The rapid evolution of VR technology is another limitation of accessing the same. As newer updates and versions quickly become available, the cost of updating the systems, replacing outdated equipment and hardware and upgrading to the latest software poses a concern on rehabilitation facilities. The cost incurred to meet the expectations of new-age technology requires a regular stream of investment which adds a recurring financial burden for small-scale and limited budget-clinics and rehabilitation centers. The additional costs incurred in training healthcare professionals to operate these advanced technologies and the expectation on practitioners to become well-equipped with advanced technical skills to operate the machine is an additional cost. A lack of trained staff can lead to underutilization of VR techniques. Efforts to address cost-related challenges in VR rehabilitation fixes various other limitations with the utilization of VR. For example, researchers are exploring ways to design affordable, portable VR systems using off-the-shelf components such as smartphone-based VR headsets. Open-source software and solutions continue to gain traction further allowing developers to create cost-effective solutions that are customizable in applications. Despite these advances, scaling VR for widespread use in rehabilitation remains a formidable challenge (Larson et al., 2014; Zanatta et al., 2022).

4.5.2 Technical Issues

Technical reliability is essential for the successful implementation of VR in neurocognitive rehabilitation. The hardware and software glitches frequently pose significant challenges. In instances where headsets mat experience tracking errors, motion control and failure to accurately register inputs can further affect visuo-auditory output in terms of latency issues. The disruptions can disrupt the immersive experience thereby reducing its effectiveness. While immersion enhances patient engagement, it increases the system sensitivity and performance issues. A delay in rendering or tracking can disorient users, especially those who have sensory impairments. Systems that are poorly designed and calibrated can produce inaccurate feedback that hinder the rehabilitation process. Poorly calibrated systems may produce inaccurate feedback, frustrating patients and potentially hindering their progress. Addressing these issues requires regular calibration, updates, and troubleshooting further straining the resources of healthcare providers (Bang et al., 2024; Quan et al., 2024; Zanatta et al., 2022).

Another technical hurdle involves compatibility across different VR systems and platforms. Clinics often use a mix of devices, which may not always integrate seamlessly. Inconsistent user experiences can undermine the uniformity of therapeutic interventions, making it difficult to standardize treatment protocols. Moreover, the limited interoperability of VR systems creates challenges in data sharing and performance tracking, which are crucial for evaluating outcomes. To mitigate these issues, developers are focusing on improving the reliability and user-friendliness of VR systems. Advances in artificial intelligence (AI) and machine learning are being leveraged to enhance the accuracy and adaptability of VR applications. For instance, AI algorithms can predict and correct tracking errors in real time, ensuring a smoother and more consistent user experience. As these technologies mature, they promise to reduce the technical barriers currently limiting VR's clinical applications (Hao et al., 2022; Larson et al., 2014).

4.5.3 Usability

Despite its potential, VR technology can be daunting for certain patient populations, particularly elderly individuals or those with cognitive impairments. The complexity of operating VR systems, from headsets to manipulating controllers can lead to overwhelming users without familiarity in contemporary technologies. Older adults recovering from strokes, for example, may struggle with the physical or cognitive demands of using equipment, while patients with neurodegenerative diseases may find it challenging to follow on-screen instructions or understand the mechanics of the system. Usability issues extend beyond patients to include clinicians and caregivers. Many rehabilitation professionals are untrained in VR technology, which can make integrating these tools into traditional therapeutic workflows challenging.

The steep learning curve for both patients and providers can lead to inconsistent application of VR in clinical practice, reducing its overall effectiveness. To address these challenges, developers are prioritizing user-centered design. Innovations such as gesture-based controls, voice commands, and simplified interfaces are being introduced to make VR systems more intuitive and accessible. For example, VR platforms now offer guided tutorials and adaptive settings that adjust difficulty levels based on the user's abilities. These features are particularly beneficial for older adults or those with severe cognitive impairments, as they minimize the cognitive and physical load required to interact with the technology. Moreover, incorporating haptic feedback and multimodal interactions can enhance the user experience, making VR systems more engaging and easier to use. For instance, touch-sensitive gloves or motion-tracking suits can provide tactile feedback, allowing users to interact with virtual environments more naturally. Such innovations not only improve usability but also enhance the therapeutic value of VR by offering a richer, more immersive experience (Bang et al., 2024; Quan et al., 2024; Wang et al., 2023).

4.5.4 Limited Evidence

While VR has shown promise in neurocognitive rehabilitation, the body of evidence supporting its efficacy remains limited. Most studies in this field suffer from small sample sizes, short follow-up periods, and a narrow focus on specific conditions, such as stroke recovery. As a result, the generalizability of these findings to diverse populations and neurological conditions remains uncertain. Longitudinal research is particularly lacking. Neurocognitive rehabilitation often requires extended periods of therapy, yet few studies track the long-term outcomes of VR interventions. Without robust data on the sustainability of VR-based improvements, it is difficult to determine whether these technologies offer lasting benefits. For example, a short-term study may report significant gains in motor function, but without follow-up, it is unclear whether these improvements translate into meaningful functional independence. Another limitation is the underrepresentation of certain populations in VR research. Most studies focus on relatively healthy and motivated participants, often excluding individuals with severe impairments, comorbidities, or socioeconomic barriers. This creates a gap in understanding how VR performs in real-world clinical settings, where patient populations are highly heterogeneous. Expanding the scope of research to include a broader range of conditions, such as Parkinson's disease, multiple sclerosis, or traumatic brain injuries, is essential for a comprehensive evaluation of VR's potential. Efforts to address these gaps are underway. Researchers are conducting larger-scale trials that incorporate diverse participant demographics and longer follow-up periods. Advances in data analytics and machine learning are also enabling more nuanced assessments of VR interventions. For instance, wearable sensors and motion-capture technologies can collect detailed data on patient performance, allowing researchers to identify patterns and optimize treatment protocols. These approaches promise to build a more robust evidence base, but achieving this goal will require sustained investment and interdisciplinary collaboration (Catania et al., 2023; Prats-Bisbe et al., 2024; Zanatta et al., 2022).

4.6 Future Directions

The potential for VR in neurocognitive rehabilitation continues to expand, driven by technological advancements:

1. *Integration with Artificial Intelligence (AI):* Artificial intelligence (AI) is redefining the potential of virtual reality (VR) in neurocognitive rehabilitation by enabling personalized, efficient and adaptive therapeutic interventions. AI-powered VR environments are used to analyze performance in real-time further allowing for dynamic adjustments to therapy sessions. These modifications ensure exercises remain challenging but achievable, promoting sustained cognitive and motor improvement. Such real-time data and feedback contributes to adaptability providing a patient-centered approach that evolves alongside their progress. AI tools, including personalized rehabilitation apps, have demonstrated efficacy in addressing various conditions such as stroke and traumatic brain injury, offering insights into patient outcomes and in refining treatment goals through predictive analytics (Khalid et al., 2024).
2. *Wearable VR Devices:* Wearable VR devices enhance accessibility and portability of neurocognitive rehabilitation. Compact and user-friendly designs allow patients to engage in therapeutic sessions at home or in community settings, minimizing the need for frequent clinical visits. These devices integrate AI to offer real-time feedback on performance and can track rehabilitation progress enabling the monitoring and adjustments of treatment remotely. By integrating gamified VR experiences with portable technology, wearable devices enhance patient motivation, ensuring consistent engagement with therapy. These solutions hold promise for populations in remote or underserved areas, providing equitable access to advanced rehabilitation tools (Khalid et al., 2024).
3. *Neurofeedback Integration:* Neurofeedback when combined with VR offers a powerful tool to monitor and influence brain activity during therapy. Technology such as EEG provide real-time insights into the patient's neural responses that can adjust interventions based on specific patterns of brain activity. The application of brain-computer interface (BCI) systems leverage cortical electrical activity to allow patients to control virtual or robotic devices offering a high precision in cognitive retraining. The integration not only enhances the personalization of therapy but also fosters patient engagement by delivering immediate feedback on their progress. Neurofeedback can be particularly beneficial for targeting motor and cognitive recovery. It directly connects brain activity with observable (Khalid et al., 2024).
4. *Tele-Rehabilitation*: Tele-rehabilitation is emerging as a critical application of VR, especially for extending the reach of neurocognitive therapies to underserved

and remote regions. Using VR-based interventions, patients can participate in immersive therapy sessions from their homes, significantly reducing barriers to care caused by distance or mobility constraints. Programs like "TeleRehab" and "Virtual Clinic" have demonstrated success, particularly in pediatric populations, by delivering structured and engaging rehabilitation programs remotely. During the COVID-19 pandemic, tele-rehabilitation proved invaluable in ensuring continuity of care, allowing patients to maintain their therapy routines while adhering to social distancing measures. The scalability of tele-rehabilitation holds potential for addressing healthcare disparities, making it a cornerstone of accessible rehabilitation services (Khalid et al., 2024).

5. *Other Emerging Technology:* In addition to VR, emerging technologies such as augmented reality (AR) and electronic health records (EHRs) are enhancing the scope of neurocognitive rehabilitation. AR creates hybrid environments where virtual elements are layered onto the physical world, making it particularly useful for functional skill training, balance exercises, and pain management. For example, AR applications can simulate real-world challenges, such as navigating busy streets, providing patients with safe and controlled environments to practice skills. EHRs also play a vital role in improving the coordination of care in rehabilitation. By consolidating patient data from various sources, EHR systems facilitate seamless communication among healthcare providers, enabling more cohesive and effective treatment plans. They also allow for longitudinal tracking of patient progress, ensuring that interventions remain aligned with evolving needs and objectives (Khalid et al., 2024).

4.7 Conclusion

Virtual reality has revolutionized the paradigm shift in neuropsychological rehabilitation by offering personalized engagement and safe interventions which can be quantified to offer innovative solutions for therapy outcomes. Its capacity to bridge the gap in creative, immersive and adaptive interventions acts as a cornerstone in future rehabilitation science. The continuous development of VR promises to speed up the recovery process and make it effective, accessible and patient-centric. The gamification of VR in rehabilitation ensures that therapy is tailored to each patient with immersive features to boost motivation and engagement. The environments enhance VR safety and break barriers in accessibility and care for patients across different strata of the society. The objective data collected by VR systems will provide valuable insights and real-time feedback for both clinicians and patients while parallelly contributing to research. As virtual reality continues to evolve the applications in rehabilitation would encompass a broader range of neurological conditions. The integration of artificial intelligence (AI) and machine learning will further enhance personalization and effectiveness of VR systems making them indispensable tools in modern medicine. Ultimately, VR integrates traditional approaches to rehabilitation with the power of technology to create immersive experiences that enhance therapy. As technology continues to advance and become more accessible, its integration into clinical practice will undoubtedly play a pivotal role in enhancing patient care and improving quality of life for individuals with diverse rehabilitation needs.

References

Aderinto, N., Olatunji, G., Abdulbasit, M. O., Edun, M., Aboderin, G., & Egbunu, E. (2023). Exploring the efficacy of virtual reality-based rehabilitation in stroke: A narrative review of current evidence. *Annals of Medicine, 55*(2), 2285907. https://doi.org/10.1080/07853890.2023.2285907

Bang, M., Kim, M. A., Kim, S. S., & Kim, H. S. (2024). Cognitive training using virtual reality: An assessment of usability and adverse effects. *Archives of Rehabilitation Research and Clinical Translation, 6*(4), 100378. https://doi.org/10.1016/j.arrct.2024.100378

Brown, J. L., Vanable, P. A., & Eriksen, M. D. (2008). Computer-assisted self-interviews: A cost effectiveness analysis. *Behavior Research Methods, 40*(1), 1–7. https://doi.org/10.3758/brm.40.1.1

Buele, J., Varela-Aldás, J. L., & Palacios-Navarro, G. (2023). Virtual reality applications based on instrumental activities of daily living (iADLs) for cognitive intervention in older adults: A systematic review. *Journal of Neuroengineering and Rehabilitation, 20*, 168. https://doi.org/10.1186/s12984-023-01292-8

Catania, V., Rundo, F., Panerai, S., & Ferri, R. (2023). Virtual reality for the rehabilitation of acquired cognitive disorders: A narrative review. *Bioengineering, 11*(1), 35. https://doi.org/10.3390/bioengineering11010035

Catania, V., Rundo, F., Panerai, S., & Ferri, R. (2024). Virtual reality for the rehabilitation of acquired cognitive disorders: A narrative review. *Bioengineering, 11*(1), 1. https://doi.org/10.3390/bioengineering11010035

Crowe, S. E., Yousefi, M., Shahri, B., Piumsomboon, T., & Hoermann, S. (2024). Interactions with virtual therapists during motor rehabilitation in immersive virtual environments: A systematic review. *Frontiers in Virtual Reality, 5*, 1284696. https://doi.org/10.3389/frvir.2024.1284696

Drigas, A., & Sideraki, A. (2024). Brain neuroplasticity leveraging virtual reality and brain–computer interface technologies. *Sensors, 24*(17), 17. https://doi.org/10.3390/s24175725

Guzmán, D. E., Rengifo, C. E., Guzmán, J. D., & Cena, G. (2024). Virtual reality games for cognitive rehabilitation of older adults: A review of adaptive games, domains and techniques. *Virtual Reality, 28*, 92. https://link.springer.com/article/10.1007/s10055-024-00968-3

Hadjipanayi, C., Banakou, D., & Michael-Grigoriou, D. (2024). Virtual reality exergames for enhancing engagement in stroke rehabilitation: A narrative review. *Heliyon, 10*(18), e37581. https://doi.org/10.1016/j.heliyon.2024.e37581

Hao, J., Haoyu, X., Kimberly, H., & Chen, Z. (2022). Effects of virtual reality intervention on neural plasticity in stroke rehabilitation: A systematic review. *Archives of Physical Medicine and Rehabilitation, 103*(11), 523–541. https://doi.org/10.1016/j.apmr.2021.06.024

Insidea. (2024, August 2). Virtual reality rehabilitation: Revolutionizing recovery. *XR Health.* https://www.xr.health/virtual-reality-rehabilitation/

Khalid, U. bint, Naeem, M., Stasolla, F., Syed, M. H., Abbas, M., & Coronato, A. (2024). Impact of AI-powered solutions in rehabilitation process: Recent improvements and future trends. *International Journal of General Medicine, 17*, 943–969. https://doi.org/10.2147/IJGM.S453903. https://www.themarginalian.org/2013/03/26/viktorfrankl-mans-search-for-meaning/

Khokale, R., Mathew, G. S., Ahmed, S., Maheen, S., Fawad, M., Bandaru, P., Zerin, A., Nazir, Z., Khawaja, I., Sharif, I., Abdin, Z. U., & Akbar, A. (2023). Virtual and augmented reality in post-stroke rehabilitation: A narrative review. *Cureus, 15*, 4. https://doi.org/10.7759/cureus.37559

Larson, E. B., Feigon, M., Jacobs, H. E., Gagliardo, P., & Dvorkin, A. Y. (2014). Virtual reality and cognitive rehabilitation: A review of current outcome research. *SAGE Journals, 34*(4), 759. https://doi.org/10.3233/NRE-141078

Lorenz, E. A., Bråten Støen, A., Lie Fridheim, M., & Alsos, O. A. (2024). Design recommendations for XR-based motor rehabilitation exergames at home. *Frontiers in Virtual Reality, 5*, 1340072. https://doi.org/10.3389/frvir.2024.1340072

Lv, Z., & Guo, J. (2022). Virtual reality neurorehabilitation. *International Journal of Mental Health Promotion, 24*(3), 287–310. Annals of Medicine. https://doi.org/10.32604/ijmhp.2022.019829

Marougkas, A., Troussas, C., Krouska, A., & Sgouropoulou, C. (2024). How personalized and effective is immersive virtual reality in education? A systematic literature review for the last decade. *Multimedia Tools and Applications, 83*(6), 18185–18233. https://doi.org/10.1007/s11042-023-15986-7

Massetti, T., da Silva, T. D., Crocetta, T. B., Guarnieri, R., de Freitas, B. L., Bianchi Lopes, P., Watson, S., Tonks, J., & de Mello Monteiro, C. B. (2018). The clinical utility of virtual reality in neurorehabilitation: A systematic review. *Journal of Central Nervous System Disease, 10*, 1179573518813541. https://doi.org/10.1177/1179573518813541

Nieto-Escamez, F., Cortes-Perez, I., Obrero-Gaitán, E., & Fusco, A. (2023). Virtual reality applications in neurorehabilitation: Current panorama and challenges. *Brain Sciences, 13*(5), 819. https://doi.org/10.3390/brainsci13050819

Prats-Bisbe, A., López-Carballo, J., García-Molina, A., Leno-Colorado, D., García-Rudolph, A., Opisso, E., & Jané, R. (2024). Virtual reality–based neurorehabilitation support tool for people with cognitive impairments resulting from an acquired brain injury: Usability and feasibility study. *JMIR Neurotechnology, 3*(1), e50538. https://doi.org/10.2196/50538

Qian, J., McDonough, D. J., & Gao, Z. (2020). The effectiveness of virtual reality exercise on individual's physiological, psychological and rehabilitative outcomes: A systematic review. *International Journal of Environmental Research and Public Health, 17*(11), 11. https://doi.org/10.3390/ijerph17114133

Quan, W., Liu, S., Cao, M., & Zhao, J. (2024). A comprehensive review of virtual reality technology for cognitive rehabilitation in patients with neurological conditions. *Applied Sciences, 14*(14), 14. https://doi.org/10.3390/app14146285

Radianti, J., Majchrzak, T. A., Fromm, J., & Wohlgenannt, I. (2020). A systematic review of immersive virtual reality applications for higher education: Design elements, lessons learned, and research agenda. *Computers & Education, 147*, 103778. https://doi.org/10.1016/j.compedu.2019.103778

Riva, G., Mancuso, V., Cavedoni, S., & Stramba-Badiale, C. (2020). Virtual reality in neurorehabilitation: A review of its effects on multiple cognitive domains. *Expert Review of Medical Devices, 17*, 1035. https://doi.org/10.1080/17434440.2020.1825939

Rojo, A., Paz, J. A. S., Sanchez-Picot, J., Raya, R., & Garcia-Carmona, R. (2022). FarmDay: A gamified virtual reality neurorehabilitation application for upper limb based on activities of daily living. *Applied Sciences, 12*(14), 7068. https://doi.org/10.3390/app12147068

Scott, H., Griffin, C., Coggins, W., Elberson, B., Abdeldayem, M., Virmani, T., Larson-Prior, L. J., & Petersen, E. (2022). Virtual reality in the neurosciences: Current practice and future directions. *Frontiers in Surgery, 8*, 807195. https://doi.org/10.3389/fsurg.2021.807195

Stefan, H., Mortimer, M., & Horan, B. (2023). Evaluating the effectiveness of virtual reality for safety-relevant training: A systematic review. *Virtual Reality, 27*(4), 2839–2869. https://doi.org/10.1007/s10055-023-00843-7

Teo, W.-P., Muthalib, M., Yamin, S., Hendy, A. M., Bramstedt, K., Kotsopoulos, E., Perrey, S., & Ayaz, H. (2016). Does a combination of virtual reality, neuromodulation and neuroimaging provide a comprehensive platform for neurorehabilitation?—A narrative review of the literature. *Frontiers in Human Neuroscience, 10*, 284. https://doi.org/10.3389/fnhum.2016.00284

Wang, C., Kong, J., & Qi, H. (2023). Areas of research focus and trends in the research on the application of VR in rehabilitation medicine. *Healthcare, 11*(14), 2056. https://doi.org/10.3390/healthcare11142056

Zanatta, F., Giardini, A., Pierobon, A., D'Addario, M., & Steca, P. (2022). A systematic review on the usability of robotic and virtual reality devices in neuromotor rehabilitation: Patients' and healthcare professionals' perspective. *BMC Health Services Research, 22*(1), 523. https://doi.org/10.1186/s12913-022-07821-w

Chapter 5
Technological Innovations in Rehabilitation: Artificial Intelligence

5.1 Introduction

Artificial intelligence (AI) refers to the application of machine learning and other advanced computational techniques to enhance diagnosis, plan treatment, and increase efficacy in the rehabilitation outcomes in neuropsychological rehabilitation (Calderone et al., 2024). AI has transformed rehabilitation by redirecting the traditional approaches by clinicians and researchers to an approach toward recovery from cognitive and neurological impairments. Through the integration of advanced machine learning in neuroscience, psychology, and computer science, AI provides tools that provide precise, adaptive, and personalized interventions. These technologies have demonstrated significant potential in increasing the possibility of recovery through the application of theoretical frameworks of neuroplasticity—which introduces the foundational concept of rehabilitation that contradicts prior beliefs of that the brain is incapable of recovering lost functions to the possibility that since functions are not localized to specific parts of the brain, it can reorganize and compensate for lost functions through its ability to adapt to the injury (Akbar et al., 2024).

AI is applicable to a wide range of neuropsychological conditions from stroke to traumatic brain injury (TBI). In case of individuals recovering from stroke, AI systems such as virtual reality (VR) platforms and robotic exoskeletons offer interactive simulations that promote motor and cognitive recovery. The systems have the capacity to tailor to each patient and further on the basis of their current levels of functioning. The systems retain progress while challenging and intensifying the program to meet the rehabilitation goals. Studies have indicated that such adaptive interventions outperform traditional methods and promote neuroplasticity by improving functional outcomes. AI has enabled the creation of highly tailored cognitive retraining programs that simulate real-world scenarios to target cognitive deficits and to adjust to the levels of task difficulty based on individual performance.

K. J. Reddy, *Innovations in Neurocognitive Rehabilitation*,
https://doi.org/10.1007/978-3-031-88117-6_5

These programs have proven to be effective and successful in sustaining the engagement and progress in patients with TBI (Khalid et al., 2024; Ying et al., 2024).

In cases of Parkinson's disease (PD), AI has revolutionized ways to analyze motor movements and motor behavior management. The wearable devices are equipped with sensors to capture detailed data on movement patterns that are further analyzed to detect abnormalities and monitor disease progress. These AI systems provide valuable insights that guide clinicians in designing the interventions to reduce risk of falls and enhance mobility. Similar applications of AI in autism spectrum disorders (ASDs) and post-traumatic stress disorder (PTSD) indicate that AI-powered VR platforms are effectively managing the symptoms. The systems enable patients to practice social interactions or confront trauma-related stimuli within controlled, customizable environments, fostering skill acquisition, and emotional resilience (Calderone et al., 2024; Yang et al., 2024).

AI functions beyond its practicality through research by addressing patient diversity and individuality. ML models trained on large sets of data can predict the trajectories of prognosis that allows therapists to plan and design the rehabilitation to meet the unique challenges that each patient experiences. The capability is critical in neuropsychological rehabilitation, where the variability in the nature and severity of deficits require a tailored approach. AI's ability to model complex data of the patients advances its ecological validity in therapy by bridging the gap between controlled and clinical settings and the challenges patients experience on an everyday basis (Seelye et al., 2012). As AI evolves they are integrated with other technological applications such as brain-computer interfaces (BCIs) and telemedicine platforms. BCIs have enabled new forms of communication for individuals with severe disabilities whereas telemedicine platforms are often supported by AI algorithms that allow patients to access care remotely, overcoming the geographical and logistical barriers. The innovations expand the reach of rehabilitation and reduce discrepancies within the healthcare systems (Calderone et al., 2024).

The integration of AI into neuropsychological rehabilitation has its challenges with respect to data privacy, security, and the equity in the access of technological services. The disparity in infrastructure and resources can limit the availability for underserved communities. Another aspect to consider are the ethical considerations around the use of AI in clinical decision-making that may reduce the purpose of the therapist or clinician in the process of rehabilitation. AI continues to reshape contemporary neuropsychological rehabilitation and the theoretical foundations in plasticity continues to address various challenges that the AI can base itself off.

5.2 Theoretical Foundations of Artificial Intelligence in Rehabilitation

Neuroplasticity serves as a fundamental theoretical framework for understanding and guiding neuropsychological rehabilitation. Neuroplasticity is the concept that refers to the brain's ability to reorganize and adapt its structure and function in response to internal and external stimuli, further changing across molecular, cellular, and behavioral levels. The concept of neuroplasticity is key to rehabilitation as it creates the scientific basis for interventions aimed at compensating for lost and declining functions by further restoring lost functions through adaptive training. Neuroplasticity is described as the brain's ability to change and has two primary forms: structural and functional plasticity. Structural plasticity involves physical changes at the synaptic formation and dendritic remodeling. Functional plasticity refers to the reorganization of neural circuitry to support new functions and compensate for impairments. These changes occur across the lifespan that makes neuroplasticity relevant to individuals of all ages undergoing neuropsychological rehabilitation (Role of Neuroplasticity in Neuro-Rehabilitation, n.d.).

Several mechanisms drive neuroplastic alterations including synaptic plasticity, dendritic remodeling, axonal reorganization, and molecular change. Enhanced expression of growth-related genes, proteins, and neurotrophic factors such as brain-derived neurotrophic factor (BDNF), promotes neural survival and adaptability. Rehabilitation exercises provide crucial stimuli that drive plasticity. Targeted interventions such as constraint-induced movement therapy for stroke patients or cognitive training for persons with TBI, activate neuroplastic mechanisms to rewire damaged areas and to improve functionality. These activities enhance and strengthen synaptic activities and neural pathways. Intrinsic neuroplastic processes play a significant role in spontaneous recovery of injured regions. The brain's innate capacity to restore function without external intervention is a limited phenomenon. Rehabilitation focuses on spontaneous processes by accelerating and guiding recovery through structured activities. While the brain displays peak plasticity during childhood, studies indicate that cortical plasticity persists through adulthood. Adults can undergo major structural and functional changes in response to learning, injury or rehabilitation. Studies with stroke patients indicate improved motor function through therapy that has applied neuroplastic changes (Berlucchi & Buchtel, 2009; Maier et al., 2019; Puderbaugh & Emmady, 2024).

Advanced neuroimaging techniques such as functional magnetic resonance imaging (fMRI), positron emission tomography (PET), and diffusion tensor imaging (DTI), provide direct evidence of neuroplastic changes. These tools allow clinicians to visualize alterations in brain structure, monitor functional connectivity, and evaluate the efficacy of rehabilitation interventions. For example, fMRI studies have documented enhanced activity in brain regions associated with motor control following targeted exercises. Research in skill acquisition highlights that neuroplasticity has the capacity to induce significant brain changes. Learning new skills, such as literacy or playing a musical instrument, demonstrates that consistent practice reshapes neural pathways among adults. Rehabilitation programs tailored to specific cognitive or motor deficits utilize this capacity to facilitate recovery. The principles of neuroplasticity inform the design of rehabilitation protocols. Targeted interventions aim to stimulate brain regions associated with the affected functions, enabling reorganization and compensation. For instance, repetitive transcranial magnetic stimulation (rTMS) and virtual reality-based therapies are innovative approaches that enhance neuroplasticity to promote functional recovery in conditions like stroke, brain injury, or neurodegenerative diseases (Prigatano et al., n.d.).

5.2.1 Integration of AI-Driven Tools in Neuroplasticity-Based Rehabilitation

Contemporary advancements in AI have complemented the neuroplasticity framework by amplifying neuroplastic changes by tailoring therapeutic interventions to individual needs to ensure that dynamic adaptation is possible throughout the recovery process. AI complements external stimuli by personalization and dynamic adjustment of therapeutic tasks based on the progress of the patient. AI enhances external stimuli through customization and dynamically adjusting therapeutic tasks based on patient progress. AI-powered systems, such as robotic exoskeletons or gamified rehabilitation platforms, support repetition. The tools used can provide real-time data, optimizing the intensity and complexity of tasks to stimulate targeted brain regions, thereby fostering synaptic plasticity and functional reorganization. AI enhances ecological validity in rehabilitation by stimulating real-world environments and activities. These realistic simulations, tailored to patients' everyday contexts, bridge the gap between clinical training and real-life functionality. Studies on TBI patients have indicated that AI systems help recover social and communication skills in controlled, interactive settings, which strengthens neural circuits associated with memory, attention and executive functioning (Berlucchi & Buchtel, 2009; Ying et al., 2024).

Feedback mechanisms used by AI uses real-time data analysis from wearable sensors or motion trackers providing immediate corrective feedback. AI-driven feedback systems have been shown to improve gait patterns by identifying improper movements and offering adaptive cues to optimize motor function. In neurocognitive rehabilitation, AI-powered tools provide personalized training programs to restore executive functions such as memory and problem-solving. Incorporating neurofeedback systems such as technologies like electroencephalogram (EEG) enable patients to actively regulate their brain activity thereby reinforcing plasticity for higher cognitive functions (Zotey et al., n.d.; Avila et al., 2004; Mennella et al., 2023). By integrating neuroplasticity into neuropsychological rehabilitation, clinicians can design evidence-based programs that harness the brain's potential for reorganization and healing. This approach highlights the dynamic interplay between therapy, neural adaptation, and behavioral improvement, offering a robust pathway to recovery and enhanced well-being for individuals with neurological conditions.

5.3 Applications of AI in Neurocognitive Rehabilitation

The advent of artificial intelligence (AI) in neurocognitive rehabilitation has transformed therapeutic approaches, offering tailored, adaptive, and effective solutions for patients with neurological and cognitive impairments. AI's ability to process large datasets and provide real-time feedback allows for personalized care that addresses specific deficits across conditions such as stroke, traumatic brain injury (TBI), Parkinson's disease, autism spectrum disorder (ASD), and post-traumatic stress disorder (PTSD). This chapter explores the role of AI in advancing neurocognitive rehabilitation practices, drawing insights from recent studies and applications.

5.3.1 AI in Stroke Rehabilitation

Stroke survivors struggle with motor impairments, cognitive deficits, and emotional difficulties that can have debilitating effects on daily functioning. AI systems are increasingly applied in the rehabilitation of lost functions or to introduce compensatory strategies.

Motor Rehabilitation AI-powered rehabilitation systems excel at facilitating motor recovery in stroke patients through gamified exercises and robotic-assisted therapies. Gamification of AI creates environments that improve coordination, balance, and fine motor skills. In cases of AI-based therapies to include activities of daily living, AI-based VR therapy that simulates daily tasks such as cooking, driving, and navigating fine motor skills to promote independence and in enhancing motor recovery and rehabilitation (Khalid et al., 2024).

Combining AI with robotics to train motor skills provides real-time feedback. The systemic analysis of patient movements and dynamic adjustments of the intensity of exercises allows customizability of treatment plans. Research published in Neuroinformatics highlights the use of AI-based exoskeletons to improve muscle activation and to improve motor rehabilitation, especially with respect to coordination in patients with stroke. Interventions with these contemporary strategies have proven to have significantly beneficial outcomes in improving mobility and in reducing recovery time when compared to traditional approaches (Javed et al., 2023; Stefano, 2023).

Cognitive Rehabilitation Stroke patients often experience cognitive impairments such as memory loss, attention deficits, and reduced problem-solving abilities. AI platforms integrate adaptive learning algorithms into cognitive training programs to address these challenges. A study by the National Institutes of Health found that AI-powered tools significantly enhanced cognitive function by tailoring tasks to patient-specific deficits. Interactive AI systems also provide emotional support during rehabilitation. Chatbots and conversational agents use natural language processing (NLP) to engage patients, reducing feelings of isolation and improving motivation. These technologies offer a holistic approach to recovery, addressing both cognitive and emotional dimensions (Javed et al., 2023; Lv & Guo, 2022).

5.3.2 AI in Traumatic Brain Injury (TBI) Rehabilitation

TBI rehabilitation is complex, requiring a combination of cognitive, motor, and emotional interventions. AI technologies have become indispensable in designing personalized rehabilitation plans for TBI patients.

Cognitive Rehabilitation AI excels in delivering cognitive therapies tailored to the unique needs of TBI patients. Systems equipped with machine learning algorithms analyze cognitive deficits and create customized training modules. For instance, research demonstrated that AI-based cognitive programs improved executive functioning and processing speed more effectively than standard therapies (Templeton et al., 2020).

Recent advancements include virtual reality-based cognitive simulations. A study in Frontiers in Psychology explored how AI-powered VR environments allowed TBI patients to practice tasks such as navigating a busy street or managing household chores. These simulations improved cognitive flexibility and problem-solving skills, enabling patients to regain autonomy in their daily lives (Calvillo & Irimia, 2020).

Motor Rehabilitation AI tools like robotic systems and motion-sensing technologies are used to improve motor functions in TBI patients. These tools monitor patient movements in real time, providing feedback to correct posture and gait irregularities. Research published in Cognitive Computation emphasized the role of AI in optimizing motor rehabilitation through predictive modeling, which helps clinicians develop more effective exercise regimens (Javed et al., 2023).

1. Traumatic Brain Injury (TBI):

AI has emerged as a transformative tool in the rehabilitation of patients with traumatic brain injury (TBI). These systems play a critical role in personalizing and adapting cognitive training modules to target-specific cognitive deficits resulting from injury. The development of personalized AI-driven rehabilitation programs relies on a comprehensive framework that integrates clinical insights with advanced machine learning (ML) techniques. The personalization process begins with comprehensive cognitive assessments, such as thorough neuropsychological assessments (NPAs), to identify deficits across multiple cognitive domains, including memory, attention, and executive function. These assessments form the foundation for mapping out a patient's cognitive profile. Using ML algorithms, raw results from cognitive assessments are aggregated and recalibrated to align with established neuropsychological standards, ensuring that the cognitive profile is both accurate and actionable (Abelmann et al., 2023).

AI systems generate tailored training modules based on identified deficits. For example, a patient struggling with working memory might receive tasks requiring tracking and manipulating information over short intervals. These created training tasks (CTTs) align with the patient's current abilities and gradually increase in complexity to foster cognitive growth. Task difficulty is dynamically adjusted based on patient performance—if a patient excels at a task, the AI increases the challenge; if they struggle, the complexity is reduced. This adaptive mechanism ensures both engagement and progress. A dynamic profile adaptation process, rooted in belief revision (BR) theory, continuously refines the patient's cognitive profile. This allows the system to respond to changes in the patient's capabilities over time, ensuring sustained therapeutic relevance. Additionally, AI training modules cover multiple cognitive domains, integrating tasks to improve memory, attention, and problem-solving in a holistic approach to neurorehabilitation (Oberholzer & Müri, 2019).

The transferability of skills to real-world tasks is a critical component of AI training modules. These programs simulate ecological scenarios, such as navigating a crowded shopping mall or managing workplace responsibilities, fostering skills directly applicable to daily life. Gamified elements are also incorporated to enhance engagement, with features like scoring, rewards, and progress tracking helping to maintain motivation during repetitive cognitive exercises. Despite the central role of AI, clinical oversight remains vital. Clinicians analyze results, adjust protocols, and provide emotional support, ensuring the technology complements human expertise rather than replacing it. Real-time feedback is another integral feature, enabling patients and clinicians to monitor performance and progress, facilitating timely adjustments to training modules and strategies. AI-driven rehabilitation programs are tailored to enhance cognitive recovery and support reintegration into society. For example, patients struggling with attentional deficits can use training modules focused on multitasking in complex scenarios, such as managing virtual customer service interactions. These systems are particularly effective in improving executive functioning and processing speed, which are often impaired in TBI patients. Furthermore, AI simulations of real-world environments allow patients to rehearse critical life skills, such as managing finances or navigating public transportation, promoting functional independence and improving quality of life (Calvillo & Irimia, 2020; Yang et al., 2024).

5.3.3 AI in Parkinson's Disease Rehabilitation

Parkinson's disease, characterized by progressive motor and cognitive decline, presents unique rehabilitation challenges. AI has proven effective in managing both motor symptoms and cognitive deficits.

Motor Symptom Management AI-driven robotic systems have revolutionized gait training for Parkinson's patients. These systems analyze gait patterns, identify abnormalities, and provide corrective feedback in real time. Fasano et al. (2018) found that AI-powered gait training programs improved walking speed and reduced freezing of gait episodes, outperforming traditional physical therapy (Khalid et al., 2024). Advanced AI applications also incorporate movement therapies such as dance or yoga into rehabilitation programs. These interventions not only improve motor coordination but also promote emotional well-being. AI algorithms continuously adapt the intensity of exercises based on patient progress, ensuring optimal outcomes.

Cognitive and Emotional Rehabilitation Cognitive deficits in Parkinson's patients, including memory loss and reduced problem-solving abilities, are addressed through AI-based training platforms. These systems incorporate puzzles, memory exercises, and problem-solving tasks that adapt to the patient's cognitive abilities. A study in Psychological Research highlighted that such interventions significantly improved cognitive flexibility and attention spans (Deal et al., 2019).

AI tools also monitor emotional states using sentiment analysis and facial recognition technologies. By detecting signs of anxiety or depression, these systems enable timely interventions, improving the overall mental health of Parkinson's patients. AI-based platforms offer adaptive training using puzzles, memory exercises, and problem-solving tasks, significantly enhancing cognitive flexibility and attention spans. Emotional monitoring is achieved through sentiment analysis and facial recognition technologies, enabling timely interventions for anxiety or depression. In motor rehabilitation, AI-powered systems analyze gait patterns using data collected from wearable or nonwearable devices. These systems process gait data, extract features like stride length and cadence, and use predictive algorithms to identify and manage motor anomalies such as freezing of gait (FOG). Advanced tools benchmark performance against clinical standards like UPDRS III, while real-time analysis supports rapid decision-making. Integration with robotics, such as adaptive exoskeletons, aids in symptom relief and mobility improvement. Personalization and multidimensional analysis ensure tailored interventions, enhancing accuracy and detecting subtle motor changes. Studies show AI can reduce FOG episodes, improve walking speed, and enable earlier diagnosis and management of PD progression (Iseki et al., 2023; MS, 2022).

5.3.4 AI in Autism Spectrum Disorder (ASD) Rehabilitation

AI applications in ASD therapy focus on enhancing social communication skills and managing sensory sensitivities. These tools create engaging, nonthreatening environments for learning, making therapy more accessible and effective.

Social Skills Development AI-powered systems simulate social scenarios to teach communication and interaction skills. Virtual reality platforms allow individuals with ASD to practice conversations, job interviews, or group activities in a controlled setting. Didden et al. (2016) demonstrated that such interventions significantly improved social anxiety and communication skills compared to traditional therapies (Khalid et al., 2024).

Chatbots are another innovative tool in ASD therapy. These AI-driven conversational agents adapt their behavior to the user's responses, providing a personalized and stress-free interaction experience. Studies have shown that practicing with chatbots increases conversational confidence and reduces social anxiety in individuals with ASD (Catania et al., 2024).

Sensory Integration Therapy AI tools address sensory sensitivities through controlled exposure to stimuli. For example, Ben-Sasson et al. (2013) used AI-driven sound therapy to reduce sensory over-responsivity in children with autism, leading to improved emotional regulation and behavior (Catania et al., 2024). AI systems can also create personalized sensory environments, gradually desensitizing individuals to triggers like loud noises or bright lights (Javed et al., 2023).

5.3.5 AI in Post-Traumatic Stress Disorder (PTSD) Rehabilitation

PTSD rehabilitation has benefited immensely from AI technologies, particularly in the areas of exposure therapy and emotional regulation.

Exposure Therapy Virtual reality exposure therapy (VRET) is a widely used AI application in PTSD treatment. Patients are gradually exposed to trauma-related stimuli within a safe, simulated environment, allowing them to process and desensitize traumatic memories. Rothbaum et al. (2014) found that VRET significantly reduced PTSD symptoms in veterans compared to traditional exposure therapy. AI systems in VRET adjust the intensity of stimuli based on the patient's physiological responses, ensuring a personalized and effective treatment experience (Templeton et al., 2020).

Emotional Monitoring and Behavioral Insights AI-powered tools monitor emotional and physiological responses during therapy, providing valuable insights for clinicians. For example, NLP algorithms analyze speech patterns to detect emotional distress, while wearable sensors track heart rate and skin conductivity. These data points help therapists adjust treatment plans in real time, optimizing outcomes (Lv & Guo, 2022).

Recent research in Cognitive Computation highlighted the use of AI-driven sentiment analysis in PTSD therapy. This approach enables early detection of emotional dysregulation, allowing for timely interventions and reducing the risk of relapse (Scott et al., 2022).

5.4 Key Benefits of AI in Rehabilitation

Artificial intelligence (AI) is reshaping the landscape of rehabilitation by addressing individual patient needs, enhancing motivation, improving safety, and providing quantifiable data for progress tracking. Below, we delve into how AI delivers these benefits and its potential to transform outcomes for patients in various rehabilitation contexts.

1. *Personalization*: AI systems allow customization to meet individual patient needs. One of the most significant advantages of AI in rehabilitation is its ability to tailor interventions to individual needs. This level of personalization is achieved through advanced data analysis and adaptive algorithms:

    - Tailored Treatment Plans: AI analyzes vast amounts of patient data, including medical history, current health status, and progress metrics, to develop personalized treatment plans. This ensures that the interventions are precisely aligned with each patient's specific deficits and rehabilitation goals (Das et al., 2022; Solo-Step, 2023).
    - Customized Rehabilitation Exercises: AI-powered tools create and recommend exercises designed to address the unique capabilities and limitations of each patient. These exercises adapt dynamically as the patient's performance improves, ensuring ongoing relevance and effectiveness (Das et al., 2022; Khalid et al., 2024).
    - Reinforcement Learning Agents: Advanced AI systems, such as reinforcement learning agents, interact with patients during therapy sessions. These agents employ algorithms that continuously adapt to patient responses, providing real-time adjustments that enhance the effectiveness of training interventions (Das et al., 2022).

    For example, a stroke survivor with limited motor function may receive AI-driven guidance through exercises designed to improve arm strength and coordination, with difficulty levels adapting to their progress over time. This personalized approach not only maximizes recovery potential but also ensures efficient use of time and resources.
2. *Motivation and Engagement:* Gamification and interactive features sustain interest and adherence. Adherence to rehabilitation programs can be a challenge for

many patients. AI addresses this issue by integrating engaging elements into therapy:

- Immersive Technologies: Augmented reality (AR) and virtual reality (VR) technologies create interactive and immersive environments that captivate patients. These environments transform mundane exercises into stimulating experiences, helping maintain patient interest and motivation (Maleki Varnosfaderani & Forouzanfar, 2024, p. 1). For instance, a VR game may require a patient to reach for virtual objects, subtly encouraging movements that support motor skill recovery.
- Visual Feedback: Tools like motion capture systems and video biofeedback provide real-time visualizations of a patient's movements. This immediate feedback enables patients to understand and correct their actions, making therapy sessions more engaging and productive.
- Mobile Accessibility: AI-powered mobile applications allow patients to access therapy exercises remotely, ensuring continuity of care. Patients can engage in guided sessions at home, reducing the inconvenience of traveling to therapy centers while maintaining consistent progress (Shomo, 2024).
- Gamification: Incorporating scoring systems, rewards, and progress tracking into therapy modules encourages adherence and provides patients with a sense of achievement. Studies show that gamified elements improve both short-term participation and long-term commitment to rehabilitation programs (Sumner et al., 2023).

3. *Safety and Accessibility*: Simulated environments reduce risks associated with real-world practice. AI technology enhances the safety and accessibility of rehabilitation, enabling patients to practice essential skills in risk-free environments:

- Controlled Virtual Spaces: Virtual reality allows patients to simulate real-world activities, such as crossing a busy street or navigating a grocery store, within a safe and controlled environment. This reduces the risks associated with practicing these tasks in real-life settings (Solo-Step, 2023).
- Supportive Mobility Systems: Technologies like overhead track systems and robotic exoskeletons provide physical support during balance and mobility exercises. These tools minimize the risk of falls and injuries, making it safer for patients to attempt challenging movements.
- Remote Therapy Accessibility: AI-powered rehabilitation platforms facilitate therapy sessions from the comfort of a patient's home. This improves access to care for individuals with mobility limitations or those living in remote areas, ensuring that geographic location is no longer a barrier to receiving quality rehabilitation (Artificial Intelligence (AI) In Health Care and Rehabilitation, n.d.).

For example, a patient recovering from a traumatic brain injury may use a virtual kitchen simulation to practice cooking. In this safe virtual setting, they can regain independence without the risks associated with handling real appliances or sharp objects.

4. *Quantifiable Progress*: Objective data collection allows clinicians to track improvements and refine interventions. The ability to measure and monitor progress is crucial for effective rehabilitation. AI systems excel in collecting and analyzing data to provide actionable insights:
   - Objective Movement Analysis: Motion capture technology and video biofeedback generate precise measurements of movement patterns, such as joint angles, stride length, and balance. These metrics allow clinicians to assess improvements over time and adjust therapy plans accordingly (Khalid et al., 2024).
   - Diagnostic Accuracy: AI tools offer objective assessments of patient abilities, reducing reliance on subjective evaluations. For example, an AI algorithm may analyze gait data to detect subtle motor deficits in a Parkinson's disease patient that might otherwise go unnoticed (Khalid et al., 2024).
   - Early Issue Detection: By processing large datasets, AI can identify early warning signs of complications or stagnation in recovery. This enables clinicians to intervene promptly and refine therapeutic strategies to address emerging challenges (Maleki Varnosfaderani & Forouzanfar, 2024).
   - Progress Visualization: AI tools provide visual representations of progress, such as graphs or animations, helping patients and clinicians better understand the impact of interventions. This transparency fosters a collaborative approach to rehabilitation, where patients actively participate in setting and achieving recovery goals.

   For instance, a patient undergoing rehabilitation for a spinal cord injury may use a motion capture system to track improvements in walking gait. The data not only quantifies their progress but also highlights areas requiring additional focus, ensuring a targeted and efficient recovery plan.

5.5 Challenges and Limitations

Despite its advantages, AI in neurocognitive rehabilitation faces challenges: The integration of artificial intelligence (AI) in neurocognitive rehabilitation holds great promise in improving treatment outcomes for individuals with neurocognitive disorders, including dementia, stroke, and traumatic brain injury. However, as with any emerging technology, several challenges must be addressed before AI can be fully integrated into rehabilitation practices. The following are key limitations and challenges of AI in neurocognitive rehabilitation:

1. Cost:

One of the primary barriers to the widespread adoption of AI in neurocognitive rehabilitation is the high initial cost. Implementing AI systems requires significant investment in both hardware and software, as well as the infrastructure necessary to support these technologies. AI solutions often involve advanced machine learning models, specialized sensors, and devices that can track patients' movements and cognitive states. These technologies come with a hefty price tag, which can be prohibitive for many healthcare facilities, especially those operating on limited budgets.

Moreover, many healthcare systems and institutions in lower-resource areas may struggle to afford these innovations, leading to disparities in access to care. While the long-term benefits of AI, such as improved rehabilitation outcomes and cost savings from more efficient therapy, may justify the initial expenditure, the cost remains a significant obstacle. Furthermore, insurance companies may not cover AI-driven rehabilitation services, creating a financial barrier for patients who would otherwise benefit from these treatments (Prats-Bisbe et al., 2024).

2. Technical Issues:

AI-based systems are often highly complex, integrating multiple technologies and software solutions. As a result, technical issues such as hardware malfunctions, software glitches, and system failures can significantly disrupt therapy and compromise patient outcomes. Technical failures during rehabilitation sessions could lead to loss of critical therapeutic data, delay in treatment, or even harm if AI systems provide incorrect guidance or recommendations due to errors.

Ensuring the reliability and robustness of AI systems is a challenging task. Unlike traditional rehabilitation tools, AI systems require constant monitoring and maintenance to prevent technical failures. Additionally, these systems must be adaptable to various rehabilitation needs and clinical settings. Developing AI technology that is both resilient to malfunctions and capable of adapting to different patient conditions requires extensive testing and quality control measures, which can be resource-intensive (Alsobhi et al., 2022).

3. Usability:

Another significant challenge in AI neurocognitive rehabilitation is the usability of AI systems, particularly for elderly individuals and patients with cognitive impairments. Older adults, who are often the primary recipients of neurocognitive rehabilitation, may find it difficult to interact with AI-based tools, especially if they have limited experience with technology. Cognitive decline can further complicate the learning and effective use of AI systems. For example, patients may struggle to navigate complex interfaces or may not understand how to engage with interactive therapy tools.

User-friendly design is essential to ensuring that AI systems are accessible to all patient populations. For patients with cognitive impairments, interfaces need to be intuitive, with clear instructions and simple navigation. Additionally, AI systems must be designed to account for the varying degrees of cognitive decline, providing personalized support and adjustments as needed. Involving patients and healthcare professionals in the design and testing phases of AI systems can help ensure these systems are truly user-centered and accessible (Maggio et al., 2023).

4. Limited Evidence of Efficacy:

While there is growing interest in the potential of AI for neurocognitive rehabilitation, there is still limited evidence regarding its long-term efficacy. The majority of existing studies focus on short-term outcomes, and comprehensive, longitudinal research is required to validate the sustained effectiveness of AI interventions. Rigorous, evidence-based trials are necessary to compare AI-driven rehabilitation to traditional therapies, assessing its impact on cognitive recovery, quality of life, and other relevant metrics over extended periods.

Furthermore, standardization of AI applications across different rehabilitation settings remains an unmet need. Without standardized protocols, it is difficult to assess the effectiveness of AI systems across diverse populations, including individuals with different neurocognitive disorders or those from various cultural and socioeconomic backgrounds. Developing these protocols will be crucial in ensuring that AI interventions are not only effective but also equitable.

5.6 Additional Challenges

Aside from the issues mentioned above, several additional challenges hinder the widespread use of AI in neurocognitive rehabilitation:

1. *Interoperability*: AI systems need to integrate seamlessly with existing healthcare technologies, such as electronic health records (EHRs) and other clinical data management systems. However, many AI solutions currently lack interoperability, making it difficult to share and analyze patient data across platforms. This lack of integration can hinder the efficiency and effectiveness of AI in rehabilitation and may create barriers for clinicians in making informed decisions based on real-time data.
2. *Transparency and Trust*: One of the ongoing concerns regarding AI systems is the lack of transparency in decision-making processes. AI algorithms, particularly those that rely on deep learning techniques, can be "black boxes" that make decisions without providing clear explanations for how they arrived at those conclusions. This lack of transparency can reduce clinicians' trust in AI systems, as they may not fully understand the rationale behind AI-driven recommendations. Building trust in AI technologies will require greater transparency in how these systems function and make decisions.
3. *Ethical and Regulatory Considerations*: The use of AI in healthcare, particularly in sensitive areas like neurocognitive rehabilitation, raises ethical and regulatory concerns. Issues such as patient privacy, data security, and informed consent are critical in ensuring that AI is used responsibly. Additionally, regulatory bodies must establish clear guidelines for the use of AI in medical applications, ensuring that these technologies meet safety and efficacy standards. Addressing these ethical and regulatory challenges will be essential for the responsible and ethical deployment of AI in healthcare.

4. *Clinical Actionability*: Finally, making the insights provided by AI systems actionable for clinicians remains a challenge. AI can generate vast amounts of data and analysis, but clinicians may find it difficult to interpret this information in a way that directly informs treatment decisions. It is essential that AI tools are designed not only to collect and analyze data but also to present it in a clear, actionable format that helps clinicians provide personalized care to patients (Akbar et al., 2024; Khalid et al., 2024; Lanotte et al., 2023; Sumner et al., 2023).

5.7 Future Directions and Innovations

The potential for AI in neurocognitive rehabilitation continues to expand, driven by technological advancements:

1. *Integration with Artificial Intelligence (AI)*: AI can personalize AI environments by analyzing patient progress and adapting exercises in real time. Artificial intelligence (AI) has the potential to revolutionize neurocognitive rehabilitation by making therapies more personalized and adaptive. AI systems can analyze a patient's progress in real time, adjusting exercises and tasks to match their individual recovery needs. For example, if a patient shows improvement in specific motor skills, the AI can increase the difficulty of exercises to ensure continued growth. Conversely, if a patient is struggling, the system can simplify tasks or provide targeted feedback to address their challenges. This real-time adaptability fosters a more dynamic and engaging therapy experience, reducing frustration and maximizing recovery outcomes. By continuously learning from each interaction, AI can tailor rehabilitation environments to suit individual preferences and cognitive abilities, offering a more effective and patient-centered approach.
2. *Wearable AI Devices*: Compact, portable AI solutions will enhance accessibility and usability. The development of compact and portable AI-powered devices, such as smartwatches, neural headbands, or sensor-equipped wearables, is making rehabilitation more accessible and convenient. These devices can monitor a patient's movements, provide real-time feedback, and even detect subtle changes in performance or behavior that may indicate progress or setbacks. For example, an AI-powered wearable might guide a stroke patient through a series of arm exercises, offering immediate corrections if improper form is detected. Such devices enable patients to practice therapy outside clinical settings, integrating rehabilitation seamlessly into their daily lives. This approach empowers patients to take a more active role in their recovery while also reducing their reliance on in-person sessions, which can be costly and time-consuming.
3. *Neurofeedback Integration*: Combining AI with neuroimaging tools like EEG can provide insights into brain activity during therapy. Combining AI with neuroimaging tools like EEG (electroencephalography) is unlocking new possibilities for neurofeedback in rehabilitation. By analyzing real-time brain activity, AI can provide deeper insights into how different therapeutic activities affect a

patient's cognitive and neural functions. For instance, during a memory exercise, an AI-integrated EEG system can detect patterns of brain activity associated with focus or fatigue. Based on this data, the AI can adjust the session—either by modifying the task or suggesting a break—to ensure that the therapy remains both effective and sustainable. This integration enables clinicians to create highly targeted interventions, accelerating recovery, and enhancing the overall effectiveness of treatment.

4. *Tele-Rehabilitation*: AI-based interventions delivered remotely will increase access for patients in underserved regions. Tele-rehabilitation, powered by AI, is breaking down geographical barriers to care, offering new opportunities for patients in underserved or rural regions. AI-based tele-rehabilitation platforms allow patients to access personalized therapy from the comfort of their homes. These platforms can monitor progress remotely, provide interactive exercises, and even offer real-time adjustments based on performance. For example, a patient recovering from a traumatic brain injury might engage with an AI-driven virtual therapist that tracks their cognitive improvements and suggests daily exercises. Clinicians, in turn, can remotely review detailed progress reports generated by the AI, ensuring continuity of care without the need for frequent in-person visits. This approach not only enhances accessibility but also provides a cost-effective way to deliver high-quality rehabilitation to a larger population.

Emerging AI technologies hold immense promise for the future of rehabilitation. Reinforcement learning, for example, enables systems to learn optimal strategies through trial and error, improving therapy recommendations over time.

Generative AI, another cutting-edge development, creates personalized simulations and exercises tailored to patient preferences. For instance, an AI system could design virtual environments that mimic a patient's favorite outdoor activities, making therapy both enjoyable and effective.

Collaborative AI, where multiple systems integrate data from various modalities such as motion tracking, speech analysis, and physiological monitoring, offers a holistic approach to rehabilitation. These systems combine insights from diverse sources to deliver comprehensive and precise care.

5.8 Conclusion

Artificial intelligence represents a paradigm shift in neurocognitive rehabilitation, offering innovative solutions to enhance therapy outcomes. Its ability to create immersive, engaging, and adaptive interventions positions it as a cornerstone in the future of rehabilitation sciences. As technology continues to evolve, the integration of AI with other cutting-edge tools promises to revolutionize the landscape of neurocognitive rehabilitation, making recovery more effective, accessible, and patient-centric. AI is revolutionizing rehabilitation by enhancing personalization, precision, and accessibility. From real-time feedback and predictive analytics to robotic assistance and cognitive training, AI addresses diverse patient needs across physical and cognitive domains. As technology continues to evolve, its potential to empower patients and clinicians alike grows exponentially. However, ethical and accessibility challenges must be addressed to ensure that these advancements benefit all.

To overcome challenges, researchers, developers, clinicians, and policymakers must collaborate to improve AI systems and their implementation in neurocognitive rehabilitation. This includes refining AI technologies for greater usability, improving evidence through long-term studies, enhancing system reliability, and developing standardized protocols. Furthermore, fostering interdisciplinary collaboration will be crucial for addressing technical, ethical, and regulatory issues and ensuring that AI benefits are accessible to all patients, regardless of their socioeconomic status or geographic location.

Ultimately, while AI in neurocognitive rehabilitation presents several challenges, it also holds the potential to revolutionize the field, offering more personalized, efficient, and accessible therapies for individuals with neurocognitive disorders. With ongoing research, innovation, and collaboration, the barriers to adoption can be mitigated, unlocking AI's full potential in transforming neurocognitive rehabilitation. The future of AI in neurocognitive rehabilitation is incredibly bright. By integrating with advanced technologies, enabling remote care, and tailoring therapies to individual needs, AI has the potential to transform lives. However, realizing this vision will require collaboration among researchers, clinicians, developers, and policymakers to address challenges and ensure ethical, equitable implementation.

With thoughtful innovation and continued investment, AI can move from being a promising tool to an integral part of rehabilitation, helping patients regain their independence and improve their quality of life.

References

Abelmann, A. C., Kessels, R. P. C., Brazil, I. A., Fasotti, L., & Bertens, D. (2023). Game-supported cognitive strategy training for slowed information processing speed after acquired brain injury: Study protocol for a randomised controlled trial. *BMJ Open, 13*(9), e067108. https://doi.org/10.1136/bmjopen-2022-067108

Akbar, K., Passaro, A., Di Gioia, M., Martini, E., Dragone, M., Zullo, A., & Stasolla, F. (2024). Reviewing the horizon: The future of extended reality and artificial intelligence in neurorehabilitation for brain injury recovery. *Information, 15*(8), 8. https://doi.org/10.3390/info15080501

Alsobhi, M., Sachdev, H. S., Chevidikunnan, M. F., Basuodan, R., Ku, D. K., & Khan, F. (2022). Facilitators and barriers of artificial intelligence applications in rehabilitation: A mixed-method approach. *International Journal of Environmental Research and Public Health, 19*(23), 23. https://doi.org/10.3390/ijerph192315919

Artificial Intelligence (AI) In Health Care and Rehabilitation. (n.d.). Physiopedia. Retrieved December 15, 2024, from https://www.physio-pedia.com/Artificial_Intelligence_(AI)_In_Health_Care_and_Rehabilitation

Avila, R., Bottino, C. M. C., Carvalho, I. A. M., Santos, C. B., Seral, C., & Miotto, E. C. (2004). Neuropsychological rehabilitation of memory deficits and activities of daily living in patients with Alzheimer's disease: A pilot study. *Brazilian Journal of Medical and Biological Research = Revista Brasileira De Pesquisas Medicas E Biologicas, 37*(11), 1721–1729. https://doi.org/10.1590/s0100-879x2004001100018

Ben-Sasson, A., Soto, T. W., Martínez-Pedraza, F., & Carter, A. S. (2013). Early sensory over-responsivity in toddlers with autism spectrum disorders as a predictor of family impairment and parenting stress. *Journal of Child Psychology and Psychiatry, 54*(8), 846–853. https://doi.org/10.1111/jcpp.12035

Berlucchi, G., & Buchtel, H. A. (2009). Neuronal plasticity: Historical roots and evolution of meaning. *Experimental Brain Research, 192*(3), 307–319. https://doi.org/10.1007/s00221-008-1611-6

Calderone, A., Latella, D., Bonanno, M., Quartarone, A., Mojdehdehbaher, S., Celesti, A., & Calabrò, R. S. (2024). Towards transforming neurorehabilitation: The impact of artificial intelligence on diagnosis and treatment of neurological disorders. *Biomedicine, 12*(10), 10. https://doi.org/10.3390/biomedicines12102415

Calvillo, M., & Irimia, A. (2020). Neuroimaging and psychometric assessment of mild cognitive impairment after traumatic brain injury. *Frontiers in Psychology, 11*, 1423. https://doi.org/10.3389/fpsyg.2020.01423

Catania, V., Rundo, F., Panerai, S., & Ferri, R. (2024). Virtual reality for the rehabilitation of acquired cognitive disorders: A narrative review. *Bioengineering, 11*(1), 1. https://doi.org/10.3390/bioengineering11010035

Das, R., Paul, S., Mourya, G. K., Kumar, N., & Hussain, M. (2022). Recent trends and practices toward assessment and rehabilitation of neurodegenerative disorders: Insights from human gait. *Frontiers in Neuroscience, 16*, 859298. https://doi.org/10.3389/fnins.2022.859298

Deal, L. S., Flood, E., Myers, D. E., Devine, J., & Gray, D. L. (2019). The Parkinson's disease activities of daily living, interference, and dependence instrument. *Movement Disorders Clinical Practice, 6*(8), 678. https://doi.org/10.1002/mdc3.12833

Didden, R., Korzilius, H., van Zonneveld, K., & Curfs, L. M. G. (2016). The impact of sensory over-responsivity on daily functioning in children with autism spectrum disorder. *Research in Autism Spectrum Disorders, 29*, 1–10. https://doi.org/10.1016/j.rasd.2016.05.004

Fasano, A., De Vloo, P., Llinas, M., Hlasny, E., Kucharczyk, W., Hamani, C., & Lozano, A. M. (2018). Magnetic resonance imaging-guided focused ultrasound thalamotomy in Parkinson tremor: Reoperation after benefit decay. Movement Disorders: *Official Journal of the Movement Disorder Society, 33*(5), 848–849. https://doi.org/10.1002/mds.27348

Iseki, C., Hayasaka, T., Yanagawa, H., Komoriya, Y., Kondo, T., Hoshi, M., Fukami, T., Kobayashi, Y., Ueda, S., Kawamae, K., Ishikawa, M., Yamada, S., Aoyagi, Y., & Ohta, Y. (2023). Artificial intelligence distinguishes pathological gait: The analysis of markerless motion capture gait data acquired by an iOS application (TDPT-GT). *Sensors, 23*(13), 13. https://doi.org/10.3390/s23136217

Javed, A. R., Saadia, A., Mughal, H., Gadekallu, T. R., Rizwan, M., Maddikunta, P. K. R., Mahmud, M., Liyanage, M., & Hussain, A. (2023). Artificial intelligence for cognitive health assessment: State-of-the-art, open challenges and future directions. *Cognitive Computation, 15*(6), 1767–1812. https://doi.org/10.1007/s12559-023-10153-4

Khalid, U. B., Naeem, M., Stasolla, F., Syed, M. H., Abbas, M., & Coronato, A. (2024). Impact of AI-powered solutions in rehabilitation process: Recent improvements and future trends. *International Journal of General Medicine, 17*, 943–969. https://doi.org/10.2147/IJGM.S453903

Lanotte, F., O'Brien, M. K., & Jayaraman, A. (2023). AI in rehabilitation medicine: Opportunities and challenges. *Annals of Rehabilitation Medicine, 47*(6), 444–458. https://doi.org/10.5535/arm.23131

Lv, Z., & Guo, J. (2022). Virtual reality neurorehabilitation. *International Journal of Mental Health Promotion, 24*(3), 287–310. https://doi.org/10.32604/ijmhp.2022.019829

Maggio, M. G., De Bartolo, D., Calabrò, R. S., Ciancarelli, I., Cerasa, A., Tonin, P., Di Iulio, F., Paolucci, S., Antonucci, G., Morone, G., & Iosa, M. (2023). Computer-assisted cognitive rehabilitation in neurological patients: State-of-art and future perspectives. *Frontiers in Neurology, 14*, 1255319. https://doi.org/10.3389/fneur.2023.1255319

Maier, M., Ballester, B. R., & Verschure, P. F. M. J. (2019). Principles of neurorehabilitation after stroke based on motor learning and brain plasticity mechanisms. *Frontiers in Systems Neuroscience, 13*, 74. https://doi.org/10.3389/fnsys.2019.00074

Maleki Varnosfaderani, S., & Forouzanfar, M. (2024). The role of AI in hospitals and clinics: Transforming healthcare in the 21st century. *Bioengineering, 11*(4), 337. https://doi.org/10.3390/bioengineering11040337

Mennella, C., Maniscalco, U., De Pietro, G., & Esposito, M. (2023). The role of artificial intelligence in future rehabilitation services: A systematic literature review. *IEEE Access, 11*, 11024–11043. https://doi.org/10.1109/ACCESS.2023.3236084

MS, M. W. (2022, November 30). *Machine learning could diagnose Parkinson's by analyzing gait | Parkinson's News Today*. https://parkinsonsnewstoday.com/news/machine-learning-could-diagnose-parkinsons-analyzing-gait/

Oberholzer, M., & Müri, R. M. (2019). Neurorehabilitation of traumatic brain injury (TBI): A clinical review. *Medical Science, 7*(3), 47. https://doi.org/10.3390/medsci7030047

Prats-Bisbe, A., López-Carballo, J., García-Molina, A., Leno-Colorado, D., García-Rudolph, A., Opisso, E., & Jané, R. (2024). Virtual reality–based neurorehabilitation support tool for people with cognitive impairments resulting from an acquired brain injury: Usability and feasibility study. *JMIR Neurotechnology, 3*(1), e50538. https://doi.org/10.2196/50538

Prigatano, G. P., Braga, L. W., Johnson, S. F., & Souza, L. M. N. (n.d.). Neuropsychological rehabilitation, neuroimaging and neuroplasticity: A clinical commentary. *NeuroRehabilitation, 49*(2), 255–265. https://doi.org/10.3233/NRE-218024

Puderbaugh, M., & Emmady, P. D. (2024). Neuroplasticity. In *StatPearls*. StatPearls Publishing. http://www.ncbi.nlm.nih.gov/books/NBK557811/

Role of Neuroplasticity in Neuro-rehabilitation. (n.d.). Physiopedia. Retrieved December 14, 2024, from https://www.physio-pedia.com/Role_of_Neuroplasticity_in_Neuro-rehabilitation

Rothbaum, B. O., Price, M., Jovanovic, T., Norrholm, S. D., Gerardi, M., Dunlop, B., Davis, M., Bradley, B., Duncan, E., Rizzo, A., & Ressler, K. J. (2014). A randomized, double-blind evaluation of D-cycloserine or alprazolam combined with virtual reality exposure therapy for posttraumatic stress disorder (PTSD) in Iraq and Afghanistan War Veterans. *American Journal of Psychiatry, 171*(6), 640–648. https://doi.org/10.1176/appi.ajp.2014.13121625

Scott, H., Griffin, C., Coggins, W., Elberson, B., Abdeldayem, M., Virmani, T., Larson-Prior, L. J., & Petersen, E. (2022). Virtual reality in the neurosciences: Current practice and future directions. *Frontiers in Surgery, 8*, 807195. https://doi.org/10.3389/fsurg.2021.807195

Seelye, A. M., Schmitter-Edgecombe, M., Das, B., & Cook, D. J. (2012). Application of cognitive rehabilitation theory to the development of smart prompting technologies. *IEEE Reviews in Biomedical Engineering, 5*, 29–44. https://doi.org/10.1109/RBME.2012.2196691

Shomo. (2024, October 6). How AI is revolutionizing personalized healthcare: Insights, benefits, and challenges of tailored.... *Medium*. https://medium.com/@ys.soumya.1/how-ai-is-revolutionizing-personalized-healthcare-insights-benefits-and-challenges-of-tailored-8a1386a6b4e4

Solo-Step. (2023, November 30). 6 Rehabilitation technologies transforming healthcare in 2024. *Solo-Step*. https://solostep.com/rehabilitation-technologies/

Stefano, G. B. (2023). Artificial intelligence as a tool for the diagnosis and treatment of neurodegenerative diseases. *Brain Sciences, 13*(6), 6. https://doi.org/10.3390/brainsci13060938

Sumner, J., Lim, H. W., Chong, L. S., Bundele, A., Mukhopadhyay, A., & Kayambu, G. (2023). Artificial intelligence in physical rehabilitation: A systematic review. *Artificial Intelligence in Medicine, 146*, 102693. https://doi.org/10.1016/j.artmed.2023.102693

Templeton, J. M., Poellabauer, C., & Schneider, S. (2020). Enhancement of neurocognitive assessments using smartphone capabilities: Systematic review. *JMIR mHealth and uHealth, 8*(6), e15517. https://doi.org/10.2196/15517

Yang, R., Yuan, Q., Zhang, W., Cai, H., & Wu, Y. (2024). Application of artificial intelligence in rehabilitation science: A scientometric investigation utilizing citespace. *SLAS Technology, 29*(4), 100162. https://doi.org/10.1016/j.slast.2024.100162

Ying, H., Pranolo, A., Nuryana, Z., & Syafitri, A. I. (2024). Emerging trends in the evolution of neuropsychology and artificial intelligence: A comprehensive analysis. *Telematics and Informatics Reports, 16*, 100171. https://doi.org/10.1016/j.teler.2024.100171

Zotey, V., Andhale, A., Shegekar, T., & Juganavar, A. (n.d.). Adaptive neuroplasticity in brain injury recovery: Strategies and insights. *Cureus, 15*(9), e45873. https://doi.org/10.7759/cureus.45873

Chapter 6
Technological Innovations in Rehabilitation: Wearable Devices and Sensors

6.1 Introduction

The landscape of neuropsychological rehabilitation has been reshaped by wearable technology in the recent decades. It provides innovative solutions that enhance patient care and improves therapeutic outcomes. The electronic devices created by wearable technology are often adapted into everyday clothing or to the body-providing data on the basis of certain parameters such as heart rate, movement, temperature among other metrics. In the context of neuropsychological rehabilitation, these devices are utilized to provide real-time, actionable feedback that provides data-driven observations in areas that require modification within the rehabilitation plan. The personalization of these interventions makes the process effective in managing the neurological conditions. Through the incorporation of wearable technology in rehabilitation protocols, the field progresses to provide highly effective treatment strategies that can be applicable within clinical settings (AdornThemes, 2024; Minen & Stieglitz, 2021; The Importance of Wearable Devices in Neuro Rehabilitation, 2023).

A core benefit of wearable technology is the ability to provide continuous monitoring of a patient's physical and cognitive condition beyond the rehabilitation setting. Devices such as smartwatches, fitness trackers, and customized neural-sensors offer the potential for uninterrupted health tracking and capture alterations in the cognitive or psychical state that cannot be registered through human observation (AdornThemes, 2024; The Importance of Wearable Devices in Neuro Rehabilitation, 2023). In cases of motor rehabilitation, monitoring movement patterns can provide insights into motor impairments, heart rate variations that indicate stress or autonomic nervous system response. The continuous streaming of data enables the early detection of deteriorating symptoms and provides timely interventions in a dynamic perspective to understand patient response to rehabilitation strategies (Minen & Stieglitz, 2021; Mitchner, 2023).

K. J. Reddy, *Innovations in Neurocognitive Rehabilitation*,
https://doi.org/10.1007/978-3-031-88117-6_6

In addition to monitoring, wearable devices foster real-time feedback for both patients and healthcare providers, promoting a more responsive and interactive rehabilitation process. Through instant notifications or visual displays, these devices encourage patients to actively participate in their own care, adhering to prescribed activities or exercises. Some wearables even incorporate gamification elements, using rewards, challenges, or progress tracking to enhance motivation and adherence to rehabilitation protocols. For example, individuals recovering from stroke or traumatic brain injuries (TBIs) might use motion-sensing devices that guide and assess their physical exercises, ensuring they perform tasks correctly and effectively. Meanwhile, healthcare providers gain access to actionable insights, enabling them to tailor treatments in response to patients' progress or challenges, thus bridging the gap between clinical assessments and real-world patient experiences (Maceira-Elvira et al., 2019; Minen & Stieglitz, 2021; Mitchner, 2023; The Importance of Wearable Devices in Neuro Rehabilitation, 2023).

6.2 Types of Wearable Devices and Sensors

The integration of wearable devices and sensors into neurorehabilitative practices marks a transition from traditional views of patient-care. These technologies facilitate noninvasive, real-time data collection, enabling the monitoring of various physiological and biomechanical parameters. The five categories of wearable devices that are explored in this chapter are classified into fitness trackers, EEG, headsets, motion sensors, smart clothing, and biosensors—whose applications, capabilities, and feasibility have been tested and proved to have impact in the rehabilitation settings.

6.2.1 Fitness Trackers

Fitness trackers have risen in popularity in the mainstream consumer market via smart watches and wristbands, more recently through fitness rings. In the context of neuropsychological rehabilitation, fitness trackers have proved to be of significant impact in measuring physical activity, heart rate, and sleep patterns that provide consistent, user-friendly monitoring of critical metrics of physiological activities. Fitness trackers measure steps taken, the distance travelled, and overall levels of vital indicators that can assess mobility in patients undergoing rehabilitation. These measures enable practitioners to design and adjust treatment plans that can be tailored to individual needs. Additionally, mobility tracking is inferred through the energy exhausted and the intensity of exertion. The information is invaluable in rehabilitation planning and ensures the patient's metabolic needs are included in treatment (physiotherapy and medication). Fitness trackers are equipped with sleep monitoring features which can analyze circadian rhythms, offering critical data for neurocognitive recovery and overall health optimization (Peake et al., 2018). Recent research displays potential of wearable devices to detect falls in older adults further enhancing safety in rehabilitation contexts. Other works highlight smartwatch-based activity tracking improved exercise adherence among cardiac rehabilitation patients, illustrating the dual benefits of motivation and monitoring. They underscore the versatility and impact of fitness trackers in supporting recovery journeys (Kristoffersson & Lindén, 2022; Seneviratne et al., 2017).

6.2.2 EEG Headsets

Electroencephalography (EEG) headsets have emerged as a powerful tool for neurological rehabilitation that offers a window into brain activity. Portable EEG devices allow continuous monitoring of brain waves, providing real-time feedback essential for evaluating neural function. Leveraging this data researchers can assess the brain's capacity to reorganize itself to make up for or recover the lost functions, neuroplasticity, which is a cornerstone of cognitive and motor recovery (Peake et al., 2018). EEG headsets enable detailed evaluations of cognitive functions including attention, memory, and executive processing. The patients recovering from TBI and stroke heavily rely on the feedback during rehabilitation to ensure optimal recovery is achieved. Recent studies have identified that EEG-based neurofeedback has resulted in significantly enhanced cognitive performance in individuals with TBI. The combination of brain-computer interfaces (BCIs) with EEG promotes motor recovery in stroke patients indicating promising outcomes in overcoming neurological impairments and functional improvements (Wearable Sensors in Other Medical Domains with Application Potential for Orthopedic Trauma Surgery — A Narrative Review, n.d.; Sigcha et al., 2023).

6.2.3 Motion Sensors

Inertial measurement units (IMUs), accelerometers, and gyroscopes are critical components in improving motion sensor technologies for capturing exact movement data. In rehabilitation programs, these sensors are used to analyze gait patterns, motor deficits, and balance. Using movement monitoring patterns, physicians can discover irregularities and follow improvement over time, ensuring that therapies are effective (Peake et al., 2018). A notable application is monitoring impairment evaluation, in which motion sensors detect and measure movements in neurological patients. The success of these devices in upper limb movement tracking in stroke patients sheds light on the recovery program. Another application is in motor impairment assessment, where motion sensors can detect and quantify movements in patients with neurological conditions. The effectiveness of IMU-based systems in tracking upper limb movements in stroke patients, providing detailed insights into their recovery. Accelerometers have also been used to assess walking speed and stride length in people with Parkinson's disease. These examples demonstrate motion sensors' precision and adaptability in meeting a variety of rehabilitative demands. The inclusion of motion sensors aids another crucial application: gait analysis. These gadgets discover anomalies from typical walking patterns by analyzing stride length, cadence, and symmetry, which guides interventions for improvement. Furthermore, gyroscopes provide extensive information about body orientation and movement, assisting in balance assessments and fall risk evaluations—an important component for older persons and individuals with vestibular diseases (Peake et al., 2018).

6.2.4 Smart Clothing

Smart clothes—new technology for implanting sensor technology in clothes, offering a nonintrusive and unobtrusive mechanism for physiologic and biomechanical information collection. In smart clothes, sensors can be implanted for tracking muscle activity, postures, and respirations, and smart clothes can most effectively contribute to rehabilitation settings. For tracking activity in muscles, smart clothes with electromyography (EMG) sensors can record activity in muscles during exercise. For rehabilitation programs, such information can effectively target specific sets of muscles. Zhang et al., for instance, developed EMG-integrated smart clothes for diagnosing and distinguishing between motions in the arm in spinal injured subjects and providing them with specific feedback during therapy sessions (Peake et al., 2018; Walls et al., 2024).

Posture tracking is one such significant application, with accelerometers and gyroscopes in smart apparel providing real-time feedback about positioning during rehabilitation therapy. Patients with spinal trauma or musculoskeletal trauma can specifically gain a lot through such a feature. Certain smart apparel even include sensors for tracking respiration, a significant feature for pulmonary disease patients. For example, Huang et al. demonstrated that smart cloth-based technology could effectively monitor activity level and breathing in older adults, with an improvement in rehabilitation requirements in complex rehabilitation scenarios (Peake et al., 2018; Zhang & He, 2023).

6.2.5 Biosensors

Biosensors expand the horizons of wearable technology by measuring physiological biomarkers such as blood sugar, hydration, and stress chemicals. These sensors provide a deeper understanding of the body's internal processes, extending their applications beyond fitness tracking to specialized rehabilitation settings. One key application is stress monitoring, where wearable biosensors continuously track physiological responses to stress. By analyzing heart rate variability and cortisol levels, these sensors enable the development of personalized stress management strategies. Similarly, biosensors that track metabolic changes, such as blood sugar and lactate levels, offer valuable insights into the recovery process, particularly for patients undergoing high-intensity rehabilitation. Hydration monitoring is another critical function of biosensors, especially for individuals engaged in demanding rehabilitation programs. Maintaining optimal hydration and a proper electrolyte balance is essential for both performance and overall well-being. For example, Lopez et al. utilized biosensors to track patients' fluid intake and electrolyte levels, ensuring their hydration needs were met during therapy sessions (Peake et al., 2018). Biosensors have demonstrated their utility in addressing chronic conditions. Studies indicate that continuous glucose monitoring systems improved glycemic control in patients with type 1 diabetes undergoing rehabilitation. Similarly, biosensors predict fatigue levels in individuals with multiple sclerosis, enabling personalized interventions to optimize recovery. The application provides precise, real-time data on various biomarkers that enhance the ability to monitor and manage complex rehabilitation needs. Their versatility and accuracy make them a valuable addition to the toolkit of rehabilitation professionals, paving the way for more personalized and effective care (AdornThemes, 2024; The Importance of Wearable Devices in Neuro Rehabilitation, 2023; Wajid, 2024).

6.3 Applications in Neurocognitive Rehabilitation

The field of rehabilitation has experienced a paradigm shift with the advent of wearable technologies. These devices provide real-time, non-invasive monitoring, enabling tailored interventions and enhancing patient engagement. This section explores the applications of wearable devices in cognitive, motor, and emotional rehabilitation, drawing on examples and findings from a wide array of studies.

6.3.1 Cognitive Rehabilitation

- *Memory and Attention Training*: EEG headsets have become a significant tool in cognitive rehabilitation, particularly for memory and attention training. These devices detect even the slightest changes in brain activity, providing real-time feedback that enables targeted cognitive exercises. For example, EEG-based training has been shown to help individuals with attention deficit disorders maintain focus during complex tasks. Through neurofeedback, patients develop a greater awareness of their cognitive states, leading to improved self-regulation and enhanced mental performance. Virtual reality (VR) environments offer immersive experiences that simulate real-life scenarios, making them highly effective in cognitive rehabilitation. Patients engage in memory recall, attention tasks, and decision-making exercises within these virtual spaces. Brown et al. demonstrated that VR-based interventions led to significant improvements in memory and executive function compared to traditional methods. Stroke survivors undergoing VR rehabilitation showed enhanced problem-solving abilities by navigating virtual environments designed to replicate real-world challenges (Peake et al., 2018).

- *Biofeedback Techniques*: Biofeedback technology equipped with heart rate variability sensors has proven highly effective for mindfulness and stress reduction. These wearables track physiological changes, such as breathing patterns, and provide real-time feedback to promote relaxation. Wilson et al. found that patients using wearable biofeedback systems experienced reduced anxiety symptoms, attributing the improvement to increased self-awareness and better stress management techniques. Such interventions have been particularly beneficial for individuals with post-traumatic stress disorder (PTSD), who often struggle to regulate their physiological responses to stress. The emphasis on the importance of standardizing output measures in wearable device research for cognitive rehabilitation. Their study highlighted the diverse range of metrics used to assess critical constructs like memory and attention. Additionally, the authors stressed the need to adapt wearable technology protocols for pediatric patients, noting key differences in movement and cognitive patterns compared to adults (Peake et al., 2018).

6.3.2 Motor Rehabilitation

- *Gait Analysis and Training*: Wearable motion sensors, such as smart insoles and body-mounted devices, have transformed gait analysis and training. These sensors capture detailed data on stride length, gait speed, and balance stability, allowing clinicians to identify abnormalities and customize interventions. For example, Taylor et al. used motion sensors to track walking patterns in Parkinson's patients, providing real-time feedback that improved their gait stability and reduced the risk of falls (Peake et al., 2018).
- *Hand Function Recovery*: Sensor-equipped gloves have become an essential tool in rehabilitating hand function for patients recovering from neurological injuries. These devices track finger movements, grip strength, and dexterity, providing quantitative feedback on progress. For example, Jones et al. used sensor-embedded gloves to design personalized therapy sessions for stroke patients, targeting deficits in fine motor skills. The data collected by these gloves also helped therapists pinpoint areas requiring additional focus, ensuring a more comprehensive recovery. Wearable device systems have proved to be effective in tracking upper limb performance during daily activities. Their research emphasized the need for standardized metrics, such as use ratios, to accurately assess progress. These insights have contributed to more reliable assessments and improved therapeutic outcomes in motor rehabilitation (Peake et al., 2018).

6.3.3 *Emotional and Behavioral Monitoring*

- *Stress and Mood Tracking*: Wearables equipped with biosensors can detect physiological changes associated with emotional states like stress, anxiety, and depression. These devices monitor parameters such as heart rate variability, skin conductance, and even facial expressions, providing insights into a patient's emotional well-being. For instance, Martin et al. found that biosensor data helped patients recognize patterns in their stress responses, enabling them to adopt healthier coping mechanisms (Peake et al., 2018).
- *Engagement Enhancement*: Gamification elements integrated into wearables have shown significant promise in enhancing patient engagement. Features like point systems, challenges, and rewards motivate individuals to adhere to rehabilitation protocols. For example, Davis et al. reported increased participation rates in physical therapy programs incorporating wearable gamification strategies. Patients not only adhered to their exercises but also reported a sense of accomplishment, further driving their commitment to recovery. Miller and colleagues highlighted the importance of developing wearable systems that generate actionable reports within minutes. They advocated for creating consumer-friendly summaries understandable by clinicians, patients, and their families, emphasizing the need for inclusivity in healthcare technology. In integrating wearable devices into cognitive, motor, and emotional rehabilitation, researchers and clinicians have unlocked new avenues for enhancing patient care. By leveraging real-time data, personalized feedback, and engaging interfaces, these technologies continue to redefine the possibilities of recovery across diverse populations (Peake et al., 2018).

6.4 Benefits of Wearable Technologies

6.4.1 *Personalized Therapy*

Wearable technology has revolutionized rehabilitation by enabling personalized therapy tailored to individual recovery needs. These devices collect real-time data, offering a comprehensive view of a patient's progress and adjusting interventions accordingly. For example, wearable sensors can monitor key parameters such as gait patterns, joint mobility, or muscle activity, providing valuable insights that guide therapeutic adjustments. The continuous feedback loop enabled by wearables ensures that treatments remain dynamic, adapting to a patient's evolving condition in real time. This level of precision minimizes the risk of over- or under-prescribing exercises, ultimately optimizing recovery times and improving outcomes. Patients benefit from customized guidance throughout their rehabilitation journey—whether through smart insoles for gait correction or wearable exoskeletons that support motor function recovery, providing immediate feedback to refine movements.

Beyond enhancing the therapeutic experience, real-time feedback also boosts patient confidence, reinforcing adherence to prescribed rehabilitation regimens. Additionally, wearable technology makes it possible to monitor complex conditions like Parkinson's disease, where symptoms fluctuate throughout the day. By capturing and analyzing these variations, clinicians can fine-tune treatment plans to better address each patient's unique needs (Kang & Exworthy, 2022; LivaFortis, 2024; Rodgers et al., 2019; therapease, 2024).

6.4.2 Objective Monitoring

One of the greatest advantages of wearable devices in rehabilitation is their ability to provide objective monitoring, delivering reliable and quantifiable data on patient progress. This marks a significant improvement over traditional methods, which often depend on subjective patient feedback and observational assessments that may not fully capture changes over time. Wearables equipped with accelerometers and gyroscopes can accurately measure key parameters such as range of motion, activity levels, and balance, offering an evidence-based framework for evaluating the effectiveness of therapy. Objective data enables healthcare providers to detect even subtle improvements or regressions in a patient's condition; changes that might go unnoticed through conventional observational techniques. Wearable EMG (electromyography) sensors can track muscle activation patterns, allowing clinicians to assess neuromuscular recovery with high precision. Over time, the accumulation of such data supports long-term tracking, helping therapists identify trends and refine rehabilitation protocols accordingly. This level of accuracy enhances the quality of care, ensuring that interventions are both effective and efficient (Adeghe et al., 2024; Alshammari et al., 2024; Bilenkyi & Dobrytskyi, 2024).

6.4.3 Accessibility

The portability and user-friendliness of wearable devices make them ideal for expanding access to rehabilitation services. Unlike traditional equipment that often requires clinical supervision, wearables are designed for ease of use, enabling patients to continue their rehabilitation from the comfort of their homes. This is particularly beneficial for individuals living in remote areas or those with limited mobility, who might otherwise face significant barriers to accessing care. By reducing the need for frequent clinic visits, wearables make rehabilitation more convenient and cost-effective. Home-based rehabilitation facilitated by wearable technology does not compromise the quality of care. Devices such as smart braces or connected physiotherapy tools transmit real-time data to healthcare providers, ensuring that patients receive continuous monitoring and support. This model of remote care has the added advantage of promoting independence among patients, empowering them to take an active role in their recovery. By making advanced rehabilitation tools more widely accessible, wearable technology has the potential to bridge gaps in care delivery and improve outcomes for underserved populations (Bove, 2019; Rodgers et al., 2019).

6.4.4 Enhanced Patient Engagement

Wearable devices are uniquely positioned to enhance patient engagement in rehabilitation programs. By incorporating interactive features such as gamification, biofeedback, and progress tracking, these technologies motivate individuals to actively participate in their therapy. For instance, virtual reward systems and competitive leaderboards can transform repetitive exercises into enjoyable and rewarding activities, encouraging patients to stick to their prescribed routines. Biofeedback mechanisms are another critical aspect of engagement. Devices like heart rate monitors or wearable EEG systems provide real-time insights into physiological responses, helping patients understand the direct impact of their efforts. For example, during a stress-management program, wearable technology can show how deep breathing exercises influence heart rate variability, reinforcing the importance of adherence to therapeutic practices. This immediate reinforcement fosters a sense of achievement and accountability, ultimately improving compliance and outcomes (Adeghe et al., 2024; therapease, 2024).

6.5 Limitations and Challenges

6.5.1 Data Privacy and Security

Wearable healthcare tools have provided unprecedented convenience, but they also raise security and privacy concerns and security. These devices constantly collect and transmit sensitive health data, exposing users to cyber threats if adequate safeguards are not in place. Ensuring compliance with regulations such as HIPAA is critical, as unauthorized access to patient data can result in serious privacy violations. Furthermore, as cyberattacks become more sophisticated, encryption techniques and secure data storage practices must evolve to protect patient information. Transparency in how data is collected and shared is equally important. Users must trust that their health information is handled responsibly (Hrabovska et al., 2023).

6.5.2 Cost and Accessibility

Wearables have the potential to improve medical treatment, but many people cannot afford them currently. High-tech, real-time wearables are expensive, making them unaffordable for low-income populations. Even when available, such wearables cannot always be translated for use in a medical setting, as information quality varies in unanticipated ways. Consumer-grade wearables, which are less expensive, frequently fail to offer accurate results for medical applications. Closing such a difference will require cost-effective technologies and policies to ensure equitable access to medical technology (Challenges of Incorporating Wearable and At-Home Device Data into Clinical Data Management, n.d.; Canali et al., 2022).

6.5.3 Integration with Clinical Workflows

To be truly effective, wearable technology must seamlessly integrate with existing healthcare workflows. However, many healthcare providers find it difficult to incorporate wearable data into electronic health records (EHRs) due to data compatibility and standardization issues. The sheer amount of data generated by these devices can be overwhelming, making it difficult to extract meaningful insights without the use of specialized tools. Furthermore, proprietary algorithms used by different manufacturers frequently prevent interoperability, resulting in data silos that healthcare professionals can't easily access or understand. Streamlining data formats and improving interoperability will be critical in making wearable-generated health data a valuable tool for clinical decision-making (LaBoone & Marques, 2024; Smuck et al., 2021).

6.5.4 Technical Limitations

While wearable technology have great prospects, technological restrictions remain a substantial barrier. Battery life is an ongoing issue—many devices require regular charging, which can lead to gaps in monitoring and decreased patient adherence. Accuracy and device calibration are extremely difficult, as tiny reading errors might result in inaccurate diagnoses or unsuccessful treatment programs. Signal interference, ambient conditions, and even user preferences for how they use their devices can all have an impact on reliability. Furthermore, data processing differences between manufacturers result in inconsistencies, hindering the integration of wearable data into clinical applications. Overcoming these constraints needs breakthroughs in battery efficiency, sensor technology, and standardization efforts to improve accuracy and reliability (Canali et al., 2022; LaBoone & Marques, 2024; Smuck et al., 2021).

6.6 Future Directions

6.6.1 Advances in Sensor Technology

The technology for wearable disease tracking has experienced rapid development, with a considerable improvement in creating complex sensing technology with high accuracy and comfort for use. One of the key advances in such technology is flexible and implantable sensors. Unlike traditional wearable technology, such sensors produce more reliable information collection, specifically for long-term tracking of long-term disease. Due to its flexible character, it can shape its form according to one's shape, minimizing any kind of discomfort and improving accuracy in measurement. As a result, it is becoming increasingly adopted for continuous medical tracking, most prominently in medical settings, where accuracy in information is critical in diagnosing and controlling disease such as heart disease, diabetes, and neurological disease (Hughes et al., 2023).

Moreover, multimodal platforms for sensing have become increasingly common. These platforms integrate information gained through a variety of sensors, such as motion sensors and EEG headsets, in an effort to paint a fuller picture of a patient's state of overall wellness. For example, integration of information gained through both motion sensors and EEG can yield information about both brain electrical activity and a person's motion, and can be immensely useful for tracking disease progression in neurological disease, or for gauging improvement in trauma cases. By using a variety of sensors, these platforms can bypass individual weaknesses in individual sensors, and allow clinicians access to a fuller range of information to inform protocols for care. AI comes into its own when dealing with complex information gained through sophisticated sensors. AI algorithms increasingly integrate with wearables to interpret information gained through sensors and make predictive statements regarding trends in terms of disease progression, and even early warnings regarding complications in overall wellness. For example, AI can interpret heart rate variability information gained through wearables in terms of predicting a patient's level of postoperative recovery, or even in terms of predicting a patient's response to a range of interventions. Not only can such use of AI make predictive through wearables even more effective, but enable even more personalized and proactive care, with a view towards optimizing patient outcomes over a period of time (Hughes et al., 2023).

Wearable technology has proven most beneficial in following individual medical ailments, with most technology developed to follow individual physiologic markers significant for both therapy and diagnosis. For instance, triaxial accelerometers are positioned at the hip to monitor activity level, particularly in cardiac rehabilitation programs. Triaxial accelerometers can monitor intensity and frequency of activity and provide immediate feedback to clinicians and patients. Clinicians can then individualize rehabilitation programs and follow that patients have safe activity levels during heart disease rehabilitation and postoperative rehabilitation (Hughes et al., 2023).

Other wearables, such as specialized, strap-on sensors, monitor a variety of vital signs, including heart and respiratory rate, ECG, and accelerometry. During exercise, these sensors monitor how a body is enduring physical strain. Such types of wearables are most useful for rehabilitation post-heart attack and in long-term respiratory disease, allowing clinicians to monitor improvement in a patient during activity and at rest. Integration of accelerometry in such a device also provides valuable information about motion patterns, important for estimating patient functional capacity and planning therapy interventions. In cardiovascular disease, smartwatches have become a significant tool for AF detection, a common arrhythmia. Wearables use photoplethysmography (PPG) sensors, tracking blood volume variation and capable of identifying aberrant heart rhythm indicative of AF. AF at an early stage detected through such smartwatches enables early intervention, possibly preventing dire complications such as stroke. Long-term arrhythmia tracking is also supported through continuous ECG patches, adhered to the skin. Patients with intermittent arrhythmic events stand to gain most from such a patch, with continuous recordings taken without having to report to a clinic for follow-up (Bayoumy et al., 2021; Hughes et al., 2023).

6.6.2 AI and Machine Learning Integration

Artificial intelligence (AI) and machine learning (ML) technologies are at the forefront of transforming wearable health monitoring. One of the most significant advancements in this area is the application of deep neural networks (DNNs) to detect specific health biomarkers, including individual particles and molecular markers that indicate disease presence or progression. These AI models are capable of processing complex data, such as those generated by biosensors or wearables, to detect patterns that may be undetectable by human analysts. For example, deep learning algorithms can be used to detect the presence of disease markers in sweat, blood, or even exhaled air, providing clinicians with critical insights for early diagnosis and monitoring. AI is also being used to enhance the accuracy of multimodal sensing systems, particularly in isolating individual signals from complex datasets. Wearable devices often collect data from multiple sources, such as motion sensors, ECG sensors, and temperature sensors, which can result in signal overlap or cross-sensitivity. AI models are being developed to isolate individual signals from these complex data streams, enabling more accurate analysis and reducing the risk of false positives or missed diagnoses. This capability is particularly useful in monitoring patients with multiple health conditions, as it ensures that each aspect of their health is accurately assessed without interference from other variables (Shajari et al., 2023).

Machine learning algorithms are also being applied to accelerometer data to predict a patient's risk of mortality. For example, wearable accelerometers can monitor a person's movement patterns, activity levels, and even gait, all of which are critical indicators of overall health and longevity. Machine learning models can analyze these data points and identify patterns that may indicate an increased risk of mortality, allowing for early intervention and more personalized care strategies (Bayoumy et al., 2021). The ability to use these algorithms to predict health outcomes based on wearable data is an exciting development in precision medicine, offering new ways to tailor treatments to individual patients and enhance preventive care efforts.

6.6.3 Multimodal Systems

Combining data from multiple wearable devices, such as EEG headsets and motion sensors, provides a comprehensive view of patient progress. Multimodal systems leverage data integration to capture and analyze diverse physiological and behavioral markers simultaneously. For instance, EEG headsets can monitor brain activity, while motion sensors track mobility and balance, offering insights into the neurological and physical aspects of recovery. Such systems are particularly beneficial for conditions like stroke rehabilitation, where understanding both cognitive and motor recovery is crucial. By synthesizing data from different sensors, clinicians can create more personalized treatment plans and monitor progress more effectively (Hughes et al., 2023).

Furthermore, these systems address limitations associated with single-device monitoring. Cross-referencing data from various devices enhances accuracy, reduces noise, and provides a multidimensional perspective on a patient's health status. For example, integrating heart rate data with motion tracking allows for a more nuanced understanding of exertion levels during physical therapy. This holistic approach not only improves diagnostics but also enables more targeted and efficient interventions.

6.6.3.1 AI/Machine Learning Applications

Artificial intelligence (AI) and machine learning (ML) technologies are at the forefront of transforming wearable health monitoring. One of the most significant advancements in this area is the application of deep neural networks (DNNs) to detect specific health biomarkers, including individual particles and molecular markers that indicate disease presence or progression. These AI models are capable of processing complex data, such as those generated by biosensors or wearables, to detect patterns that may be undetectable by human analysts. For example, deep learning algorithms can be used to detect the presence of disease markers in sweat, blood, or even exhaled air, providing clinicians with critical insights for early diagnosis and monitoring. AI is also being used to enhance the accuracy of multimodal sensing systems, particularly in isolating individual signals from complex data sets. Wearable devices often collect data from multiple sources, such as motion sensors, ECG sensors, and temperature sensors, which can result in signal overlap or cross-sensitivity. AI models are being developed to isolate individual signals from these complex data streams, enabling more accurate analysis and reducing the risk of false positives or missed diagnoses. This capability is particularly useful in monitoring patients with multiple health conditions, as it ensures that each aspect of their health is accurately assessed without interference from other variables.

Machine learning algorithms are also being applied to accelerometer data to predict a patient's risk of mortality. For example, wearable accelerometers can monitor a person's movement patterns, activity levels, and even gait, all of which are critical indicators of overall health and longevity. Machine learning models can analyze these data points and identify patterns that may indicate an increased risk of mortality, allowing for early intervention and more personalized care strategies (Bayoumy et al., 2021; Shajari et al., 2023). The ability to use these algorithms to predict health outcomes based on wearable data is an exciting development in precision medicine, offering new ways to tailor treatments to individual patients and enhance preventive care efforts.

6.6.4 Remote Monitoring and Tele-Rehabilitation

The integration of wearable devices with telehealth platforms has greatly expanded the potential for remote patient monitoring. Wearable health devices that track vital signs and other physiological data can be seamlessly connected to telemedicine systems, allowing healthcare providers to monitor patients in real time, even if they are located in different geographic areas. This development has been particularly important during the COVID-19 pandemic, as remote monitoring allowed clinicians to continue providing care while minimizing the risk of exposure to the virus. Telemedicine platforms integrated with wearable devices enable patients to receive personalized care from the comfort of their homes, making healthcare more accessible and efficient. Smartphone apps that work with wearable devices are also playing an important role in delivering personalized care. These apps can use data from wearables to provide individualized coaching and feedback, helping patients manage their health on a daily basis. For example, a smartphone app might use heart rate and activity data from a smartwatch to recommend exercises, offer reminders to take medications, or alert the user to abnormal health data. This direct feedback loop can empower patients to take a more active role in managing their health, improving adherence to treatment plans and overall health outcomes (Hughes et al., 2023).

AI-enabled systems are also revolutionizing the way healthcare providers can respond to abnormal health data. For example, AI algorithms can automatically analyze data from wearables, such as smartwatches, to detect abnormal rhythms or other irregularities that might signal a health emergency. When an abnormality is detected, the system can immediately alert the patient and their healthcare provider, enabling swift intervention. This capability is particularly valuable for managing conditions like atrial fibrillation, where timely treatment can prevent more serious health complications (Bayoumy et al., 2021).

6.7 Conclusion

Wearable devices and sensors represent a significant leap forward in neurocognitive rehabilitation, offering innovative solutions to longstanding challenges in the field. By providing continuous, objective, and personalized data, these technologies enhance the efficacy of therapeutic interventions and empower patients to take an active role in their recovery. Despite current limitations, advancements in technology, integration, and accessibility hold promise for a future where wearable devices and sensors become integral to neurocognitive rehabilitation practices.

The advancements in wearable technology for health monitoring are paving the way for more personalized, efficient, and proactive healthcare. By incorporating advanced sensing technologies, AI algorithms, and telehealth platforms, wearable devices are becoming more accurate, comprehensive, and accessible. These developments offer tremendous potential for improving patient outcomes, particularly for those with chronic conditions or those undergoing rehabilitation. As wearable technology continues to evolve, it is likely to play an increasingly central role in both preventive and therapeutic healthcare, offering new opportunities for remote care, early detection, and personalized treatment strategies.

References

Adeghe, E. P., Okolo, C. A., & Ojeyinka, O. T. (2024). A review of wearable technology in healthcare: Monitoring patient health and enhancing outcomes. *Open Access Research Journal of Multidisciplinary Studies, 7*(1), 142–148. https://doi.org/10.53022/oarjms.2024.7.1.0019

AdornThemes. (2024, November 28). *The use of wearable technology for rehabilitation*. TheTouchPoint Solution™. https://thetouchpointsolution.com/blogs/touchpoints-blog/the-use-of-wearable-technology-for-rehabilitation

Alshammari, B., Alanazi, N. F., Kreedi, F., Alshammari, F., Alkubati, S. A., Alrasheeday, A., Madkhali, N., Alshara, A., Bakthavatchaalam, V., Al-Masaeed, M., Alshammari, S. K., Alshammari, N. K., Ansari, M., Hussain, A., & Al-Sadi, A. K. (2024). Exposure to secondary traumatic stress and its related factors among emergency nurses in Saudi Arabia: A mixed method study. *BMC Nursing, 23*(1), 337. https://doi.org/10.1186/s12912-024-02018-4

Bayoumy, K., Gaber, M., Elshafeey, A., Mhaimeed, O., Dineen, E. H., Marvel, F. A., Martin, S. S., Muse, E. D., Turakhia, M. P., Tarakji, K. G., & Elshazly, M. B. (2021). Smart wearable devices in cardiovascular care: Where we are and how to move forward. *Nature Reviews Cardiology, 18*(8), 581–599. https://doi.org/10.1038/s41569-021-00522-7

Bilenkyi, S., & Dobrytskyi, D. (2024). *Mind studios*. Blog - Mind Studios. https://themindstudios.com/blog/digital-rehabilitation-technology/

Bove, L. A. (2019). Increasing patient engagement through the use of wearable technology. *The Journal for Nurse Practitioners, 15*(8), 535–539. https://doi.org/10.1016/j.nurpra.2019.03.018

Canali, S., Schiaffonati, V., & Aliverti, A. (2022). Challenges and recommendations for wearable devices in digital health: Data quality, interoperability, health equity, fairness. *PLOS Digital Health, 1*(10), e0000104. https://doi.org/10.1371/journal.pdig.0000104

Challenges of Incorporating Wearable and At-Home Device Data into Clinical Data Management. (n.d.). *MaxisIT*. Retrieved January 30, 2025, from https://maxisit.com/blog/challenges-of-incorporating-wearable-and-at-home-device-data-into-clinical-data-management/

Hrabovska, N., Kajati, E., & Zolotova, I. (2023). A validation study to confirm the accuracy of wearable devices based on health data analysis. *Electronics, 12*(11), 11. https://doi.org/10.3390/electronics12112536

Hughes, A., Shandhi, M. M. H., Master, H., Dunn, J., & Brittain, E. (2023). Wearable devices in cardiovascular medicine. *Circulation Research, 132*(5), 652–670. https://doi.org/10.1161/CIRCRESAHA.122.322389

Kang, H. S., & Exworthy, M. (2022). Wearing the future—Wearables to empower users to take greater responsibility for their health and care: Scoping review. *JMIR mHealth and uHealth, 10*(7), e35684. https://doi.org/10.2196/35684

Kristoffersson, A., & Lindén, M. (2022). A systematic review of wearable sensors for monitoring physical activity. *Sensors, 22*(2), 2. https://doi.org/10.3390/s22020573

LaBoone, P. A., & Marques, O. (2024). Overview of the future impact of wearables and artificial intelligence in healthcare workflows and technology. *International Journal of Information Management Data Insights, 4*(2), 100294. https://doi.org/10.1016/j.jjimei.2024.100294

LivaFortis. (2024, September 19). *5 reasons to try digital health physical therapy.* https://static.wixstatic.com/media/7173f9_4147e6cad47e450f98514fd409654d6f~mv2.jpg/v1/fill/w_1000,h_563,al_c,q_85,usm_0.66_1.00_0.01/7173f9_4147e6cad47e450f98514fd409654d6f~mv2.jpg. LivaFortis Website. https://www.livafortis.us/post/5-reasons-to-try-digital-health-physical-therapy

Maceira-Elvira, P., Popa, T., Schmid, A.-C., & Hummel, F. C. (2019). Wearable technology in stroke rehabilitation: Towards improved diagnosis and treatment of upper-limb motor impairment. *Journal of Neuroengineering and Rehabilitation, 16*(1), 142. https://doi.org/10.1186/s12984-019-0612-y

Minen, M. T., & Stieglitz, E. J. (2021). Wearables for neurologic conditions. *Neurology: Clinical Practice, 11*(4), e537–e543. https://doi.org/10.1212/CPJ.0000000000000971

Mitchner, A. (2023, May 31). *Beyond the lab: Wearable technology for mental health.* Labfront. https://www.labfront.com/article/wearable-technology-in-psychological-research

Peake, J. M., Kerr, G., & Sullivan, J. P. (2018). A critical review of consumer wearables, mobile applications, and equipment for providing biofeedback, monitoring stress, and sleep in physically active populations. *Frontiers in Physiology, 9*, 743. https://doi.org/10.3389/fphys.2018.00743

Rodgers, M. M., Alon, G., Pai, V. M., & Conroy, R. S. (2019). Wearable technologies for active living and rehabilitation: Current research challenges and future opportunities. *Journal of Rehabilitation and Assistive Technologies Engineering, 6*, 2055668319839607. https://doi.org/10.1177/2055668319839607

Seneviratne, S., Hu, Y., Nguyen, T., & Lan, G. (2017). A survey of wearable devices and challenges. *IEEE Communications Surveys & Tutorials, 99*(1), 1. https://doi.org/10.1109/COMST.2017.2731979

Shajari, S., Kuruvinashetti, K., Komeili, A., & Sundararaj, U. (2023). The emergence of AI-based wearable sensors for digital health technology: A review. *Sensors (Basel, Switzerland), 23*(23), 9498. https://doi.org/10.3390/s23239498

Sigcha, L., Borzì, L., Amato, F., Rechichi, I., Ramos-Romero, C., Cárdenas, A., Gascó, L., & Olmo, G. (2023). Deep learning and wearable sensors for the diagnosis and monitoring of Parkinson's disease: A systematic review. *Expert Systems with Applications, 229*, 120541. https://doi.org/10.1016/j.eswa.2023.120541

Smuck, M., Odonkor, C. A., Wilt, J. K., Schmidt, N., & Swiernik, M. A. (2021). The emerging clinical role of wearables: Factors for successful implementation in healthcare. *NPJ Digital Medicine, 4*(1), 1–8. https://doi.org/10.1038/s41746-021-00418-3

The Importance of Wearable Devices in Neuro Rehabilitation. (2023, August 9). Lusio Rehab. https://lusiorehab.com/blogs/news/the-importance-of-wearable-devices-in-neuro-rehabilitation

therapease. (2024, March 10). *Harnessing the power of patient controlled rehabilitation technology for patient outcomes—TherapEase.* https://therapease.com/2024/03/10/harnessing-the-power-of-wearable-rehabilitation-technology-for-patient-outcomes/

Wajid, M. (2024, January 23). *What are wearable technology used in rehabilitation?* https://www.icliniq.com/articles/physiotherapy-physiohealth/usage-of-wearable-technology-in-rehabilitation

Walls, H. J., Dausch, D. E., Temple, D. S., Hegarty-Craver, M. S., Davis-Wilson, H. C., Gaur, P., Hegarty-Craver, M., Davis-Wilson, H. C., Gaur, P., Walls, H. J., Dausch, D. E., & Temple, D. S. (2024). *Wearable sensors for service members and first responders: Considerations for using commercially available sensors in continuous monitoring*. RTI Press. https://doi.org/10.3768/rtipress.2024.op.0090.2402

Wearable Sensors in Other Medical Domains with Application Potential for Orthopedic Trauma Surgery—A Narrative Review. (n.d.). Retrieved December 19, 2024, from https://www.mdpi.com/2077-0383/13/11/3134

Zhang, J., & He, C. (2023). Evidence-based rehabilitation medicine: Definition, foundation, practice and development. *Medical Review, 4*(1), 42–54. https://doi.org/10.1515/mr-2023-0027

Chapter 7
Patient Perspectives and Outcomes

7.1 Introduction

Neuropsychological rehabilitation is a complex and evolving process designed to help individuals regain cognitive, emotional, and functional abilities after a brain injury or neurological condition. Within this intricate framework, integrating patient perspectives has become essential for developing effective interventions and supporting meaningful recovery. Patient feedback not only enriches our understanding of individual rehabilitation experiences but also plays a key role in shaping care strategies that address diverse challenges and evolving needs. By identifying gaps in care and unmet needs, patient input enables clinicians to fine-tune treatment plans, ensuring interventions remain aligned with personal priorities. For instance, individuals recovering from brain injuries often face difficulties with cognitive function, emotional regulation, and managing their medical care. Engaging patients in discussions about their experiences allows clinicians to assess subjective improvements in these areas and adjust rehabilitation strategies accordingly. Patients offer invaluable insights into their recovery journeys, shedding light on the complexities of navigating rehabilitation. These firsthand perspectives help clinicians understand the nuances of each individual's challenges, motivations, and perceptions of treatment effectiveness. Moreover, actively seeking patient input fosters a collaborative approach to rehabilitation, strengthening the therapeutic alliance—a fundamental element of successful outcomes. When patients feel heard and valued, they are more likely to engage in their treatment, adhere to rehabilitation protocols, and experience greater satisfaction with their recovery process (Wain et al., 2008).

Patient-centered care has become a cornerstone of modern healthcare, emphasizing the patient's voice in decision-making and treatment planning. In neuropsychological rehabilitation, this approach highlights the importance of assessing care quality from the patient's perspective, rather than relying solely on traditional provider-evaluated outcomes. While clinicians often focus on measurable

K. J. Reddy, *Innovations in Neurocognitive Rehabilitation*,
https://doi.org/10.1007/978-3-031-88117-6_7

improvements—such as enhanced motor skills or higher cognitive test scores—these metrics don't always capture the full scope of a patient's recovery.

Patient-reported outcome measures provide a more holistic view by considering personal experiences and perceptions of progress. For many individuals, regaining the ability to participate in social activities or manage daily responsibilities may be more meaningful than improvements reflected in standardized cognitive tests. Recognizing and balancing both provider and patient perspectives is essential for delivering truly comprehensive care. Throughout rehabilitation, patients encounter various barriers and facilitators that shape their experiences. For instance, brain injury survivors frequently cite communication as both a help and a hindrance to accessing care. Clear, effective communication with healthcare providers fosters trust and allows patients to express their needs, while poor communication can lead to frustration, misunderstandings, and unmet expectations. Additionally, compensatory strategies—such as memory aids and assistive technologies—are often identified as key tools for promoting independence and improving medical self-management. These resources enable patients to navigate their conditions more effectively and regain a sense of autonomy. Despite these advancements, knowledge gaps among healthcare providers regarding brain injuries and their long-term effects remain a significant challenge. Strengthening provider education and training in this area is crucial to enhancing care quality and ensuring that interventions are both relevant and effective. Research consistently shows that when patients feel their unique needs and insights are acknowledged, they report higher satisfaction with their treatment—reinforcing the value of a truly collaborative approach to rehabilitation (Domensino et al., 2022; Torregrossa et al., 2023).

A growing body of research supports the integration of holistic approaches in neuropsychological rehabilitation, recognizing the deep connection between cognitive and emotional factors in shaping recovery. As Prigatano emphasized, treating cognitive disturbances in isolation is insufficient—emotional well-being plays a crucial role in influencing thought processes and behavior. A key component of this approach is fostering a therapeutic milieu, an environment that encourages healing, self-discovery, and psychological growth. Rehabilitation is not just about overcoming physical or cognitive impairments; it also involves navigating the emotional and existential challenges that arise from neurological trauma. Psychotherapy has proven to be a valuable tool within this framework, providing patients with a space to process the impact of their injuries and work toward self-acceptance. By addressing the emotional dimensions of recovery, therapy helps individuals redefine their sense of identity and purpose, ultimately strengthening their resilience. Research supports this comprehensive model, showing that emotional well-being often serves as a foundation for cognitive improvement. The shift toward patient-centered rehabilitation further reinforces the importance of a holistic approach. Collaborative care models prioritize patient perspectives, encouraging individuals to take an active role in their recovery. Shared decision-making—where patients and clinicians work together to set goals, explore treatment options, and develop personalized rehabilitation plans—ensures that interventions are both evidence-based and relevant to the patient's daily life. For example, a patient struggling with memory impairments may benefit more from practical strategies like reminders and organizational tools rather than generic cognitive exercises. Similarly, those dealing with emotional distress may require targeted support through counseling or peer support groups to address feelings of isolation and loss. Beyond individual treatment, patient insights are also invaluable for identifying systemic challenges in healthcare, such as accessibility barriers and the need for improved communication within multidisciplinary teams. By integrating holistic care, psychotherapy, and patient-driven insights, neuropsychological rehabilitation becomes more than just a clinical process—it evolves into a dynamic, patient-empowered journey toward meaningful recovery (Nair & Lincoln, 2013).

Integrating patient perspectives into neuropsychological rehabilitation is not just a best practice—it is an ethical obligation that upholds the autonomy and dignity of individuals navigating their recovery. By actively valuing patient feedback, adopting patient-centered approaches, addressing barriers to care, and embracing holistic treatment models, clinicians can provide rehabilitation that is both compassionate and effective. For instance, stroke survivors have reported significantly better recovery experiences when their emotional well-being is prioritized alongside physical therapies. Similarly, research on brain injury rehabilitation highlights the profound impact of bridging healthcare provider knowledge gaps and utilizing compensatory tools to support cognitive and functional recovery. These insights reinforce the need for a rehabilitation approach that acknowledges the intricate connection between cognitive and emotional health.

Ultimately, recognizing recovery as both a scientific process and a deeply human experience allows for more meaningful, patient-driven outcomes. By embracing this perspective, clinicians can create rehabilitation programs that are not only evidence-based but also truly responsive to the lived experiences of those they serve (Rea et al., 2023).

This chapter aims to illuminate the transformative potential of patient perspectives, advocating for a rehabilitation paradigm that places the individual's voice at the heart of the healing process. As healthcare systems evolve, recognizing and embedding the wisdom of patients within clinical frameworks will remain a cornerstone of effective neuropsychological rehabilitation. By advancing these principles, clinicians and researchers can contribute to a more inclusive and responsive model of care, ultimately improving the lives of individuals navigating the profound challenges of recovery.

7.2 The Importance of Patient Perspectives

Patients provide unique insights into their rehabilitation experiences, including challenges, motivations, and the perceived effectiveness of interventions. Their feedback helps clinicians:

- *Identify Unmet Needs and Adapt Treatment Plans:* Patients often highlight challenges that clinical assessments might miss—things like struggles with daily routines, emotional stress, or specific social barriers. For example, someone recovering from a stroke might find it difficult to manage household tasks or return to work, but these concerns may not come up in a standard evaluation. Addressing these real-world difficulties helps clinicians create more personalized rehabilitation plans. Research shows that incorporating patient feedback leads to more effective and relevant interventions, especially for individuals with complex medical histories or diverse needs. Additionally, by truly listening to patients, clinicians can tackle systemic barriers—like unreliable transportation to therapy or a lack of culturally appropriate care—ensuring that rehabilitation is both fair and accessible (Bridger et al., 2021; Prigatano et al., 2021).
- *Assess Subjective Improvements in Cognitive and Emotional Domains:* Standardized tests and clinical metrics often fail to capture the nuanced, subjective improvements reported by patients. For instance, beyond measurable cognitive gains, patients might report feeling more confident in social situations, experiencing fewer episodes of anxiety, or regaining a sense of independence. These subjective outcomes are critical for understanding the holistic impact of rehabilitation. Research highlights the importance of patient-reported outcomes, noting that such feedback provides a deeper understanding of how rehabilitation affects quality of life. For example, a patient participating in cognitive rehabilitation might express relief at being able to recall names during conversations—a significant milestone that traditional assessments might overlook (Jesus et al., 2024).
- *Foster a Collaborative Approach, Enhancing Patient Engagement:* Engaging patients in the decision-making process ensures that their priorities and goals are integrated into the rehabilitation plan, fostering a sense of ownership and motivation. When patients feel that their voices are heard and valued, they are more likely to remain committed to therapy and actively participate in the process. Collaborative approaches, such as shared goal-setting, have been shown to improve adherence and satisfaction. For example, a patient recovering from a traumatic brain injury might prioritize regaining the ability to drive, while another might focus on returning to work (Kennard et al., 2024; Winegardner & Coetzer, 2017). By aligning therapy goals with these priorities, clinicians can create personalized plans that resonate with each individual's aspirations.

7.2.1 Capturing Patient Feedback

Tools and techniques used for gathering patient perspectives include:

- *Structured Interviews:* Structured interviews involve a predefined set of questions, enabling researchers to gather consistent information across multiple patients. These interviews focus on specific aspects of the rehabilitation journey, such as physical recovery, emotional well-being, or perceived barriers to progress. Structured interviews are particularly valuable for identifying common challenges and areas for improvement in rehabilitation programs (Domensino et al., 2022).

- *Semi-Structured Interviews:* Semi-structured interviews combine predefined questions with the flexibility to explore topics in greater depth based on patients' responses. This method allows for a more personalized approach, capturing nuanced experiences and perspectives that may not emerge in structured formats. For example, semi-structured interviews have been used to understand the long-term challenges faced by patients recovering from neurological injuries, such as difficulties in maintaining social relationships or returning to work (Guàrdia-Olmos et al., 2015; Nair & Lincoln, 2013).
- *Surveys and Questionnaires:* Surveys and questionnaires are standardized tools designed to capture patient feedback on a larger scale. Instruments like the Patient-Reported Outcomes Measurement Information System (PROMIS) provide validated measures of physical, mental, and social health. These tools are particularly useful for tracking changes over time and comparing outcomes across different patient populations. For example, PROMIS has been employed to assess the effectiveness of cognitive rehabilitation programs, highlighting areas where patients report significant improvements or ongoing difficulties (Bennett, 2001; Dür et al., 2022).
- *Focus Groups:* Focus groups involve guided discussions among small groups of patients, providing an opportunity to explore shared experiences and diverse viewpoints. This method is particularly effective for identifying common themes and generating ideas for improving rehabilitation practices. For instance, focus groups with stroke survivors have revealed insights into the social and emotional challenges of recovery, such as feelings of isolation or frustration with the pace of progress. These findings have informed the development of support groups and community-based rehabilitation initiatives (Bennett, 2001).

7.3 Measuring Outcomes in Neurocognitive Rehabilitation

Cognitive tests like the MoCA and WMS might reveal deficits in these areas, but physiological measures can provide a richer picture. fMRI might show reduced activity in the dorsolateral prefrontal cortex and hippocampus—regions associated with attention and memory, respectively. This information doesn't just confirm what the cognitive tests are showing; it adds a layer of detail that can inform treatment. If neurofeedback training is introduced to specifically target these areas, subsequent fMRI scans might show increased activity, providing tangible evidence that the intervention is working. Similarly, EEG can track how the brain's electrical activity changes over time. In the early stages of rehabilitation, a patient might show disrupted connectivity between different brain regions, as evidenced by low coherence in their EEG patterns. With therapy, these patterns might become more synchronized, reflecting improved communication within the brain. For clinicians, this kind of data is invaluable—it provides a clear, measurable indicator of progress that can be used to refine and adapt treatment plans. The combination of cognitive tests and physiological measures is what makes modern neuropsychological rehabilitation so powerful. Together, they offer a comprehensive view of recovery that is both broad and deep. Cognitive tests capture the practical, real-world implications of someone's cognitive challenges—how they affect their ability to function day-to-day.

Physiological measures, meanwhile, delve into the neural mechanisms underlying these challenges, providing a scientific basis for understanding what's happening in the brain. No two patients are the same, and their rehabilitation shouldn't be either. By combining data from cognitive tests and physiological measures, clinicians can create tailored interventions that address both the behavioral and neural aspects of recovery. For example, if a patient's cognitive tests show significant improvement but their fMRI scans reveal lingering deficits in connectivity, therapy might continue to focus on strengthening those neural pathways. Conversely, if the brain appears to be recovering well but cognitive tests reveal ongoing challenges, the focus might shift to strategies for compensating or adapting. Ultimately, these tools are not just about measuring recovery—they are about facilitating it. By providing a detailed, nuanced understanding of both cognitive function and brain activity, they empower clinicians to deliver the kind of personalized, evidence-based care that makes a real difference in people's lives. In the world of neuropsychological rehabilitation, where progress can sometimes feel slow and uncertain, this combination of cognitive and physiological insights offers a powerful reminder of the brain's incredible capacity for change and growth.

7.3.1 Objective Metrics

- *Cognitive Tests:* Cognitive tests have long been a cornerstone of assessing and tracking the outcomes of neuropsychological rehabilitation. Standardized tools like the Montreal Cognitive Assessment (MoCA) and the Wechsler Memory Scale (WMS) are widely used to evaluate an individual's cognitive abilities in a structured and replicable way. These tests offer more than just numbers on a page—they provide a window into how someone is navigating their world post-injury or illness, shedding light on the areas where they may be struggling and where they are succeeding. The MoCA, for example, is a versatile tool that has gained significant traction in clinical settings due to its sensitivity in identifying neurocognitive disorders (NCDs). What sets the MoCA apart is its ability to assess multiple cognitive domains, including memory, executive function, visuospatial skills, and attention. These domains are integral to daily life, from remembering to take medication to problem-solving at work or navigating a complex social interaction (Julayanont et al., n.d.).

- The MoCA doesn't just look at the big picture of cognition; it gets into the details, helping clinicians and researchers understand where interventions might be most effective. One of the exciting advancements in the use of MoCA is the development of MoCA domain scores (MDS). These scores, which are rooted in robust neuropsychological and neuroimaging research, allow for a more nuanced interpretation of test results. Studies have shown that MDS correlates strongly with traditional neuropsychological tests, demonstrating their validity ($r = 0.15$–0.43, ps < 0.001). This means that MDS not only offers a quick way to understand cognitive performance but also align well with more detailed, time-consuming assessments. For clinicians, this is invaluable—it allows them to gather high-quality data efficiently, which is especially important when working with individuals who may fatigue easily or struggle to focus during testing. The clinical applications of MDS go even further. By examining the patterns within these domain scores, clinicians can differentiate between the causes of early cognitive decline. For instance, someone with Alzheimer's disease might show different deficits compared to someone with vascular dementia or frontotemporal lobar degeneration. This ability to tease apart the specific nature of someone's cognitive challenges means that interventions can be tailored in a way that feels personal and relevant to the individual. If a patient is struggling primarily with memory, for example, the focus can be on strategies to enhance recall (Dautzenberg et al., 2020).

- Alternatively, if executive function is the primary issue, interventions might center on planning and organization. The Wechsler Memory Scale (WMS) complements the MoCA by offering a deep dive into memory function specifically. Memory is one of the most frequently affected domains in conditions like traumatic brain injury, stroke, or neurodegenerative diseases, and the WMS is designed to measure it from multiple angles. Whether it's immediate recall (like remembering a phone number) or delayed recall (like recalling the details of a story after a short break), the WMS provides a comprehensive profile of memory abilities. This kind of detailed assessment is critical for tracking progress over time. It allows clinicians to see whether someone is improving, plateauing, or even declining, which in turn informs how their rehabilitation plan should evolve (Ang et al., 2023).
- *Physiological Measures:* While cognitive tests like the MoCA and WMS provide invaluable behavioral data, they are only part of the story. To truly understand the outcomes of neuropsychological rehabilitation, it's important to look under the hood—at the brain itself. This is where physiological measures like functional MRI (fMRI) and EEG come into play. These tools allow clinicians and researchers to peer into the brain's inner workings, capturing the structural and functional changes that underpin cognitive recovery. Functional MRI (fMRI) is an incredible technology that maps brain activity by measuring changes in blood flow. When a particular brain region is active, it requires more oxygen, and fMRI can detect these changes with remarkable precision. In the context of rehabilitation, fMRI is particularly useful for tracking how the brain adapts over time. For

instance, it might show increased activity in the prefrontal cortex during tasks that require attention, suggesting that therapy is helping to strengthen this critical cognitive function. Similarly, it can reveal how different regions of the brain communicate with each other—information that is vital for understanding brain plasticity, or the brain's ability to reorganize itself after injury or disease. EEG, on the other hand, provides a real-time view of the brain's electrical activity (Nordström et al., 2024; de Wit et al., 2022).

- Unlike fMRI, which measures blood flow indirectly, EEG directly captures the brain's electrical signals, offering a dynamic perspective on neural function. It's particularly good at detecting abnormalities, such as disrupted communication between brain regions or unusual patterns of brain waves. For example, excessive theta wave activity might indicate drowsiness or reduced cognitive engagement, while a shift toward more normalized patterns during rehabilitation might suggest that the brain is beginning to recover. One of the most exciting aspects of using tools like fMRI and EEG is their ability to provide objective, quantifiable data about brain recovery. While cognitive tests rely on someone's performance on a given day—performance that might be influenced by fatigue, stress, or other external factors—physiological measures go deeper. They capture the underlying neural changes that are happening as a result of therapy, offering a more stable and reliable marker of progress. For example, a patient recovering from a traumatic brain injury who is struggling with attention and memory (Nordström et al., 2024; de Wit et al., 2022).

7.3.2 Subjective Metrics

Patient-reported outcomes (PROs) have become an essential aspect of modern healthcare, allowing individuals to directly share their health status and experiences through self-reported measures. These outcomes are increasingly valued for their ability to enhance clinical decision-making, boost patient satisfaction, and contribute to better health outcomes. By centering care around the patient's perspective, PROs enable healthcare providers to personalize treatments and interventions, ultimately improving overall quality of care. A key subset of PROs is quality of life (QoL) assessments, which focus on how patients perceive their health and well-being. These assessments capture vital dimensions such as physical and mental health, social interactions, and environmental influences, offering a comprehensive view of how illnesses and treatments impact daily life (Bullinger & Quitmann, 2014).

Several well-established tools exist to measure QoL, with two of the most commonly used being the WHOQOL-BREF—a concise version of the World Health Organization's Quality of Life instrument—and the SF-36, a general health survey assessing overall well-being. Additionally, condition-specific instruments provide deeper insights into particular populations, such as individuals with mental health disorders like depression or anxiety. These tools help clinicians better understand the unique challenges faced by different patient groups, enabling more targeted and effective interventions. With advancements in technology, QoL assessments are now increasingly integrated into digital platforms, streamlining data collection and analysis in clinical settings. Innovations such as computerized adaptive testing (CAT) have made assessments more efficient by tailoring questions to individual patients, reducing the number of items while maintaining accuracy and reliability (Geerards et al., 2019).

CAT-based assessments have shown promise in reducing survey length by approximately 25%, easing the burden on patients. However, shorter assessments do not always lead to better engagement. In some cases, patients may be less inclined to share results with their healthcare providers or engage fully with the process. This highlights the importance of thoughtful implementation when incorporating digital tools. One crucial factor in optimizing digital QoL assessments is the quality of feedback provided. Research indicates that graphical or tailored text-based feedback significantly enhances user experience, helping patients better understand their results and fostering greater involvement in their healthcare journey. For example, visual representations of QoL scores or comparisons to population norms can make results more accessible for both patients and clinicians. However, studies also suggest that while feedback improves user experience, it does not necessarily make CAT assessments superior to traditional fixed-length surveys (Aiyegbusi et al., 2024; Geerards et al., 2019).

Despite these challenges, digital platforms present significant opportunities for integrating QoL assessments into routine clinical practice. They allow for real-time data collection and monitoring, giving healthcare providers a clearer picture of patient progress and enabling timely adjustments to treatment plans. Additionally, incorporating QoL data into electronic health records facilitates a more comprehensive, data-driven approach to patient care. Digital tools also hold particular promise for expanding access to QoL assessments, especially for underserved or remote populations. Mobile apps and web-based platforms can bridge gaps in healthcare accessibility, ensuring that a broader range of individuals benefit from personalized, patient-centered care (Aiyegbusi et al., 2024).

For digital QoL assessments to be truly effective, key factors such as platform usability, meaningful feedback, and patient engagement must be carefully considered. While innovations like CAT and graphical feedback enhance efficiency, their implementation requires careful planning to maximize their impact. Future research should explore how factors like question relevance, ease of use, and feedback clarity influence patient engagement. Additionally, integrating QoL assessments seamlessly into clinical workflows will be crucial for their long-term adoption and success (Bullinger & Quitmann, 2014).

Digital platforms have revolutionized the administration of these assessments, offering innovative solutions like computerized adaptive testing and graphical feedback to enhance efficiency and user engagement. However, the success of these tools depends on careful consideration of factors such as assessment length, feedback design, and the broader context of their use in clinical practice. By addressing these challenges, healthcare providers can harness the full potential of PROs and

QoL assessments to deliver more personalized, patient-centered care that improves outcomes and satisfaction for diverse populations.

7.3.3 Technology-Driven Evaluations

Wearable devices have seen a surge in popularity due to their noninvasive nature and ease of use, allowing individuals to monitor their health continuously. These devices can track a variety of metrics, including heart rate, blood pressure, oxygen levels, physical activity, sleep patterns, and even biochemical markers like glucose levels. Among the most widely used wearables are fitness trackers, smartwatches, accelerometer-based monitors, and ECG chest patches, with wrist-worn devices being the most common (73%), followed by hip-mounted (13.2%) and chest-worn (11.1%) options (Huhn et al., 2022; Shajari et al., 2023).

Their adaptability makes them seamlessly integrable into daily routines, making health tracking more accessible and convenient for a broad audience. The integration of AI and machine learning has significantly expanded the capabilities of wearable devices. These technologies analyze vast amounts of collected data, transforming raw numbers into actionable insights that help identify patterns, detect health risks, and even predict potential medical conditions. For example, AI-enabled wearables can detect abnormal heart rhythms, providing early warnings that allow for timely intervention ("AI-Powered Wearable Devices & Real-Time Monitoring," 2023).

Beyond improving diagnostics, AI-powered wearables also facilitate predictive analytics, helping both users and healthcare professionals address health concerns before they become severe. One of the major advantages of AI in wearable technology is its ability to enhance data accuracy and reliability. Wearable devices, particularly those monitoring biochemical signals, often struggle with inaccuracies due to environmental factors or device limitations. AI algorithms help mitigate these issues by correcting errors and minimizing cross-sensitivity in multimodal sensing, ensuring that the collected data is more precise and meaningful (Shajari et al., 2023). This is particularly crucial in medical applications, where even small deviations in data can lead to significant clinical consequences. Beyond accuracy, personalization is another key strength of AI-powered wearables. By analyzing individual health trends, these devices can provide tailored recommendations for lifestyle adjustments, exercise plans, and dietary improvements, encouraging better health outcomes. Additionally, wearable technology serves as an extension of clinical care, giving healthcare providers real-time access to patient data. This integration supports informed decision-making, facilitates early interventions, and enhances the management of chronic conditions ("AI-Powered Wearable Devices & Real-Time Monitoring," 2023).

For instance, individuals with diabetes can benefit from continuous glucose monitoring wearables that provide immediate feedback, helping them manage their condition more effectively. Wearable devices with AI capabilities are proving invaluable in various healthcare applications. Cardiovascular monitoring is one of the most significant areas, as AI-powered wearables can detect early signs of heart conditions like atrial fibrillation—potentially preventing serious complications. Additionally, they play a crucial role in tracking physical activity and sleep patterns, helping to prevent lifestyle-related diseases such as obesity and diabetes. The ability to continuously analyze these metrics empowers users to make informed decisions about their well-being. Beyond personal health management, wearables have also demonstrated public health benefits. For instance, during the COVID-19 pandemic, they were used to track symptoms such as changes in heart rate or oxygen saturation, enabling early detection and isolation of potentially infected individuals. These examples highlight the broader potential of wearables in addressing global health challenges and promoting disease prevention on a larger scale. Furthermore, as wearable sensors continue to advance, devices are being developed to track more complex health metrics, such as stress levels and mental health indicators, paving the way for a more holistic approach to health monitoring ("AI-Powered Wearable Devices & Real-Time Monitoring," 2023; Shajari et al., 2023).

Despite their many advantages, AI-powered wearables face several challenges that need to be addressed to maximize their impact. One of the most pressing concerns is data privacy and security. Since these devices continuously collect and transmit sensitive health data, ensuring strong encryption and compliance with privacy regulations is essential for building trust among users and healthcare providers. Additionally, while AI enhances accuracy, its algorithms must be rigorously validated to ensure they are reliable and effective in real-world applications. Overcoming these hurdles is critical to widespread adoption and success in healthcare. Cost and accessibility are also key considerations. Although wearable technology is becoming more affordable, advanced AI-powered models remain expensive, potentially limiting access for some populations. Bridging this gap will require technological innovations, strategic partnerships, and policy support to make these devices more widely available (Huhn et al., 2022).

Another challenge is user engagement and adherence—wearables are only effective if users wear them consistently and interact with the data they provide. Encouraging long-term engagement through intuitive designs, personalized feedback, and seamless integration into daily life will be crucial for maximizing their impact. Overall, wearable devices, especially those enhanced with AI, are revolutionizing health monitoring by providing real-time insights, improving diagnostic accuracy, and promoting preventive healthcare. As technology continues to evolve, addressing challenges related to data privacy, accessibility, and user engagement will be key to unlocking their full potential. With thoughtful implementation, AI-powered wearables can significantly contribute to personalized, data-driven healthcare, improving outcomes for individuals and populations alike.

7.4 The Role of Technology in Enhancing Patient Engagement

Technological advancements have revolutionized the way patients interact with rehabilitation programs. Virtual reality (VR), mobile applications, and telehealth platforms are at the forefront of these innovations, transforming cognitive rehabilitation by improving accessibility, engagement, and outcomes.

7.4.1 Virtual Reality (VR)

Virtual reality (VR) has emerged as a transformative tool in cognitive rehabilitation, offering immersive and interactive environments that simulate real-world tasks. VR is particularly effective in improving cognitive function and alleviating symptoms in patients with neurological conditions such as Alzheimer's disease, traumatic brain injuries (TBIs), and stroke. By simulating daily activities, VR enables patients to practice and improve their cognitive abilities in a controlled, safe environment. Studies have demonstrated that VR-based cognitive training can lead to significant improvements in attention, memory, executive functions, and processing speed compared to traditional paper-based or computer-based methods. For instance, in stroke survivors, VR has been shown to enhance spatial navigation, problem-solving, and working memory through task repetition and adaptive difficulty levels (Abedi et al., 2024).

Similarly, Alzheimer's disease patients have experienced reductions in cognitive decline and an enhanced quality of life through VR exercises that simulate memory-based tasks. The immersive nature of VR provides sensory feedback and fosters motivation, ensuring higher levels of engagement compared to traditional methods. One of VR's unique strengths is its ability to provide immediate feedback and adapt to patient performance. This adaptability ensures that the training remains challenging yet achievable, fostering continuous improvement in cognitive function. Moreover, VR systems can integrate with other tools and devices, such as electroencephalography (EEG) or physiological activity monitoring tools, allowing therapists to track patient progress and physiological responses during training sessions (Abedi et al., 2024).

This integration enhances the effectiveness of rehabilitation programs by offering insights into the patient's neurological and physiological states. In addition to its cognitive benefits, VR is highly engaging and enjoyable, which contributes to sustained participation in rehabilitation programs. Patients often perceive VR exercises as less monotonous and more entertaining, leading to better adherence to treatment regimens. This feature is particularly important in long-term rehabilitation programs, where maintaining motivation can be challenging. Overall, VR holds immense potential to revolutionize cognitive rehabilitation by providing immersive, personalized, and effective training solutions (Abedi et al., 2024).

7.4.2 Mobile Applications

The widespread adoption of smartphones has paved the way for mobile applications to become valuable tools in cognitive rehabilitation. These apps offer an accessible, cost-effective, and scalable approach to cognitive training, allowing individuals to engage in rehabilitation exercises tailored to their specific needs. Their portability eliminates the need for frequent clinic visits, making it easier for patients to participate in therapy from the comfort of their homes. A systematic review of mobile-based cognitive training programs has shown moderate improvements in key cognitive areas, including memory, attention, executive function, and processing speed. Many of these apps incorporate gamification elements, such as rewards, levels, and interactive challenges, to enhance engagement and motivation. For example, memory-based games, problem-solving puzzles, and attention-enhancing tasks are commonly integrated into these platforms to make cognitive training more dynamic and enjoyable. This game-like experience helps ensure greater adherence to rehabilitation programs by making the process feel less like a chore and more like an interactive learning opportunity. One of the major advantages of mobile apps in cognitive rehabilitation is their suitability for home-based training. Unlike traditional clinic-based interventions, mobile apps offer patients the flexibility to practice exercises at their own pace and on their own schedule. This is particularly beneficial for individuals facing mobility issues, transportation barriers, or those living in remote areas with limited access to specialized healthcare services. By overcoming these logistical challenges, mobile technology significantly broadens the reach of cognitive rehabilitation, ensuring that more people can benefit from these interventions (Mantovani et al., 2020).

In addition to their accessibility, mobile apps leverage artificial intelligence (AI) and machine learning to deliver personalized training experiences. By continuously analyzing a user's performance, AI-driven algorithms can adjust the difficulty level, select appropriate exercises, and provide tailored feedback to optimize cognitive progress. For instance, if a patient consistently performs well in memory exercises, the app can dynamically increase task complexity to ensure continued cognitive growth. This level of customization keeps the training both challenging and engaging, ultimately enhancing the effectiveness of the rehabilitation process.

Beyond cognitive exercises, mobile apps also serve as educational tools, helping users better understand their condition and the goals of their rehabilitation program. Many applications offer progress tracking, goal setting, and exercise logs, allowing patients to monitor their improvements over time. These features empower individuals to take an active role in their rehabilitation journey, fostering a sense of ownership and accountability. By integrating real-time progress tracking and interactive feedback, mobile apps create a structured yet adaptable environment that supports long-term cognitive health. As technology continues to evolve, mobile applications are poised to play an increasingly central role in cognitive rehabilitation. Their ability to provide flexible, engaging, and AI-driven training makes them a powerful tool for improving cognitive function, particularly for those who may struggle to access traditional in-person therapy. By combining scientific principles with modern technology, these apps not only enhance cognitive performance but also promote greater independence and patient involvement in the rehabilitation process.

7.4.3 Telehealth Platforms

Telehealth platforms have become indispensable in delivering cognitive rehabilitation, especially in the wake of the COVID-19 pandemic, which restricted access to traditional in-person care. Telehealth leverages digital communication technologies to provide remote consultations, therapy sessions, and cognitive training programs. This approach has been found to be feasible, well-accepted, and effective in providing rehabilitation to patients with chronic neurological conditions. One of the primary advantages of telehealth platforms is their ability to increase participation and continuity of care in natural environments. Patients can receive therapy in their homes, where they feel most comfortable, reducing the stress and logistical challenges associated with clinic visits. This convenience encourages consistent engagement with rehabilitation programs, leading to better outcomes. Moreover, telehealth allows therapists to remotely monitor patient progress, assess the execution of exercises, and modify treatments as needed. This level of oversight ensures that the rehabilitation program remains aligned with the patient's evolving needs. Studies have shown that telehealth interventions are not inferior to conventional face-to-face approaches in terms of efficacy, validity, reliability, and patient satisfaction. For instance, telehealth platforms that offer video conferencing, virtual assessments, and guided exercises have been found to deliver comparable outcomes to in-person therapy sessions. Patients and caregivers often express high levels of satisfaction with telehealth services, citing convenience, accessibility, and the ability to maintain regular contact with healthcare professionals as key benefits. Telehealth platforms also address critical barriers to care, such as geographic and socioeconomic disparities. By eliminating the need for physical presence, telehealth makes cognitive rehabilitation accessible to patients in underserved areas or those with mobility limitations. Additionally, telehealth reduces costs associated with travel, childcare, and time off work, making rehabilitation more affordable and feasible for patients and their families. The adaptability of telehealth platforms extends beyond cognitive training to include multidisciplinary care. For example, a telehealth program might combine cognitive exercises with physical therapy, speech therapy, or mental health counseling, providing a holistic approach to rehabilitation. This integration ensures that patients receive comprehensive care tailored to their unique needs (Mantovani et al., 2020; Matamala-Gomez et al., 2021).

Virtual reality, mobile applications, and telehealth platforms collectively represent a paradigm shift in cognitive rehabilitation. By enabling real-time engagement, personalized interventions, and continuity of care, these technologies address critical gaps in traditional rehabilitation methods. While challenges such as data privacy, user adherence, and cost remain, ongoing advancements are likely to mitigate these issues and further establish these tools as indispensable components of modern healthcare. Through their combined strengths, these technologies are poised to improve patient outcomes, enhance accessibility, and empower individuals on their rehabilitation journey.

7.5 Case Studies

7.5.1 Stroke Rehabilitation

Case Study: 62-Year-Old Stroke Survivor Using VR-Based Rehabilitation
A 62-year-old male stroke survivor participated in a VR-based rehabilitation program aimed at improving motor function and cognitive engagement. The program included tasks such as virtual cooking and shopping simulations, which allowed him to practice daily living activities in a safe and controlled environment. Over 6 months, the patient's attention, memory, and executive function improved significantly compared to traditional therapy participants. This aligns with studies indicating that VR-based interventions enhance motivation and cognitive recovery by providing immersive, real-world tasks (Aderinto et al., 2023; *Virtual Reality for Individuals Affected by Stroke*, n.d.).

- Improvement in Cognitive Functions Using VR: Research demonstrates that VR-based rehabilitation improves cognitive functions like attention, memory, and executive function. For instance, a study found that stroke survivors engaging in VR cognitive tasks showed 25% greater improvement in memory recall than those undergoing standard therapies (Hussain et al., 2024). Another study highlighted that VR increased processing speed by 20%, attributed to the active engagement and repetition involved in VR-based tasks (Ku & Kang, 2018; Manivannan et al., 2019).
- Real-World Simulation for Skill Training: VR environments offer a unique advantage by replicating real-world scenarios. For example, a study demonstrated how VR was used for gait training, allowing stroke patients to simulate walking through different environments like parks or crowded malls. This approach improves balance and mobility and enables patients to practice skills that may be difficult or unsafe in real settings (*Virtual Reality for Individuals Affected by Stroke*, n.d.; Zanier et al., 2018).
- Personalized Rehabilitation Through VR: VR platforms such as the Oculus Rift or HTC Vive allow customization of tasks to cater to individual patient deficits. A study showcased how a VR program tailored to improve upper limb function significantly outperformed generic tasks, as patients were more engaged and saw faster progress in movement range and precision (Ku & Kang, 2018).
- Limitations and Future Research: Despite these advances, the long-term effectiveness of VR rehabilitation remains underexplored. A systematic review

emphasized the need for randomized controlled trials with larger sample sizes to understand the full potential of VR in stroke recovery beyond motor improvements (Aderinto et al., 2023).

7.5.2 Traumatic Brain Injury (TBI)

Case Study: Young Adult Using Mobile App for Cognitive Recovery
A 25-year-old recovering from a moderate TBI used a gamified mobile application called "NeuroFit" for 6 months. The app focused on memory and executive function improvement through daily exercises like pattern matching, logic puzzles, and timed recall tasks. Results showed a 30% improvement in attention span and a 20% increase in short-term memory recall compared to baseline testing. Gamification, such as earning rewards and badges, kept the individual engaged and motivated (Manivannan et al., 2019; Quan et al., 2024).

- Effects of Mobile Applications on Cognitive Function: Mobile apps have demonstrated moderate success in improving memory, attention, and executive functions. A study evaluated a group of TBI patients using a memory-enhancing app and found a 35% increase in memory task performance after 8 weeks. The portability of mobile devices allows patients to engage in rehabilitation consistently, even outside clinical settings (Catania et al., 2023).
- Home-Based Cognitive Training: Mobile apps are transforming rehabilitation by enabling patients to undergo therapy at home. A review published found that home-based cognitive training programs led to 40% higher adherence rates compared to clinic-based programs. This is especially beneficial for TBI patients in rural or underserved areas (Hussain et al., 2024; Quan et al., 2024).
- Personalized Cognitive Training: Many apps, such as "BrainHQ" and "Lumosity," are designed to target-specific cognitive deficits. In one study, patients recovering from mild TBI showed substantial improvement in processing speed when the app adapted to their progress by increasing task difficulty as they improved (Catania et al., 2023; Hussain et al., 2024).
- Engagement Through Gamification: Gamified features, including rewards, progress tracking, and challenges, make mobile cognitive training more appealing. A study reported that patients using gamified rehabilitation apps completed 90% of their prescribed sessions, compared to 65% for non-gamified tools. The study suggests that making the exercises enjoyable significantly improves adherence and outcomes (Manivannan et al., 2019).

7.6 Challenges and Ethical Considerations

Technology has brought remarkable advancements to rehabilitation, but its integration also presents a range of challenges and ethical dilemmas. The following sections explore these issues, highlighting both practical and moral complexities.

7.6.1 Challenges

The use of technology in rehabilitation offers opportunities to expand access and improve care, but it also introduces significant obstacles that must be addressed. One of the most pressing challenges is ensuring accessibility and affordability. Advanced rehabilitation technologies often come with high costs, making them inaccessible to many, particularly those in underserved or economically disadvantaged communities. For example, patients in rural areas face geographic barriers that can limit access to the necessary equipment or reliable Internet connectivity. This uneven distribution of resources raises concerns about reinforcing existing socioeconomic inequalities, with those who can afford advanced technology receiving better care than others.

Another key issue is the diversity of patient needs. Rehabilitation is rarely a one-size-fits-all process, yet many technological solutions fail to adapt to the unique conditions and preferences of individual patients. For instance, older adults or those with limited digital literacy may struggle to engage with tech-driven platforms, creating additional barriers to effective treatment. Without adequate support and education, such disparities can hinder progress, leaving certain groups at risk of being left behind.

The availability of healthcare professionals adds another layer of complexity. Even with cutting-edge technology, the uneven distribution of trained providers—especially in remote regions—limits the capacity to deliver consistent, high-quality care. While tele-rehabilitation can help bridge this gap, it is not a complete solution. Technical constraints, such as unstable Internet connections or incompatible devices, can still impede its effectiveness, leaving some patients with suboptimal experiences. Finally, while technology offers potential solutions, it is important to recognize its limitations. For many patients, rehabilitation is as much about human connection as it is about physical or cognitive progress. The introduction of digital tools risks eroding this personal element, replacing face-to-face interactions with screens and data-driven algorithms. Addressing these challenges requires a commitment to making technology more inclusive, adaptable, and human-centered.

7.6.2 Ethical Considerations

The integration of technology into rehabilitation care raises complex ethical questions, requiring thoughtful consideration to ensure fairness and respect for all involved. One fundamental concern is safeguarding patient privacy. Technology often relies on collecting and storing sensitive health data, which introduces risks of breaches and unauthorized access. Patients must feel confident that their information is secure, and healthcare providers have a duty to implement robust data protection measures. Without this trust, the adoption of technology-based solutions will remain limited. Another ethical issue is the need to respect patient autonomy. Rehabilitation technologies must not only work effectively but also empower patients to make informed decisions about their care. This includes providing clear, transparent information about the capabilities and limitations of such tools. Patients should never feel pressured into using technologies they do not fully understand or trust. The caregiver-patient relationship is also at stake. Technology has the potential to depersonalize care, reducing the warmth and empathy that come with human interaction. Rehabilitation is often an emotional journey as much as a physical one, and the absence of personal connection can negatively impact a patient's recovery. Healthcare providers must find ways to integrate human touch into digital solutions, ensuring that patients feel seen, heard, and supported (Guy et al., n.d.; Veras et al., 2024).

Bias in technology poses another ethical challenge. Algorithms and data-driven tools used in tele-rehabilitation can unintentionally reinforce inequalities if not designed with care. For example, a device calibrated for certain demographic groups may produce less accurate results for others, creating disparities in care quality. Ethical responsibility demands that healthcare providers and developers critically evaluate these tools to minimize bias and ensure fairness. The rapid pace of technological advancement also places a burden on healthcare professionals. Staying up-to-date with new tools and platforms requires continuous training and adaptation, which can strain already overstretched providers. Ethical guidelines and support systems are essential to help professionals navigate this new landscape while maintaining their commitment to patient welfare (Rabanifar & Abdi, 2021).

In response to these ethical challenges, public authorities, scientific bodies, and professional organizations must take proactive steps. Public institutions can play a leadership role by ensuring that underserved communities have equitable access to technology and infrastructure. At the same time, professional organizations must adapt their codes of ethics to reflect the realities of digital care, offering guidance to providers on navigating new and complex scenarios. Tele-rehabilitation, while transformative, should never be viewed as a replacement for in-person care. Instead, it must be seen as a complementary tool, designed to enhance traditional methods without eroding their core values. Striking this balance is critical to preserving the dignity and humanity of rehabilitation care. Ultimately, the adoption of technology in rehabilitation demands ongoing ethical reflection and dialogue among all stakeholders. Patients, providers, developers, and policymakers must work together to ensure that these tools are used responsibly, with fairness, transparency, and empathy at the forefront. By addressing these challenges and ethical considerations thoughtfully, we can unlock the potential of technology to improve lives while honoring the human experience at the heart of rehabilitation (Nicolas et al., 2024).

7.7 Future Directions

In the rapidly evolving field of neuropsychological rehabilitation, the drive toward a patient-centric approach is reshaping the way interventions are designed, implemented, and evaluated. By integrating advanced technologies with an acute awareness of individual needs, researchers and clinicians are reshaping rehabilitation into something deeply human. It is about seeing the person behind the patient and tailoring care to honor their unique story. This includes leveraging innovations while respecting diverse perspectives. This focus on personalization is grounded in the belief that every patient's journey is unique, shaped by their cognitive abilities, cultural background, personal goals, and access to resources. By addressing these factors, we can create rehabilitation experiences that are as diverse and multifaceted as the patients themselves (Khalid et al., 2024; Lanotte et al., 2023).

One crucial area of development is the creation of inclusive tools that consider the wide array of patient demographics. Rehabilitation cannot be one-size-fits-all, and understanding the perspectives of individuals across different age groups, cognitive abilities, and cultural contexts is essential. For instance, older adults recovering from neurological injuries may have different needs compared to younger individuals. Similarly, cultural values and norms can influence how patients perceive and engage with rehabilitation efforts. Designing tools that are adaptable to such variations requires intentionality. User-friendly interfaces are paramount, particularly when considering patients with limited digital literacy. By integrating multimodal input methods—such as voice commands, text options, and gesture-based controls—these tools can accommodate diverse communication styles and abilities, ensuring no patient is left behind.

The emergence of artificial intelligence (AI) and machine learning offers transformative potential in personalizing rehabilitation. Imagine an AI algorithm as a compassionate observer, learning from patterns in patient data to craft interventions that feel almost hand-designed. For example, these systems can adapt in real time to each person's journey, making every adjustment feel like it's meeting the patient where they are. An individual recovering from a traumatic brain injury might benefit from a rehabilitation plan informed by AI that adapts in real time, based on their progress and response to treatment. Machine learning models can predict likely outcomes, allowing clinicians to adjust strategies proactively. This predictive capacity extends beyond immediate care, assisting with long-term planning by identifying potential challenges and areas of focus for ongoing rehabilitation. AI's ability to personalize care fosters a sense of empowerment for patients, who can see their unique circumstances and progress reflected in their treatment plans (Calderone et al., 2024a; Khalid et al., 2024).

While these technological advancements hold promise, they also highlight the importance of accessibility. Integrating technology-driven interventions into underserved communities is critical to ensuring equity in healthcare. Many patients face barriers such as geographic isolation, financial constraints, and limited access to specialized care. Home-based rehabilitation systems bring care closer to the heart of patients' lives. With accessible technologies, they offer a way to overcome barriers like distance and cost, making recovery a shared experience within the comfort of familiar surroundings. AI-driven tele-rehabilitation, for example, enables patients in remote areas to access high-quality care without the need for frequent travel. These innovations not only expand access but also allow patients to engage with rehabilitation on their own terms, fostering a sense of autonomy and participation in their recovery journey (Rasa, 2024; Yang et al., 2024).

In parallel with these efforts, the development of advanced tools and technologies must be accompanied by rigorous validation. To ensure these innovations truly serve patients, large-scale clinical trials act as the proving ground. They validate both the safety and the transformative potential of AI-driven tools, reinforcing trust among patients and caregivers alike. Such trials provide the evidence base needed to integrate these innovations into standard healthcare practices. At the same time, developers and researchers must address critical ethical considerations. Data privacy is a paramount concern, particularly given the sensitive nature of health information. Ensuring algorithm transparency is equally important, as it builds trust among patients and clinicians by providing insight into how decisions are made. Moreover, tackling potential biases in AI models is crucial to prevent the perpetuation of inequities in care (Calderone et al., 2024b).

Beyond traditional approaches, emerging technologies such as virtual reality (VR) and augmented reality (AR) are opening new frontiers in neuropsychological rehabilitation. Immersive tools like VR invite patients into worlds where healing feels natural. These interactive environments offer not just cognitive stimulation but moments of triumph that remind patients of their resilience. A patient recovering from a stroke, for instance, might use VR to simulate real-world tasks, rebuilding both confidence and functional abilities. Similarly, wearable devices are becoming increasingly sophisticated, enabling continuous monitoring of a patient's condition. These devices provide real-time data on metrics such as movement, heart rate, and cognitive engagement, allowing clinicians to adjust interventions dynamically. By integrating such tools into rehabilitation programs, clinicians can create more responsive and adaptive care plans (Lanotte et al., 2023).

The role of technology in rehabilitation extends to supporting clinicians themselves. Decision support systems, powered by AI, are helping healthcare providers design tailored rehabilitation programs. These systems synthesize complex data, offering evidence-based recommendations that align with a patient's unique profile. By streamlining the decision-making process, clinicians can focus more on patient interaction and less on administrative tasks, ultimately enhancing the quality of care. Despite the promising advancements in technology, it is important to remember that rehabilitation is fundamentally a human-centered endeavor. Tools and systems must be designed with empathy, keeping the patient's experience at the forefront. Patients navigating the challenges of neuropsychological recovery often face feelings of frustration, isolation, and vulnerability. Technologies that prioritize usability and accessibility can alleviate some of these burdens, creating a sense of connection and support. For example, incorporating patient feedback into the design of rehabilitation tools ensures that their needs and preferences are respected. This collaborative approach fosters trust and engagement, key elements of successful rehabilitation (Akbar et al., 2024).

Moreover, addressing the broader social determinants of health is essential in achieving equitable outcomes. Expanding access to rehabilitation services involves not only technological innovation but also policy changes and community engagement. Advocacy for funding and resources to support underserved populations is crucial. Community-based interventions, supported by technology, can provide culturally sensitive and contextually relevant care. These efforts reflect a commitment to inclusivity, ensuring that the benefits of innovation are felt by all, regardless of their background or circumstances. As we look to the future of neuropsychological rehabilitation, the convergence of technology and human-centered care offers exciting possibilities. By embracing advancements in AI, machine learning, VR, and wearable devices, while remaining mindful of ethical considerations and accessibility challenges, we can redefine what it means to provide patient-centric care. These efforts are not just about improving clinical outcomes but also about enhancing the overall quality of life for patients. By empowering individuals to take an active role in their recovery and creating systems that adapt to their unique needs, we can build a more inclusive and effective approach to neuropsychological rehabilitation. The journey ahead is both challenging and inspiring, as we strive to transform rehabilitation into a truly personalized and equitable experience for all.

7.8 Conclusion

Neuropsychological rehabilitation is undergoing a transformation through patient-centered approaches and innovative technologies. At its core, it is about recognizing the individuality of each patient and designing interventions that honor their unique needs and stories. By integrating advanced tools and prioritizing patient perspectives, researchers and clinicians are crafting care that is both effective and deeply personal. Incorporating patient perspectives and leveraging technology to measure outcomes is essential for advancing neurocognitive rehabilitation. This approach ensures that interventions are not only clinically effective but also meaningful and impactful from the patient's viewpoint. Inclusive tools play a crucial role in addressing diverse patient demographics. Rehabilitation cannot adopt a one-size-fits-all model. Tools must accommodate variations in age, cognitive abilities, and cultural backgrounds, offering user-friendly interfaces that bridge gaps in digital literacy. Multimodal input methods, such as voice commands or gesture-based controls, allow patients to engage in ways that feel natural to them. Artificial intelligence (AI) and machine learning offer new avenues for personalized rehabilitation. AI systems analyze patient data to tailor interventions, adapting dynamically to individual progress and needs. These tools provide clinicians with insights that can predict outcomes and refine strategies, creating a care plan that evolves alongside the patient. Accessibility remains a cornerstone of these innovations. Home-based rehabilitation systems and AI-driven tele-rehabilitation bring high-quality care to underserved communities, overcoming barriers like distance and cost. By enabling patients to engage with rehabilitation in their own environments, these systems foster autonomy and make recovery more accessible.

Emerging technologies such as virtual reality (VR) and wearable devices are also redefining rehabilitation. VR immerses patients in interactive environments that support cognitive and physical recovery, while wearable devices monitor progress in real time, allowing clinicians to adjust treatments as needed. Despite the promise of these advancements, rehabilitation must remain a fundamentally human endeavor. Empathy and collaboration are essential to building trust and engagement. Incorporating patient feedback into tool development ensures that interventions align with their needs and experiences, fostering a sense of partnership in the recovery journey. Ultimately, neuropsychological rehabilitation is about resilience and connection. Each patient's journey is unique, and the integration of technology should enhance, not replace, the human connection at the heart of care. By balancing innovation with empathy, we can create a future where recovery is both clinically effective and profoundly meaningful, empowering patients to reclaim their lives with dignity and hope.

References

Abedi, A., Colella, T. J. F., Pakosh, M., & Khan, S. S. (2024). Artificial intelligence-driven virtual rehabilitation for people living in the community: A scoping review. *NPJ Digital Medicine, 7*(1), 1–13. https://doi.org/10.1038/s41746-024-00998-w

Aderinto, N., Olatunji, G., Abdulbasit, M. O., Edun, M., Aboderin, G., & Egbunu, E. (2023). Exploring the efficacy of virtual reality-based rehabilitation in stroke: A narrative review of current evidence. *Annals of Medicine, 55*(2), 2285907. https://doi.org/10.1080/07853890.2023.2285907

AI-Powered Wearable Devices & Real-Time Monitoring. (2023, December 6). *Minttihealth.* https://minttihealth.com/integrating-artificial-intelligence-with-wearable-devices-and-real-time-monitoring-for-enhanced-healthcare-outcomes/

Aiyegbusi, O. L., Cruz Rivera, S., Roydhouse, J., Kamudoni, P., Alder, Y., Anderson, N., Baldwin, R. M., Bhatnagar, V., Black, J., Bottomley, A., Brundage, M., Cella, D., Collis, P., Davies, E.-H., Denniston, A. K., Efficace, F., Gardner, A., Gnanasakthy, A., Golub, R. M., et al. (2024). Recommendations to address respondent burden associated with patient-reported outcome assessment. *Nature Medicine, 30*(3), 650–659. https://doi.org/10.1038/s41591-024-02827-9

Akbar, K., Passaro, A., Di Gioia, M., Martini, E., Dragone, M., Zullo, A., & Stasolla, F. (2024). Reviewing the horizon: The future of extended reality and artificial intelligence in neurorehabilitation for brain injury recovery. *Information, 15*(8), 8. https://doi.org/10.3390/info15080501

Ang, L. C., Yap, P., Tay, S. Y., Koay, W. I., & Liew, T. M. (2023). Examining the validity and utility of Montreal cognitive assessment (MoCA) domain scores for early neurocognitive disorders. *Journal of the American Medical Directors Association, 24*(3), 314–320.e2. https://doi.org/10.1016/j.jamda.2022.12.028

Bennett, T. L. (2001). Neuropsychological evaluation in rehabilitation planning and evaluation of functional skills. *Archives of Clinical Neuropsychology, 16*(3), 237–253. https://doi.org/10.1016/S0887-6177(00)00082-2

Blennow Nordström, E., Evald, L., Mion, M., Segerström, M., Vestberg, S., Ullén, S., Heimburg, K., Gregersen Oestergaard, L., Grejs, A. M., Keeble, T. R., Kirkegaard, H., Rylander, C., Wise, M. P., & Lilja, G. (2024). Combined use of the Montreal Cognitive Assessment and Symbol Digit Modalities Test improves neurocognitive screening accuracy after cardiac arrest: A validation sub-study of the TTM2 trial. *Resuscitation, 202*, 110361. https://doi.org/10.1016/j.resuscitation.2024.110361

Bridger, K., Kellezi, B., Kendrick, D., Radford, K., Timmons, S., Rennoldson, M., Jones, T., Kettlewell, J., & On Behalf Of The Rowtate Team. (2021). Patient perspectives on key outcomes for vocational rehabilitation interventions following traumatic injury. *International Journal of Environmental Research and Public Health, 18*(4), 2035. https://doi.org/10.3390/ijerph18042035

Bullinger, M., & Quitmann, J. (2014). Quality of life as patient-reported outcomes: Principles of assessment. *Dialogues in Clinical Neuroscience, 16*(2), 137–145.

Calderone, A., Latella, D., Bonanno, M., Quartarone, A., Mojdehdehbaher, S., Celesti, A., & Calabrò, R. S. (2024a). Towards transforming neurorehabilitation: The impact of artificial intelligence on diagnosis and treatment of neurological disorders. *Biomedicine, 12*(10), 2415. https://doi.org/10.3390/biomedicines12102415

Calderone, A., Latella, D., Bonanno, M., Quartarone, A., Mojdehdehbaher, S., Celesti, A., & Calabrò, R. S. (2024b). Towards transforming neurorehabilitation: The impact of artificial intelligence on diagnosis and treatment of neurological disorders. *Biomedicine, 12*(10), 10. https://doi.org/10.3390/biomedicines12102415

Catania, V., Rundo, F., Panerai, S., & Ferri, R. (2023). Virtual reality for the rehabilitation of acquired cognitive disorders: A narrative review. *Bioengineering, 11*(1), 35. https://doi.org/10.3390/bioengineering11010035

Dautzenberg, G., Lijmer, J., & Beekman, A. (2020). Diagnostic accuracy of the Montreal Cognitive Assessment (MoCA) for cognitive screening in old age psychiatry: Determining cutoff scores in clinical practice. Avoiding spectrum bias caused by healthy controls. *International Journal of Geriatric Psychiatry, 35*(3), 261–269. https://doi.org/10.1002/gps.5227

de Wit, L., Goldstein, F. C., & Loring, D. W. (2022). Clinical value of the Montreal Cognitive Assessment free recall condition alone versus cued recall and recognition conditions to detect true memory impairment. *Applied Neuropsychology: Adult*. https://doi.org/10.1080/23279095.2022.2161050

Domensino, A.-F., Verberne, D., Prince, L., Fish, J., Winegardner, J., Bateman, A., Wilson, B., Ponds, R., & van Heugten, C. (2022). Client experiences with holistic neuropsychological rehabilitation: "It is an ongoing process". *Neuropsychological Rehabilitation, 32*(8), 2147–2169. https://doi.org/10.1080/09602011.2021.1976222

Dür, M., Wenzel, C., Simon, P., & Tucek, G. (2022). Patients' and professionals' perspectives on the consideration of patients' convenient therapy periods as part of personalised rehabilitation: A focus group study with patients and therapists from inpatient neurological rehabilitation. *BMC Health Services Research, 22*(1), 372. https://doi.org/10.1186/s12913-022-07755-3

Geerards, D., Pusic, A., Hoogbergen, M., van der Hulst, R., & Sidey-Gibbons, C. (2019). Computerized quality of life assessment: A randomized experiment to determine the impact of individualized feedback on assessment experience. *Journal of Medical Internet Research, 21*(7), e12212. https://doi.org/10.2196/12212

Guàrdia-Olmos, J., Peró-Cebollero, M., & Gudayol-Ferré, E. (2015). Neuropsychological rehabilitation and quality of life: A meta-analysis. *Revista Iberoamericana de Psicología y Salud, 6*(1), 11–18. https://doi.org/10.1016/S2171-2069(15)70002-5

Guy, M., Blary, A., Ladner, J., & Gilliaux, M. (n.d.). Ethical issues linked to the development of telerehabilitation: A qualitative study. *International Journal of Telerehabilitation, 13*(1), e6367. https://doi.org/10.5195/ijt.2021.6367

Huhn, S., Axt, M., Gunga, H.-C., Maggioni, M. A., Munga, S., Obor, D., Sié, A., Boudo, V., Bunker, A., Sauerborn, R., Bärnighausen, T., & Barteit, S. (2022). The impact of wearable technologies in health research: Scoping review. *JMIR mHealth and uHealth, 10*(1), e34384. https://doi.org/10.2196/34384

Hussain, M. A., Waris, A., Gilani, S. O., Mushtaq, S., Pujari, A. N., Khan, N. B., Jameel, M., Daminova, G., & Khan, M. I. (2024). Virtual reality as a non-conventional rehabilitation for stroke: A comprehensive review. *Journal of Neurorestoratology, 12*(3), 100135. https://doi.org/10.1016/j.jnrt.2024.100135

Jesus, T. S., Buschbacher, J., Struhar, J., Walters, T., Lopez, C., Fernandez, A., Gracz, K., & Colby, K. (2024). What is so great about inpatient rehabilitation from the patient experience perspective: Qualitative content analysis of an appreciative inquiry during a bedside experience rounding. *Healthcare, 12*(17), 17. https://doi.org/10.3390/healthcare12171711

Julayanont, N., Phillips, N., Chertkow, H., & Nasreddine, Z. (n.d.). The Montreal Cognitive Assessment (MoCA): Concept and clinical review. In *Cognitive screening instruments: A practical approach* (pp. 111–152). Springer-Verlag. Retrieved December 27, 2024, from https://www.researchgate.net/figure/Estimated-Marginal-Means-for-total-scores-on-the-Wechsler-Memory-Scale-WMS-and_fig2_273149959

Kennard, A. L., Rainsford, S., Hamilton, K. L., Glasgow, N. J., Pumpa, K. L., Douglas, A. M., & Talaulikar, G. S. (2024). Patient perspectives and preferences for rehabilitation among people living with frailty and chronic kidney disease: A qualitative evaluation. *BMC Nephrology, 25*(1), 304. https://doi.org/10.1186/s12882-024-03740-6

Khalid, U. B., Naeem, M., Stasolla, F., Syed, M. H., Abbas, M., & Coronato, A. (2024). Impact of AI-powered solutions in rehabilitation process: Recent improvements and future trends. *International Journal of General Medicine, 17*, 943–969. https://doi.org/10.2147/IJGM.S453903

Ku, J., & Kang, Y. J. (2018). Novel virtual reality application in the field of neurorehabilitation. *Brain & Neurorehabilitation, 11*(1), e5. https://doi.org/10.12786/bn.2018.11.e5

Lanotte, F., O'Brien, M. K., & Jayaraman, A. (2023). AI in rehabilitation medicine: Opportunities and challenges. *Annals of Rehabilitation Medicine, 47*(6), 444–458. https://doi.org/10.5535/arm.23131

Manivannan, S., Al-Amri, M., Postans, M., Westacott, L. J., Gray, W., & Zaben, M. (2019). The effectiveness of virtual reality interventions for improvement of neurocognitive performance after traumatic brain injury: A systematic review. *Journal of Head Trauma Rehabilitation, 34*(2), E52–E65. https://doi.org/10.1097/HTR.0000000000000412

Mantovani, E., Zucchella, C., Bottiroli, S., Federico, A., Giugno, R., Sandrini, G., Chiamulera, C., & Tamburin, S. (2020). Telemedicine and virtual reality for cognitive rehabilitation: A roadmap for the COVID-19 pandemic. *Frontiers in Neurology, 11*, 926. https://doi.org/10.3389/fneur.2020.00926

Matamala-Gomez, M., Bottiroli, S., Realdon, O., Riva, G., Galvagni, L., Platz, T., Sandrini, G., De Icco, R., & Tassorelli, C. (2021). Telemedicine and virtual reality at time of COVID-19 pandemic: An overview for future perspectives in neurorehabilitation. *Frontiers in Neurology, 12*, 646902. https://doi.org/10.3389/fneur.2021.646902

Nair, R. D., & Lincoln, N. B. (2013). The effectiveness of memory rehabilitation following neurological disabilities: A qualitative inquiry of patient perspectives. *Neuropsychological Rehabilitation, 23*(4), 528. https://doi.org/10.1080/09602011.2013.792290

Nicolas, B., Leblong, E., Fraudet, B., Gallien, P., & Piette, P. (2024). Telerehabilitation solutions in patient pathways: An overview of systematic reviews. *Digital Health, 10*, 20552076241294110. https://doi.org/10.1177/20552076241294110

Prigatano, G. P., Braga, L. W., Johnson, S. F., & Souza, L. M. N. (2021). Neuropsychological rehabilitation, neuroimaging and neuroplasticity: A clinical commentary. *NeuroRehabilitation, 49*(2), 255–265. https://doi.org/10.3233/NRE-218024

Quan, W., Liu, S., Cao, M., & Zhao, J. (2024). A comprehensive review of virtual reality technology for cognitive rehabilitation in patients with neurological conditions. *Applied Sciences, 14*(14), Article 14. https://doi.org/10.3390/app14146285

Rabanifar, N., & Abdi, K. (2021). Barriers and challenges of implementing telerehabilitation: A systematic review. *Iranian Rehabilitation Journal, 19*(2), 121–128. https://doi.org/10.32598/irj.19.2.1404.1

Rasa, A. R. (2024). Artificial intelligence and its revolutionary role in physical and mental rehabilitation: A review of recent advancements. *BioMed Research International, 2024*(1), 9554590. https://doi.org/10.1155/bmri/9554590

Rea, N. M. A., Neubeck, L., Kydonaki, K., & McHale, S. (2023). Patients' experiences and perspectives of post-hospital follow-up care to improve physical recovery for intensive care survivors: A systematic review of qualitative research. *International Journal of Nursing Studies Advances, 5*, 100168. https://doi.org/10.1016/j.ijnsa.2023.100168

Shajari, S., Kuruvinashetti, K., Komeili, A., & Sundararaj, U. (2023). The emergence of AI-based wearable sensors for digital health technology: A review. *Sensors (Basel, Switzerland), 23*(23), 9498. https://doi.org/10.3390/s23239498

Torregrossa, W., Torrisi, M., De Luca, R., Casella, C., Rifici, C., Bonanno, M., & Calabrò, R. S. (2023). Neuropsychological assessment in patients with traumatic brain injury: A comprehensive review with clinical recommendations—PMC. *Biomedicine, 11*(7), 1991. https://doi.org/10.3390/biomedicines11071991

Veras, M., Sigouin, J., Auger, C., Auger, L.-P., Ahmed, S., Boychuck, Z., Cavallo, S., Lévesque, M., Lovo, S., Miller, W. C., Nelson, M., Norouzi-Gheidari, N., O'Neil, J., Perreault, K., Urbanowski, R., Sheehy, L., Singh, H., Vincent, C., Wang, R., et al. (2024). A rapid review protocol of physiotherapy and occupational therapy telerehabilitation to inform ethical and equity concerns. *Digital Health, 10*, 20552076241260367. https://doi.org/10.1177/20552076241260367

Virtual Reality for Individuals Affected by Stroke. (n.d.). Physiopedia. Retrieved December 28, 2024, from https://www.physio-pedia.com/Virtual_Reality_for_Individuals_Affected_by_Stroke

Wain, H. R., Kneebone, I. I., & Billings, J. (2008). Patient experience of neurologic rehabilitation: A qualitative investigation. *Archives of Physical Medicine and Rehabilitation, 89*(7), 1366–1371. https://doi.org/10.1016/j.apmr.2007.11.050

Winegardner, J., & Coetzer, R. (2017). A global perspective on neuropsychological rehabilitation: Maximising outcomes when funds are short. In *Neuropsychological rehabilitation: The international handbook* (pp. 480–491). Routledge/Taylor & Francis Group.

Yang, R., Yuan, Q., Zhang, W., Cai, H., & Wu, Y. (2024). Application of Artificial Intelligence in rehabilitation science: A scientometric investigation utilizing citespace. *SLAS Technology, 29*(4), 100162. https://doi.org/10.1016/j.slast.2024.100162

Zanier, E. R., Zoerle, T., Di Lernia, D., & Riva, G. (2018). Virtual reality for traumatic brain injury. *Frontiers in Neurology, 9*, 345. https://doi.org/10.3389/fneur.2018.00345

Chapter 8
Neuroplasticity and Technology

8.1 Introduction

Neuroplasticity, the brain's remarkable ability to adapt, reorganize, and rewire itself in response to experiences, injuries, and environmental changes, has revolutionized how we understand brain function. Once believed to be fixed and unchanging after early development, the brain is now seen as a dynamic and malleable organ capable of lifelong adaptation. This discovery has redefined the possibilities for recovery and rehabilitation in neuropsychology, offering hope to individuals who face neurological challenges such as stroke, traumatic brain injury (TBI), and cognitive impairments. Neuroplasticity forms the foundation of modern rehabilitation approaches, allowing the brain to repair itself, compensate for lost functions, and even strengthen existing capabilities through targeted interventions (Galetto & Sacco, 2017; Marzola et al., 2023; Zotey et al., 2023).

Research has consistently demonstrated that neuroplasticity plays a central role in recovery. Neuroimaging studies, for instance, have revealed significant changes in brain activity and structure during rehabilitation. Activities such as physical therapy, speech therapy, and cognitive exercises can foster the reorganization of neural networks, with improvements in brain activation patterns often correlating with better recovery outcomes. This ability to rewire itself is driven by processes like synaptic growth, dendritic remodeling, and even the formation of new neurons and connections. The brain is also capable of recruiting other regions, including those not directly connected to the injured areas, to take over lost functions. These mechanisms highlight the brain's resilience and adaptability, even in the face of significant challenges. Harnessing this potential has transformed the field of neuropsychological rehabilitation. Traditional therapies like speech and language therapy are designed to retrain brain regions responsible for communication, while occupational therapy targets memory, attention, and problem-solving skills. Sensory retraining focuses on improving sensory-motor integration, while memory and

K. J. Reddy, *Innovations in Neurocognitive Rehabilitation*,
https://doi.org/10.1007/978-3-031-88117-6_8

attention training help consolidate neural pathways and strengthen cognitive circuits. Each of these interventions relies on the principle that repeated, purposeful activity can reshape the brain and create new opportunities for learning and recovery. What makes neuroplasticity particularly exciting is its ability to offer tailored solutions for individuals based on their unique needs, strengths, and challenges (Prigatano et al., n.d.; Pub, n.d.).

Advances in technology have expanded the ways we can leverage neuroplasticity, bringing cutting-edge tools into rehabilitation settings. Virtual reality (VR) and augmented reality (AR) are two such innovations that create immersive environments to engage the brain in ways that mimic real-life scenarios. These tools provide stimulating and adaptive platforms for motor and cognitive recovery, making therapy more engaging and effective. Brain-computer interfaces (BCIs) represent another revolutionary development, enabling direct communication between the brain and external devices. BCIs have been used to assist individuals in regaining motor function or controlling assistive technologies, offering a glimpse into the future of neuroplastic learning. Similarly, non-invasive brain stimulation techniques like transcranial magnetic stimulation (TMS) and transcranial direct current stimulation (tDCS) can modulate brain activity in targeted areas, supporting recovery in conditions such as depression, anxiety, and stroke ([1], [6]). Neurofeedback, which allows individuals to monitor and regulate their own brain activity in real time, further exemplifies how technology can enhance the natural processes of neuroplasticity (Galetto & Sacco, 2017; Neuroplasticity in Stroke and Brain Injury, n.d.; Zotey et al., 2023).

One of the most promising aspects of neuroplasticity-based rehabilitation is its increasing focus on personalization. Tailored approaches that consider a person's cognitive profile, age, injury severity, and comorbidities can maximize the benefits of neuroplasticity. Advanced imaging techniques and sensor technologies now make it possible to guide these interventions with precision, offering detailed insights into a person's brain activity and progress. This level of personalization ensures that therapies remain adaptable, effective, and aligned with individual needs, paving the way for optimized recovery outcomes. The implications of neuroplasticity extend beyond recovery to reshaping how we approach rehabilitation itself. By understanding and harnessing the brain's capacity for change, we are not only able to restore lost functions but also to create new pathways for growth and improvement. The integration of technological innovations like artificial intelligence, real-time neuroimaging, and virtual environments further amplifies the potential of neuroplasticity, allowing us to push the boundaries of what is possible in rehabilitation. These advancements hold immense promise for improving the lives of individuals recovering from neurological challenges while emphasizing the importance of equitable access to these transformative tools (Galetto & Sacco, 2017; Marzola et al., 2023; Puderbaugh & Emmady, 2024).

Neuroplasticity has redefined hope for recovery in neuropsychology. It is the driving force behind therapies that enable individuals to regain independence and improve their quality of life. As our understanding of neuroplasticity continues to evolve, so too will our ability to leverage it through innovative technologies and personalized approaches. With sustained research and a commitment to ethical practices, we can ensure that the potential of neuroplasticity benefits all those in need, transforming rehabilitation and offering new possibilities for recovery and resilience (Puderbaugh & Emmady, 2024).

8.2 Understanding Neuroplasticity

8.2.1 Definition and Mechanisms

Neuroplasticity is the brain's remarkable ability to adapt, reorganize, and change its structure and function in response to internal and external stimuli. This property allows the brain to modify itself throughout life, facilitating processes such as learning, memory formation, recovery from injury, and adaptation to new experiences or environments. Once considered a rigid and unchangeable organ after early development, the brain is now recognized as a highly dynamic system capable of constant reorganization. This adaptability is evident not only during critical periods of early childhood but also in adulthood and even old age, making neuroplasticity a cornerstone of neuroscience and rehabilitation practices. At its core, neuroplasticity reflects the brain's capacity to rewire itself by forming, strengthening, weakening, or eliminating neural connections based on the demands placed on it. This process occurs through several distinct but interconnected mechanisms, each serving a unique role in the brain's ability to adapt. Synaptic plasticity modifies the strength of synaptic connections between neurons. Long-term potentiation (LTP), one key mechanism, increases synaptic strength and is critical for learning and memory. Conversely, long-term depression (LTD) decreases synaptic strength and allows the brain to refine its networks, ensuring that only the most efficient connections are maintained. These synaptic changes demonstrate how neuroplasticity operates at the microscopic level to optimize the brain's functionality in response to environmental stimuli and experiences.

Another vital aspect of neuroplasticity is structural plasticity, which involves the formation or elimination of neural pathways. This process encompasses neurogenesis, the growth of new neurons, and synaptogenesis, the formation of new synapses. Both processes are essential for brain development, learning, and recovery from damage. Structural plasticity can also include physical changes to existing neurons, such as the growth of dendrites or axonal remodeling, which enhance the brain's ability to process and transmit information effectively. This reorganization enables the brain to maintain its functionality and adaptability even under challenging conditions, such as injury or disease. Functional plasticity, on the other hand, reflects the brain's ability to reassign specific functions to different regions. For instance, after an injury that impairs one part of the brain, other regions can compensate by taking over the lost functions. This reallocation of tasks can involve mechanisms such as homologous area adaptation, where functions are shifted to corresponding areas in the opposite hemisphere, or cross-modal reassignment, where a region deprived of its usual sensory input becomes responsive to stimulation from other modalities. These adaptive changes highlight the brain's incredible ability to compensate for damage and maintain its core functions, even under adverse circumstances (Neuroplasticity, n.d.; Neuroplasticity | Different Types, Facts, & Research, n.d.; Tonti et al., 2021).

The mechanisms underlying neuroplasticity operate at both microscopic and macroscopic levels. At the cellular level, changes occur in synaptic activity, ion channel density, neurotransmitter release, and postsynaptic receptor sensitivity. These adjustments are fundamental for modifying the strength and efficiency of neural connections. At the network level, neuroplasticity encompasses the reorganization of entire brain regions and functional systems. Studies have indicated that cross-modal plasticity allows areas deprived of sensory input, such as the visual cortex in blind individuals, to become responsive to other types of input, such as auditory or tactile stimuli (Appelbaum et al., 2023; Tonti et al., 2021). This ability to repurpose brain regions underscores the flexibility of neural networks and their capacity to adapt to new challenges and demands. The biological basis of neuroplasticity is deeply rooted in the brain's ability to generate and modify neural circuits. Neurogenesis plays a critical role in this process, particularly in areas like the hippocampus, which is involved in memory and learning. The dynamic activity of glial cells and the extracellular matrix further supports neuroplasticity by modulating neural circuits and maintaining an environment conducive to adaptation and growth (Role of Neuroplasticity in Neuro-Rehabilitation, n.d.; Tonti et al., 2021). These processes ensure that the brain remains a responsive and adaptable organ, capable of meeting the ever-changing demands of its environment.

The importance of neuroplasticity extends to virtually every aspect of brain function and development. During early childhood, neuroplasticity is particularly pronounced, facilitating the rapid acquisition of language, motor skills, and cognitive abilities. This developmental plasticity ensures that children can learn and adapt quickly during critical periods of growth. In adulthood, neuroplasticity supports lifelong learning, memory consolidation, and the ability to adapt to new experiences or challenges. Even in old age, the brain retains its capacity for plasticity, allowing individuals to compensate for age-related changes and maintain cognitive function. Neuroplasticity is also critical for recovery and rehabilitation following injury or disease. After a stroke, for instance, neuroplastic mechanisms can facilitate the reorganization of brain networks, allowing individuals to regain lost motor or cognitive functions. This recovery often involves the recruitment of perilesional areas or remote brain regions to compensate for damaged networks. Similarly, neuroplasticity plays a central role in therapies designed to treat conditions such as TBI, sensory deprivation, or limb loss, enabling individuals to adapt and recover despite significant challenges (Kania et al., 2017; Tonti et al., 2021).

The capacity for neuroplasticity also underpins learning and memory, as it enables the brain to form and reinforce new neural connections based on experience. This adaptability is what allows individuals to acquire new skills, adapt to changing environments, and recover from setbacks. Neuroplasticity ensures that the brain remains a resilient and versatile organ, capable of meeting the demands of both internal and external challenges. From microscopic changes at the synaptic level to large-scale reorganization of neural networks, the mechanisms of neuroplasticity enable the brain to learn, recover, and maintain its functionality under a wide range of conditions. Its role in development, learning, recovery, and adaptation highlights its significance in shaping the brain's capacity to navigate an ever-changing world. As research continues to uncover the intricacies of neuroplasticity, it holds promise for advancing therapies and interventions that harness the brain's incredible potential for change (Appelbaum et al., 2023; Kania et al., 2017; Tonti et al., 2021).

8.2.2 Types of Neuroplasticity

Experience-Dependent Plasticity Experience-dependent plasticity refers to the brain's ability to adapt and change in response to learning and new experiences, driving structural and functional modifications in neural circuits. Evidence supporting this form of plasticity can be found across a variety of learning contexts. Physical activities such as aerobic exercise have been shown to promote angiogenesis (the growth of new blood vessels) in critical areas of the brain, such as the hippocampus and cerebellum, which are associated with memory and motor control. In addition to angiogenesis, exercise enhances hippocampal neurogenesis, demonstrating the direct link between physical activity and the brain's ability to grow new neurons. Similarly, motor skill learning is associated with synaptogenesis (formation of new synapses) and dendritic spine plasticity in the motor cortex and cerebellum. These changes are rapid, with dendritic spine turnover occurring shortly after skill acquisition and persisting long after the motor skill has been established. Environments that encourage exploration and social interaction have also been found to foster plasticity. Complex environments can lead to dendritic growth, synaptogenesis, and angiogenesis in regions such as the visual cortex and other sensory and associative brain areas. Fear conditioning, a form of associative learning where a neutral stimulus becomes linked with an aversive event, provides another example. In this case, synaptic changes occur in the amygdala, a brain region critical for processing emotions and memory. These changes are evidence of how the brain adapts to encode new experiences and adjust behavioral responses (Neuroplasticity—an Overview | ScienceDirect Topics, n.d.-a, n.d.-b; Caeyenberghs et al., 2018; Kerr et al., 2011).

Injury-Induced Plasticity The brain's ability to reorganize and recover following injury, such as a stroke or traumatic brain injury, is a prime example of neuroplasticity at work. Injury-induced plasticity is characterized by both degenerative and regenerative processes, enabling the brain to adapt to damage and recover lost functions. Following ischemic stroke, for instance, the initial impact includes cell death, inflammation, neurotoxicity, and vascular dysfunction in the affected region and surrounding tissues. However, over time, the brain initiates repair mechanisms such as dendritic growth, axonal sprouting, synaptogenesis, vascular remodeling, and the migration of newly generated cells to the injured area. This regenerative capacity is facilitated by the brain's inherent plasticity, which supports spontaneous recovery to some extent. In cases of recovery from diaschisis—the suppression of function in areas remote but connected to the site of injury—the brain relies on the reactivation and reorganization of neural circuits. Plasticity in perilesional regions (areas surrounding the injury) and in distant connected areas contributes significantly to this process. Behavioral experiences and targeted rehabilitation interventions, such as physical and occupational therapy, can further enhance these mechanisms, demonstrating how the brain's adaptive capacity can be guided and optimized in response to injury. Stroke recovery provides a clear example of injury-induced plasticity. Over time, damaged neural networks in the affected hemisphere can undergo reorganization, and compensatory recruitment of regions in the opposite hemisphere may also occur. This adaptive response reflects the brain's ability to harness its plasticity for recovery, underscoring the critical role of environmental stimulation and rehabilitation strategies in shaping post-injury outcomes (Cherry, 2024; Kerr et al., 2011; Zotey et al., 2023).

Developmental Plasticity Developmental plasticity refers to the brain's ability to adapt and change during periods of growth and maturation, with structural and functional changes occurring in response to environmental experiences and learning opportunities. During early childhood, the brain is particularly plastic, allowing it to adapt rapidly to sensory input and experiences. Behavioral experiences during this period shape neural function and structure, creating the foundation for cognitive, motor, and sensory development. This plasticity is driven by processes such as neurogenesis, which continues throughout childhood and adolescence, enabling the formation of new neurons and their integration into existing neural networks. Synaptic plasticity also plays a vital role during development. As children acquire new skills and knowledge, changes in synaptic efficacy strengthen or weaken specific connections, optimizing the brain's efficiency for learned tasks. This synaptic pruning ensures that only the most relevant and frequently used neural pathways are retained, while others are eliminated. This process is particularly pronounced during critical periods of development, where the brain is especially sensitive to environmental input and capable of profound reorganization (Cherry, 2024; Kerr et al., 2011; Zotey et al., 2023).

Environmental enrichment during childhood further enhances developmental plasticity. Stimulating environments that include opportunities for social interaction, play, and exploration have been found to increase dendritic growth and synaptogenesis in regions such as the hippocampus and visual cortex. These changes demonstrate how the brain's plasticity during development can be influenced by external factors, emphasizing the importance of a nurturing and stimulating environment for optimal cognitive and emotional development. In cases of early brain injury, developmental plasticity also enables the brain to compensate for lost functions. Following damage to specific brain regions, other areas may take over these functions through mechanisms such as homologous area adaptation or functional redundancy. This compensatory reorganization is more pronounced in children due to the heightened plasticity of the developing brain, allowing them to recover from injuries that might have more severe consequences in adults (Cherry, 2024; Cramer et al., 2017; Kerr et al., 2011).

These mechanisms allow the brain to adapt to new learning experiences, recover from injuries, and optimize its structure and function during development. Experience-dependent plasticity drives changes through activities such as motor skill learning, exercise, and environmental exploration, highlighting the brain's capacity to reorganize itself based on input and demand. Injury-induced plasticity underscores the brain's resilience, demonstrating its ability to repair and reorganize after trauma, with significant implications for rehabilitation and recovery. Finally, developmental plasticity illustrates the dynamic processes that shape the brain during critical periods of growth, emphasizing the importance of nurturing environments and targeted interventions during childhood and adolescence. These interconnected forms of plasticity highlight the brain's extraordinary capacity for change and adaptation, making it a central focus of neuroscience research and therapeutic innovation (Cherry, 2024; Jones, 2011; Kerr et al., 2011).

8.3 Role of Technology in Enhancing Neuroplasticity

8.3.1 Brain–Computer Interfaces (BCIs)

BCIs offer unparalleled opportunities to harness the brain's neuroplasticity for both motor recovery and cognitive enhancement. Their ability to directly link brain activity to external feedback systems creates a powerful platform for driving structural and functional changes in the brain. In the context of neurorehabilitation, BCIs have demonstrated significant success in promoting motor recovery for stroke survivors, with evidence of improved corticospinal connectivity and enhanced motor function. These systems also empower patients to regain independence, which has profound psychological and social benefits beyond the physical recovery itself. Similarly, the use of BCIs for cognitive enhancement represents a growing frontier in neuroscience. By modulating brainwave frequencies and targeting specific neural networks, neurofeedback-based BCIs can improve cognitive functions such as memory, attention, and emotional regulation. The flexibility of these systems to address a wide range of cognitive challenges, from ADHD to post-stroke speech impairments, underscores their potential as a transformative tool for personalized brain training (Caeyenberghs et al., 2018; Cherry, 2024; Cramer et al., 2011; Jones, 2011).

While the promise of BCIs is clear, it is important to acknowledge the need for continued research to fully understand their long-term effects and optimize their use. Large-scale, controlled studies are essential to validate the findings from smaller trials and refine the technology for widespread clinical application. Ethical considerations, such as ensuring accessibility and addressing potential misuse, must also be prioritized as BCIs become more integrated into rehabilitation and enhancement practices (Neuroplasticity—an Overview, n.d.-b; Cherry, 2024). BCIs facilitate direct interaction between the brain and external devices, fostering neuroplasticity through:

- *Neurorehabilitation*: Brain-computer interfaces (BCIs) have emerged as transformative tools in neurorehabilitation, particularly for individuals recovering from stroke. These systems leverage the brain's neuroplasticity to help patients regain motor functions by creating a direct pathway between neural activity and external devices. One of the key mechanisms by which BCIs enhance recovery is through real-time modulation of spinal cord circuits. By linking efferent brain signals to afferent feedback, BCIs promote integrated neuroplasticity in affected corticospinal connections, fostering voluntary motor control and coordination (Neuroplasticity—an Overview, n.d.-a; Kerr et al., 2011).
- BCI training has also been shown to increase motor-evoked potentials (MEPs) in stroke survivors, a marker of strengthened intracortical synaptic connections. This finding underscores the capacity of BCIs to facilitate both functional and structural changes in the brain. A meta-analysis of nine randomized controlled trials involving 235 post-stroke patients demonstrated that BCI training significantly improved motor recovery, as evidenced by an effect size of 0.79 in the Fugl-Meyer Assessment for upper limb functionality. These improvements highlight BCIs' potential to surpass traditional rehabilitation methods in restoring motor skills. Additionally, BCIs can modulate sensorimotor rhythms (SMRs) generated by the sensorimotor gyrus, even in patients with severe chronic motor deficits. By enabling patients to control external devices or interact with virtual environments, BCIs create opportunities for neuroplastic changes that are otherwise difficult to achieve. These systems essentially close the feedback loop between brain activity and motor function, allowing patients to "retrain" neural pathways critical for movement. Beyond clinical measures, BCIs also enhance patients' independence and confidence, which are vital for their overall recovery journey (Cramer et al., 2011; Jones, 2011; Kerr et al., 2011; Tymofiyeva & Gaschler, 2021).
- *Cognitive enhancement*: In addition to motor recovery, BCIs hold tremendous promise for cognitive enhancement. Neurofeedback-based BCI systems are increasingly being used to improve cognitive functions in individuals with conditions such as attention deficit hyperactivity disorder (ADHD), mild cognitive impairments, and traumatic brain injuries. These systems provide real-time feedback on brain activity, enabling individuals to consciously regulate neural responses and foster adaptive neuroplastic changes. Studies have reported significant improvements in cognitive domains such as concentration, visual perception, categorization, emotional regulation, and speech following neurofeedback training (Caeyenberghs et al., 2018).

- One of the mechanisms driving these cognitive benefits is the modulation of brainwave frequencies. The vast literature details the upregulation of higher-frequency oscillations (alpha and beta bands) and downregulation of lower-frequency oscillations (delta and theta bands) have been associated with enhanced cognitive performance. Specific brainwave patterns are linked to distinct cognitive functions: for instance, sensorimotor rhythm (SMR)-based neurofeedback has been shown to improve declarative memory, while theta/beta ratio training enhances inhibition and cognitive flexibility. These targeted approaches allow for precision in addressing the unique cognitive challenges faced by individuals (Caeyenberghs et al., 2018; Tymofiyeva & Gaschler, 2021).
- BCIs also demonstrate potential for improving working memory. Studies have shown that alpha-band upregulation can lead to enhanced performance in memory-related tasks. Additionally, neurofeedback training targeting specific brain regions, such as the contralesional inferior frontal gyrus, has shown promise for improving speech and language functions in stroke survivors. These findings suggest that BCIs can be tailored to address individual cognitive deficits, fostering both recovery and enhancement of brain function. Beyond clinical applications, BCIs are increasingly being explored as tools for general cognitive enhancement in healthy individuals. For instance, neurofeedback training has been shown to improve attention span and problem-solving abilities, highlighting the broader implications of these systems for optimizing brain function. The ability to directly interact with neural activity through BCIs opens up new possibilities for personalized cognitive training, where interventions are guided by real-time data and adapted to an individual's specific needs (Chen et al., 2010; Kerr et al., 2011; Zotey et al., 2023).

8.3.2 Virtual Reality (VR) and Augmented Reality (AR)

- *Replicating real-life scenarios for motor and cognitive recovery*: Virtual reality (VR) and augmented reality (AR) have demonstrated significant potential in neurorehabilitation by replicating real-life scenarios in safe and controllable environments. These technologies allow patients to engage in realistic, task-specific training tailored to their individual needs, fostering meaningful neuroplastic changes. For instance, VR provides a controlled setting where patients can practice activities such as driving, navigating public spaces, or interacting in social situations—tasks that may be too complex or unsafe to recreate in the real world during early stages of recovery (Gangemi et al., 2023; Georgiev et al., 2021).
- The ability to simulate these real-world experiences is particularly valuable for individuals recovering from stroke, traumatic brain injury (TBI), or other neurological disorders. Studies have shown that VR-based interventions can improve cognitive functions such as attention, memory, and problem-solving skills by providing repetitive and task-specific training. This type of training not only

enhances neural connections but also allows for constructive feedback, helping patients refine their skills over time (Gangemi et al., 2023; Rogers et al., 2019). By involving patients in active participation, VR can encourage greater engagement, which is often a critical factor in achieving successful rehabilitation outcomes. Additionally, VR has been associated with structural and functional brain changes. Research indicates that patients undergoing VR therapy experience increases in cortical gray matter volumes, higher concentrations of beta waves as measured through electroencephalography, and overall improvements in cognitive performance. These findings highlight the ability of immersive VR environments to drive neuroplasticity by stimulating neural networks involved in motor and cognitive functions (Drigas & Sideraki, 2024; Georgiev et al., 2021). The repetitive nature of VR-based exercises also reinforces neural pathways, supporting long-term retention of learned skills. Patients benefit from the opportunity to practice complex tasks in a risk-free environment while receiving real-time performance feedback. This feedback mechanism plays a crucial role in promoting adaptive neuroplasticity, enabling the brain to reorganize itself in response to injury or cognitive decline (Catania et al., 2023; Gangemi et al., 2023).

- *Providing stimulating and engaging therapeutic platforms*: VR and AR technologies go beyond traditional rehabilitation methods by creating stimulating and engaging therapeutic platforms that motivate patients to actively participate in their recovery journey. Unlike conventional approaches, VR environments are designed to be immersive, interactive, and multisensory, allowing patients to engage with computer-based simulations that provide immediate feedback on their performance. This immersive nature not only enhances the therapeutic experience but also stimulates specific brain regions, triggering neurochemical changes that facilitate neuroplasticity. The ability of VR to engage multiple senses simultaneously contributes to its effectiveness. By providing a richer sensory environment, VR can stimulate synaptic plasticity—an essential mechanism for learning and memory. Research has shown that immersive VR experiences lead to increased neural connectivity and synaptic density, particularly in brain areas involved in sensory processing, motor coordination, and executive functions (Catania et al., 2023; Gangemi et al., 2023; Wairimu, 2024).

- These changes are key to fostering recovery, as they enable the brain to form new pathways and strengthen existing ones in response to targeted rehabilitation activities. Furthermore, VR-based interventions have been found to be more engaging and motivating for patients compared to traditional rehabilitation methods. The interactive nature of these platforms allows for personalization, tailoring exercises to the unique needs and progress of each patient. This individualized approach not only ensures that therapy remains relevant and effective but also encourages patients to remain committed to their treatment plans, improving adherence rates and its outcomes (Gangemi et al., 2023; Georgiev et al., 2021). The motivation provided by VR is particularly important for patients who may struggle with the repetitive and monotonous nature of conventional rehabilitation exercises. VR transforms these exercises into dynamic and enjoyable experiences, which can significantly reduce dropout rates and increase the time patients spend engaging with their therapy. Recent research on stroke survivors using VR to practice motor tasks often reports higher levels of satisfaction and a greater sense of achievement, which reinforces their willingness to continue with their rehabilitation program (Aderinto et al., 2023; Rogers et al., 2019).

- The adaptability of VR also enables therapists to design experiences that match the patient's current abilities while gradually increasing the level of difficulty as they progress. This scalability ensures that patients are continually challenged, promoting sustained neuroplastic changes without causing frustration or burnout. By providing a balance between challenge and reward, VR fosters a positive feedback loop that encourages ongoing participation and improvement. In addition to its use in structured rehabilitation programs, VR is increasingly being employed as a tool for cognitive training. Immersive VR environments have been shown to enhance cognitive functions such as attention, spatial awareness, and problem-solving by targeting specific neural circuits. Studies have applied these concepts to discover that VR scenarios that require navigation through complex virtual environments can stimulate hippocampal activity, which is critical for memory and spatial reasoning (Catania et al., 2023; Gangemi et al., 2023; Wairimu, 2024).

VR and AR technologies represent a paradigm shift in neurorehabilitation. Their ability to create engaging, personalized, and scientifically validated therapeutic platforms positions them as powerful tools for enhancing neuroplasticity in both motor and cognitive recovery. The combination of immersive environments, real-time feedback, and patient-centered design ensures that these technologies can provide meaningful benefits for individuals recovering from neurological deficits while fostering long-term improvements in brain function (Gangemi et al., 2023; Shen et al., 2023). By extending the opportunities to patients to practice complex tasks in a controlled and safe setting, these tools enable meaningful structural and functional changes in the brain. The ability to stimulate synaptic plasticity, enhance neural connectivity, and increase motivation makes VR an invaluable addition to modern neurorehabilitation practices. The evidence supporting the effectiveness of VR and AR in motor and cognitive recovery highlights their potential to improve outcomes for individuals with neurological deficits. From increasing cortical gray matter volumes to providing multisensory feedback, these technologies go beyond traditional methods by engaging patients in ways that drive sustained neuroplastic changes. The interactive and adaptable nature of VR ensures that therapy is tailored to the needs of each individual, fostering a sense of achievement and encouraging long-term participation. As research continues to explore the full capabilities of VR and AR, their integration into neurorehabilitation programs is likely to expand. These technologies offer hope not only for improved recovery outcomes but also for enhancing the overall rehabilitation experience, making it more engaging, effective, and accessible. By harnessing the power of neuroplasticity, VR and AR are transforming the landscape of rehabilitation, providing new opportunities for patients to regain independence and improve their quality of life (Catania et al., 2023; Drigas & Sideraki, 2024; Georgiev et al., 2021).

8.3.3 Non-Invasive Brain Stimulation

Techniques such as Transcranial Magnetic Stimulation (TMS) and Transcranial Direct Current Stimulation (tDCS) promote neuroplasticity by:

- *Modulating neural activity in targeted brain regions*: Non-invasive brain stimulation (NIBS) techniques, such as transcranial magnetic stimulation (TMS) and transcranial direct current stimulation (tDCS), have emerged as powerful tools for modulating neural activity in targeted brain regions. By delivering controlled electromagnetic or electrical currents, these methods can induce changes in cortical excitability, which is a key mechanism underlying neuroplasticity. For instance, studies have demonstrated that TMS can influence the amplitude of motor-evoked potentials (MEPs), providing measurable evidence of changes in synaptic strength and cortical responsiveness before and after stimulation (Kricheldorff et al., 2022; Schlaug et al., 2008).
- TMS-evoked potentials (TEPs), measured through neuroimaging and electrophysiological techniques, further reveal plastic changes in areas of the brain that do not produce direct behavioral responses. This demonstrates the ability of TMS to influence neural activity at both local and distant sites, thanks to the intricate connectivity between brain regions. Repeated TMS (rTMS), in particular, has shown long-lasting effects on cortical excitability, with changes persisting for minutes or even hours after stimulation. This extended impact underscores the potential of TMS to induce meaningful neuroplastic adaptations. Additionally, TMS can produce network-wide effects that go beyond the directly stimulated area. By leveraging the connectivity between different regions of the brain, TMS can enhance or suppress neural activity in interconnected areas. A combination of TMS with neuroimaging techniques such as functional MRI (fMRI) has revealed changes in functional connectivity between targeted regions and their networks. These insights highlight the versatility of TMS as a tool for modulating not just individual regions but also broader neural circuits involved in cognition, motor function, and emotion regulation (Kricheldorff et al., 2022; Schlaug et al., 2008).
- *Supporting recovery in conditions like depression and stroke*: TMS and tDCS have proven particularly valuable in supporting recovery in conditions such as depression and stroke. TMS has received FDA approval for several psychiatric and neurological disorders, including depression, migraine with aura, obsessive-compulsive disorder, smoking cessation, and anxiety comorbid with depression. Its efficacy in these areas is largely attributed to its ability to modulate neural activity in targeted brain regions, such as the dorsolateral prefrontal cortex in depression (Gomez Palacio Schjetnan et al., 2013).

- In the context of stroke recovery, NIBS techniques such as TMS and tDCS have demonstrated encouraging results. These interventions work by promoting neuroplasticity through cellular processes of neuromodulation, counteracting the molecular and cellular mechanisms involved in the pathophysiology of cerebral ischemia. Studies have highlight that TMS and tDCS can improve motor and somatosensory functions by enhancing the excitability of perilesional areas (regions surrounding the damaged site) and promoting connectivity within motor networks. These effects are particularly beneficial during the first month post-stroke, a critical period for recovery, but they have also been observed in chronic stroke patients and even those with severe impairments (Badoiu et al., 2023; Morya et al., 2019).
- Meta-analyses further support the efficacy of these techniques in post-stroke rehabilitation. One such analysis, which included over 600 subacute and chronic stroke patients, found that combining TMS with mirror therapy significantly improved upper extremity function. The use of bilateral transcranial electric stimulation and cathodal tDCS over the contralesional hemisphere was found to be especially effective, demonstrating the importance of targeted stimulation protocols tailored to individual patients' needs. In addition to motor recovery, TMS and tDCS have shown promise in addressing post-stroke cognitive and emotional challenges. For instance, these techniques have been associated with reductions in depression scale scores and improvements in cognitive functions such as working memory, episodic memory, and attention. By enhancing neuroplasticity in regions associated with mood regulation and cognitive control, TMS and tDCS offer a holistic approach to post-stroke recovery that addresses both physical and psychological dimensions (Alloush et al., 2020; Badoiu et al., 2023; Zhao et al., 2023).
- Moreover, the versatility of TMS and tDCS allows for their application across different stages of recovery. For acute and subacute stroke patients, repetitive tDCS has been shown to accelerate improvements in motor and sensory functions. Meanwhile, chronic stroke patients have benefited from TMS protocols that focus on reactivating dormant neural pathways and enhancing network connectivity. These findings highlight the adaptability of NIBS techniques, making them suitable for a wide range of patient populations and recovery timelines. The therapeutic benefits of NIBS extend beyond stroke and depression. In migraine management, TMS has been used to alleviate aura symptoms and reduce headache frequency. Similarly, in patients with obsessive-compulsive disorder, TMS targeting the orbitofrontal cortex has shown promising results in reducing compulsive behaviors. These diverse applications underscore the broad potential of TMS and tDCS as tools for addressing a variety of neurological and psychiatric conditions (Badoiu et al., 2023; Kricheldorff et al., 2022; Sloane & Hamilton, 2024; Stagg & Johansen-Berg, 2013).

Both TMS and tDCS have demonstrated significant potential in modulating neural activity and supporting recovery across a range of conditions, including depression and stroke. By inducing targeted changes in cortical excitability and enhancing connectivity within neural networks, these techniques harness the brain's natural capacity for neuroplasticity. The ability of TMS to produce both local and network-wide effects makes it a powerful tool for addressing complex brain disorders, while the versatility of tDCS allows for personalized treatment protocols tailored to individual patient needs. In stroke recovery, NIBS techniques have proven effective in promoting motor and cognitive improvements, with benefits observed across acute, subacute, and chronic stages. Their capacity to enhance neuroplasticity in perilesional areas and strengthen functional networks offers a new avenue for rehabilitation, complementing traditional therapies and providing hope for patients with severe impairments. Similarly, in the treatment of depression, TMS has established itself as a safe and effective intervention, capable of restoring balance in neural circuits associated with mood and emotion regulation (Badoiu et al., 2023; Gomez Palacio Schjetnan et al., 2013; Kricheldorff et al., 2022).

Despite the promising outcomes, further research is needed to refine stimulation protocols and optimize their application. Large-scale, randomized controlled trials will be essential to validate the findings from smaller studies and explore the long-term effects of these interventions. Ethical considerations, such as ensuring accessibility and addressing potential side effects, must also be prioritized as these techniques become more widely adopted. Nevertheless, the current evidence highlights the transformative potential of TMS and tDCS in enhancing neuroplasticity and supporting recovery in a wide range of conditions. As our understanding of these techniques continues to grow, their integration into clinical practice is likely to expand, offering new possibilities for improving patient outcomes. By leveraging the brain's innate ability to adapt and reorganize, TMS and tDCS represent a significant step forward in the development of innovative, non-invasive therapies for neurological and psychiatric conditions (Kricheldorff et al., 2022; Morya et al., 2019; Schlaug et al., 2008; Stagg & Johansen-Berg, 2013).

8.3.4 Neurofeedback

Using Real-Time Brain Activity Monitoring (EEG, fMRI) Neurofeedback has evolved into a cutting-edge tool for brain training, leveraging real-time brain activity monitoring to provide individuals with insights into their neural processes. Techniques like electroencephalography (EEG) and functional magnetic resonance imaging (fMRI) allow researchers and clinicians to observe brain activity as it happens, facilitating targeted interventions that can improve cognition, behavior, and emotion regulation. EEG neurofeedback, for instance, has a long history of use and has been extensively validated through studies showing its ability to alter electrical brain activity and associated behaviors in both humans and animals (Saj et al., 2021; Stoeckel et al., 2014). Recent advancements in technology have paved the way for real-time functional magnetic resonance imaging (rtfMRI) neurofeedback. Unlike traditional EEG, rtfMRI provides a window into deeper brain structures and can map specific patterns of brain activity with high spatial resolution. This technology allows individuals to view their brain activity in real time while inside an MRI scanner, enabling them to learn how to regulate neural processes associated with specific cognitive or emotional states (Ciccarelli et al., 2023; Stoeckel et al., 2014). The ability to visualize and influence these patterns in real time represents a significant leap forward in our ability to modulate brain function for therapeutic and performance enhancement purposes. By using these real-time technologies, neurofeedback provides immediate feedback on the individual's neural responses, creating an interactive loop that facilitates rapid learning. The immediacy of the feedback accelerates the training process and enables patients to adjust their mental strategies in real time, resulting in more effective brain regulation and recovery (Saj et al., 2021; Treves et al., 2024).

Trains Individuals to Self-Regulate Neural Responses At its core, neurofeedback is designed to teach individuals how to consciously regulate their brain activity. This is achieved by presenting participants with real-time information about the activity in specific brain regions and encouraging them to adopt mental strategies to modulate this activity. For instance, a patient might be shown visual or auditory cues reflecting activity in a particular neural network and asked to increase or decrease the intensity of that activity through focused mental efforts (Saj et al., 2021; Stoeckel et al., 2014). Once participants have developed effective regulation strategies, they are tested to determine whether they can maintain these skills in the absence of feedback. Studies have consistently shown that neurofeedback training can lead to sustained changes in brain activity, as evidenced by alterations in EEG patterns and event-related potentials. These changes often correlate with improvements in behavioral and cognitive outcomes, such as enhanced attention, memory, and emotional regulation (Stoeckel et al., 2014). The effectiveness of neurofeedback extends beyond training sessions. Even after a single 30-min neurofeedback session, changes in fMRI responses and EEG patterns have been observed in targeted neural networks. This demonstrates the brain's ability to adapt rapidly to neurofeedback interventions, highlighting its potential to facilitate meaningful changes in both clinical and non-clinical populations. The self-regulation of neural responses has profound implications for both rehabilitation and personal development. Neurofeedback has been used to help patients with attention-deficit/hyperactivity disorder (ADHD) improve focus and self-control, while individuals recovering from stroke have used it to regain motor and cognitive functions. These applications demonstrate the versatility of neurofeedback as a tool for enhancing brain function and fostering resilience in the face of neurological challenges (Ciccarelli et al., 2023; Saj et al., 2021; Stoeckel et al., 2014).

Enhancing Adaptive Neuroplastic Changes One of the most compelling aspects of neurofeedback is its ability to enhance neuroplasticity—the brain's capacity to reorganize itself in response to experience. Real-time feedback provided by EEG or rtfMRI facilitates specific changes in brain function by accelerating and optimizing systems-level neuroplasticity. This process is theorized to occur as participants engage in repeated mental efforts to regulate their neural activity, effectively training their brains to develop new pathways and strengthen existing connections (Drigas & Sideraki, 2024; Stoeckel et al., 2014). Neuroplasticity underlies many of the behavioral and cognitive improvements observed after neurofeedback training. Skill learning through neurofeedback is thought to result from adaptive changes in both cortical and subcortical regions, leading to improvements in attention, memory, and emotional regulation. These changes are not limited to functional adaptations; neurofeedback has also been shown to induce structural changes in the brain, such as increased gray matter density in targeted areas (deCharms, 2007; Stoeckel et al., 2014). In rehabilitation settings, neurofeedback has demonstrated its ability to produce clinical improvements by enhancing neuroplasticity. For instance, patients recovering from spatial neglect—a condition often resulting from stroke—have shown significant improvements after undergoing rtfMRI neurofeedback training targeting specific brain regions. Similarly, individuals with mood disorders, such as depression, have experienced reduced symptoms through neurofeedback interventions that promote neuroplastic changes in neural circuits associated with emotion regulation (Ciccarelli et al., 2023; Treves et al., 2024). The effectiveness of neurofeedback in fostering neuroplasticity also makes it a promising tool for healthy individuals seeking to optimize their cognitive performance. Athletes and executives have used neurofeedback to improve focus, decision-making, and stress management. By training the brain to adapt more effectively to challenges, neurofeedback enables users to achieve peak performance while maintaining emotional balance. Neurofeedback's ability to produce rapid and lasting changes is a testament to the brain's remarkable capacity for adaptation. Unlike traditional approaches that rely on passive interventions, neurofeedback actively engages participants in their own recovery or development, empowering them to take control of their brain's function and potential (Drigas & Sideraki, 2024; Saj et al., 2021; Stoeckel et al., 2014).

8.4 Applications in Clinical Neuropsychology

8.4.1 Neurorehabilitation

- *Stroke recovery*: Neurorehabilitation has significantly advanced in its ability to aid stroke recovery, with innovative techniques improving both motor and cognitive outcomes. Robotic-assisted therapy, for instance, has shown tremendous potential in enhancing walking speed and distance in stroke survivors, as well as improving balance. When combined with attention and motivation-focused programs, robotic-assisted therapy demonstrated the greatest improvements in functional outcomes. Additionally, mobile game-based virtual reality (VR) programs have been particularly effective in promoting upper limb recovery in stroke patients, offering a fun and interactive way to engage in rehabilitation (Virtual Reality for Individuals Affected by Stroke, n.d.; Calderone et al., 2024).

- Brain-computer interfaces (BCIs) represent another groundbreaking intervention for stroke recovery. By connecting brain signals to external devices, BCIs help induce input–output properties of spinal cord circuits, promoting neuroplasticity in affected corticospinal connections. Stroke survivors using BCIs have shown increased motor-evoked potentials (MEPs), which indicate strengthened intracortical synaptic connections. BCIs have also demonstrated significant improvements in motor function, particularly in upper limb recovery, as measured by the Fugl-Meyer Assessment (Aderinto et al., 2023). Virtual reality (VR) and augmented reality (AR) therapies have gained traction for their ability to immerse patients in engaging and realistic rehabilitation scenarios. These tools have been shown to outperform traditional physical therapy in restoring balance and lower extremity function. Patients using VR systems have also demonstrated better improvements in motor function, cognitive abilities, and activities of daily living. Studies have highlighted significant advancements in performance on metrics like the Wolf Motor Function Test, Box and Block Test, and grip strength for patients undergoing VR-based rehabilitation compared to control groups (Aderinto et al., 2023; Stein et al., 2017; Torrisi et al., 2021).
- Action observation therapy, particularly when combined with VR, has further enhanced upper limb recovery in stroke patients. Immersive VR rehabilitation systems have been associated with notable gains in hand dexterity, shoulder flexion, and elbow pronation. AR-based systems have similarly demonstrated significant functional improvements in assessments like the Fugl-Meyer Assessment for Upper Extremity (FMA-UE), Functional Ambulation Category, and Berg Balance Scale (Gunawan et al., 2024; Stein et al., 2017; Torrisi et al., 2021).
- Rehabilitation gloves integrated with VR and biofeedback have shown promise in improving motor function and reducing the time required to complete tasks like the Nine Holes Peg Test. Specialized VR systems like PABLO have contributed to gains in hand function, shoulder and elbow movement, and overall motor performance. These interventions have even shown the ability to influence biomarkers and improve patient-reported outcomes on scales like the Stroke Impact Scale (Ku & Kang, 2018). Other innovative systems, such as the Microsoft Xbox Kinect and the Jintronix exergame system, have added supplemental benefits when used alongside conventional therapies. These platforms promote participation and enhance upper limb function in stroke survivors. Additionally, virtual reality gait training (VRT) has proven effective in restoring lower extremity function, though some studies suggest no significant differences in participation compared to non-VRT methods. Nonetheless, VRT remains a valuable tool for encouraging movement and balance recovery in stroke rehabilitation (Gunawan et al., 2024; Stein et al., 2017; Torrisi et al., 2021).
- *Traumatic brain injury (TBI)*: Although the evidence for neurorehabilitation techniques specifically targeting traumatic brain injury (TBI) is less extensive than for stroke, early findings suggest significant potential for improving executive functioning and cognitive recovery in TBI patients. VR-based training has emerged as a particularly promising intervention, enabling individuals to engage

in realistic, interactive tasks that target higher-order cognitive functions. These immersive environments allow patients to practice executive skills like decision-making, planning, and problem-solving, which are often impaired after a TBI (Saj et al., 2021).

- The principles underlying VR for TBI rehabilitation are similar to those applied in stroke recovery. By immersing patients in controlled, goal-directed environments, VR can stimulate neuroplasticity and encourage the development of new neural pathways. Patients with TBI can benefit from the repetitive, task-specific nature of VR training, which helps reinforce neural connections and improve cognitive and motor outcomes. While more research is needed to expand on these findings, the current evidence highlights the potential for VR and other neurorehabilitation technologies to address the unique challenges faced by TBI patients (Virtual Reality for Individuals Affected by Stroke, n.d.; Stein et al., 2017).

8.4.2 Cognitive Enhancement

Digital cognitive exercises and neurofeedback exemplify how innovative interventions can address cognitive challenges in diverse populations. For individuals with dementia or Alzheimer's disease, digital cognitive exercises focusing on attention training offer a structured and engaging way to combat cognitive decline. These programs not only improve specific skills but also foster broader cognitive enhancements, underscoring their value as a key component of comprehensive care strategies.

For those with ADHD, neurofeedback provides a unique opportunity to engage in real-time brain training, enabling participants to self-regulate and improve essential cognitive functions. Despite some limitations in the research, neurofeedback's personalized and non-invasive nature makes it a compelling option for individuals seeking alternative treatments. Together, these interventions highlight the power of technology in shaping the future of cognitive enhancement, providing hope for individuals and families navigating the challenges of cognitive impairment and developmental disorders (Fotuhi et al., 2023; Jiang et al., 2017; Neurofeedback For Cognitive Decline, 2023).

- *Dementia and Alzheimer's disease*: Digital cognitive exercises have emerged as a promising tool for addressing cognitive decline in individuals with dementia and Alzheimer's disease. These exercises are designed to target specific cognitive domains, with attention training being one of the most effective interventions. Attention plays a foundational role in cognitive processes, making it a crucial focus area for cognitive training programs. Research has shown that attention training can moderately improve attention and memory in older adults with cognitive impairments, providing a valuable means of enhancing overall cognitive function (Jiang et al., 2017; Marlats et al., 2020).

- Meta-analyses have further substantiated the benefits of cognitive training programs. One analysis, which reviewed 61 independent samples of adults over the age of 60, demonstrated significant improvements in performance on trained tasks, with some evidence of transfer to untrained but similar tasks. This suggests that digital cognitive exercises can help enhance not only specific skills but also broader cognitive abilities. In particular, attention training has shown promise in improving focused attention tasks, cognitive function, and working memory. While evidence of these improvements translating into everyday life remains limited, attention training has consistently outperformed other cognitive domains in terms of measurable gains (Jiang et al., 2017; Lavy et al., 2021). These findings highlight the potential of digital cognitive exercises as an intervention for delaying cognitive decline in dementia and Alzheimer's disease. The repetitive and targeted nature of these exercises fosters neural plasticity, enabling the brain to adapt and reorganize in response to training. Although more research is needed to explore the long-term impact and real-world applicability of these interventions, they represent a valuable tool in the fight against cognitive decline, offering individuals the chance to maintain higher levels of mental function and independence for longer periods (Marlats et al., 2020; Neurofeedback For Cognitive Decline, 2023).
- *ADHD*: Neurofeedback (NF) has gained significant attention as an intervention for attention deficit hyperactivity disorder (ADHD), particularly for its ability to improve attention and self-regulation. Unlike traditional cognitive training methods, neurofeedback provides real-time feedback on brain activity, enabling participants to learn how to modulate their neural responses. This process helps improve key cognitive functions such as attention and working memory, which are often impaired in individuals with ADHD (Jiang et al., 2017; Vilou et al., 2023).

- Research suggests that neurofeedback may be a more effective intervention than traditional attention training or working memory training for children with ADHD. By targeting specific neural networks associated with attention and self-regulation, NF enables participants to actively engage in their cognitive development. Studies have reported significant improvements in attention and focus following neurofeedback sessions, highlighting its potential as a valuable tool for managing ADHD symptoms (Israsena et al., 2021; Jiang et al., 2017). Despite these promising results, the evidence for neurofeedback in ADHD is not without its limitations. Meta-analyses of randomized controlled trials have shown mixed findings, with some studies failing to support NF as an effective treatment for ADHD in children and adolescents. Interestingly, significant treatment results are often observed in outcome measures that are not properly blinded, raising questions about the robustness of the findings. However, the mixed results may also reflect the variability in neurofeedback protocols, participant characteristics, and study designs, emphasizing the need for standardized approaches and further research (Jiang et al., 2017; Neurofeedback For Cognitive Decline, 2023).

- Even with these challenges, neurofeedback continues to show potential in enhancing attention and self-regulation. Its ability to provide immediate, individualized feedback allows for tailored interventions that can address the unique needs of each participant. For children with ADHD, this personalized approach can be particularly impactful, helping them develop the skills needed to succeed academically, socially, and emotionally. Neurofeedback also offers a non-invasive and medication-free alternative for managing ADHD symptoms, making it an attractive option for parents and caregivers seeking holistic treatment options (Cognitive Training Research, n.d.; Vilou et al., 2023). While more studies are needed to establish the long-term efficacy of neurofeedback in ADHD, its ability to promote neuroplasticity and improve cognitive functions positions it as a valuable complement to existing interventions. As research continues to refine neurofeedback techniques and expand our understanding of their mechanisms, the potential for this innovative approach to revolutionize ADHD treatment remains strong (Li, 2022; Neurofeedback for Cognitive Decline, 2023).

8.4.3 Psychiatric Disorders

Transcranial Direct Current Stimulation (tDCS) Transcranial direct current stimulation (tDCS) has emerged as a promising non-invasive tool for addressing depression and anxiety by modulating activity in key emotional regulation networks. A significant body of evidence highlights the role of the left dorsolateral prefrontal cortex (DLPFC) in depressive symptoms. Structural and functional differences in this region have been consistently observed in patients with major depressive disorder (MDD) compared to healthy controls. Targeting the left DLPFC with tDCS has shown significant antidepressant effects, not only in patients with MDD but across various clinical diagnoses involving emotional dysregulation (Jiang et al., 2017). The mechanisms underlying tDCS's therapeutic effects are tied to its ability to induce neuroplastic changes in the DLPFC and its associated networks. These changes include enhanced cerebral blood flow, modulation of widespread perfusion, and generation of magnetic fields at the stimulated region. By improving activity in the left DLPFC, tDCS supports better emotional regulation, counteracting the dysregulated neural activity often observed in individuals with depression and anxiety (Bielas & Michalczyk, 2021).

Clinical findings also highlight the importance of optimizing stimulation parameters for effectiveness. For instance, meta-analytic results suggest that a current intensity of 2-mA is particularly effective in reducing depressive symptoms. Intensity, rather than duration, appears to play a more critical role in achieving desired outcomes. For patients with treatment-resistant depression, more intense protocols—such as 0.1 mA/cm^2 applied twice daily—have been shown to yield positive effects (Fotuhi et al., 2023; Jiang et al., 2017). The growing evidence base for tDCS demonstrates its potential not only as a stand-alone treatment but also as part of an integrated approach for managing depression and anxiety. By leveraging neuroplasticity to enhance the functionality of emotional regulation networks, tDCS offers hope for individuals who may not respond to traditional interventions (Bielas & Michalczyk, 2021; Marlats et al., 2020).

Virtual Reality (VR)-Based Therapy Virtual reality (VR)-based therapy has rapidly gained attention as an innovative approach to treating anxiety and depression. By immersing individuals in simulated environments, VR creates realistic, controlled scenarios that allow for targeted therapeutic interventions. Studies have shown that VR can be particularly effective in reducing anxiety symptoms. For instance, research by Mehler et al. (2018) demonstrated that combining brain–computer interface (BCI)-based neurofeedback with VR significantly reduced anxiety. Participants exhibited increased alpha-wave activity and decreased beta-wave activity, indicating enhanced relaxation and reduced anxiety levels (Vilou et al., 2023).

The combination of VR and BCIs represents a powerful therapeutic tool. While VR provides immersive, anxiety-provoking scenarios, BCIs offer real-time feedback on neural responses, allowing for adaptive interventions tailored to the patient's needs. This combination has been especially effective in promoting neuroplastic changes by engaging patients in active emotional regulation. Individuals with social anxiety disorder exposed to social scenarios in VR while receiving neurofeedback experienced significant reductions in anxiety symptoms. These improvements were attributed to the ability of the VR-BCI system to help patients identify and regulate their anxiety triggers in real time. Beyond its stand-alone use, VR has also been effectively combined with tDCS to address emotional dysregulation in conditions such as post-traumatic stress disorder (PTSD). The dual approach leverages the strengths of both technologies: VR provides experiential learning opportunities, while tDCS enhances the neural plasticity required for sustained emotional regulation. Pilot studies have shown that this combined therapy not only reduces symptoms but also improves patients' ability to manage triggers in their everyday lives (Bielas & Michalczyk, 2021; Lavy et al., 2021; Vilou et al., 2023). In addition to addressing specific conditions, VR-based therapy has broader implications for fostering long-term changes in brain function. Immersive environments stimulate neuroplasticity by engaging the brain in meaningful, adaptive tasks, encouraging the development of new neural connections. This process is particularly relevant for individuals with depression and anxiety, as it helps rewire the brain's emotional regulation networks to better manage stressors and challenges (Israsena et al., 2021; Vilou et al., 2023).

8.5 Applications in Forensic Neuropsychology

8.5.1 *Behavioral Modification*

Behavioral modification in forensic neuropsychology leverages neuroplasticity-based interventions to address impulse control, emotional regulation, and decision-making, with the ultimate goal of promoting rehabilitation. The concept of use-dependent neuroplasticity, which describes the brain's ability to reorganize itself in response to specific experiences or training, has become a cornerstone for understanding how targeted interventions can lead to meaningful behavioral changes. For instance, advanced brain imaging technologies, such as functional magnetic resonance imaging (fMRI), have been used to study how neuroplastic changes occur during recovery from neurological injury. These findings have opened new avenues for designing interventions aimed at fostering self-regulation and improving executive functions, both of which are critical for rehabilitation (Levin et al., 2014). While specific evidence related to impulse control in offenders is limited in the provided sources, it is well established that neuroplasticity-based interventions often focus on enhancing executive functions, which encompass decision-making and the ability to resist impulsive behavior. By targeting these core cognitive processes, rehabilitation programs can potentially reduce the likelihood of recidivism among offenders.

Emotional regulation, a key area of focus in behavioral modification, also benefits from these interventions. Techniques like fMRI and magnetic resonance spectroscopy (MRS) have been explored to study the remodeling of cortical representations associated with cognitive and emotional abilities. These technologies could be further adapted to assess changes in emotional regulation networks during rehabilitation, providing valuable insights into the effectiveness of interventions (Drigas & Sideraki, 2024; Levin et al., 2014). One promising application of these findings is in offender rehabilitation programs. By using targeted training to promote changes in brain function, it may be possible to address underlying issues such as poor emotional regulation and impulsivity, which are often linked to criminal behavior. Although the field is still in its early stages, the integration of neuroscience into rehabilitation practices represents a significant step forward in addressing the root causes of criminal behavior and promoting long-term behavioral change (Zilverstand et al., 2016).

8.5.2 *Lie Detection and Deception Research*

Lie detection is a critical area of forensic neuropsychology, and advances in neuroscience have brought innovative approaches to understanding deception. One of the most promising techniques involves the use of P300 event-related potentials (ERPs), which are brainwave patterns elicited when an individual recognizes a significant or meaningful stimulus. The "3-stimulus protocol" (3SP) is an advanced method that uses P300 ERPs to detect deception while minimizing the impact of countermeasures. This protocol separates the probe recognition task from target-non target discrimination, enhancing its reliability and resistance to intentional manipulation (Zotey et al., 2023).

Research on the 3SP has demonstrated impressive accuracy. Studies have reported detection rates of 90–100% for guilty participants, regardless of whether they employed countermeasures to evade detection. Additionally, false positive rates for innocent participants remain low, ranging from 0–8%. These findings suggest that the 3SP is a robust and reliable method for detecting concealed information, making it a valuable tool for forensic investigations. It has been applied to scenarios such as identifying concealed autobiographical information, recognizing details of a crime, and even in anti-terrorism contexts (Putchala & Putchala, 2024; Zotey et al., 2023). The effectiveness of P300-based lie detection lies in its ability to directly measure neural responses, bypassing the subjective biases associated with traditional polygraph tests. Furthermore, researchers have continuously refined their protocols to address vulnerabilities identified through countermeasures. By combining P300 ERPs with other psychological and neuroscientific techniques, these protocols are becoming increasingly sophisticated, offering a more comprehensive understanding of deception (Luber et al., 2009).

One notable application of this technology is in the detection of malingering, where individuals feign symptoms of illness or disability for personal gain. The ability to objectively measure neural responses provides investigators with a powerful tool for distinguishing between genuine and fabricated accounts. This is particularly valuable in legal and clinical settings, where accurate assessments are critical. Despite these advancements, the use of neuroimaging and neurostimulation techniques in lie detection raises important ethical and legal questions. Concerns about privacy, consent, and the potential misuse of these technologies must be carefully addressed to ensure their responsible application in forensic contexts. While these methods hold great promise, their integration into legal systems will require rigorous validation and clear ethical guidelines (Ying et al., 2024; Zotey et al., 2023).

8.5.3 Risk Assessment

Biometric and neuroplasticity-based interventions are proving to be valuable tools in assessing and mitigating the risks associated with recidivism. By leveraging advancements in neuroscience and biometrics, these approaches enable a more comprehensive understanding of the factors contributing to criminal behavior and provide targeted solutions for addressing them. Cognitive and behavioral training programs, informed by biometric data and neuroplasticity principles, offer tailored interventions designed to reduce the likelihood of reoffending. One of the key advantages of neuroplasticity-based interventions in risk assessment is their ability to identify and address deficits in cognitive and emotional regulation. By using tools such as functional magnetic resonance imaging (fMRI) and electroencephalography (EEG), practitioners can measure activity in specific brain regions associated with decision-making, impulse control, and empathy. These insights allow for the development of personalized training programs aimed at enhancing these critical functions, which are often impaired in individuals at high risk of recidivism (Levin et al., 2014; Ying et al., 2024; Zilverstand et al., 2016).

Biometric data, including heart rate variability and galvanic skin response, further augment risk assessment processes by providing objective measures of physiological arousal and emotional states. These metrics can help identify patterns of emotional dysregulation or heightened stress responses that may contribute to impulsive or harmful behaviors. By integrating these findings with cognitive and behavioral training, practitioners can design interventions that target both the physiological and psychological factors underlying recidivism risks (INVITED SYMPOSIUM, n.d.; Ying et al., 2024).

Neuroplasticity-based training programs focus on promoting adaptive changes in brain function through repeated, targeted exercises. These programs may include activities aimed at improving executive function, emotional regulation, and social cognition, all of which are crucial for reducing recidivism. By fostering new neural connections and strengthening existing pathways, these interventions help individuals develop the skills needed to make better decisions, manage stress, and interact more positively with others. In forensic settings, these approaches offer the dual benefit of providing actionable insights for risk assessment while also equipping individuals with the tools needed to successfully reintegrate into society. The ability to combine biometric data with evidence-based neuroplasticity training represents a promising avenue for addressing recidivism at both an individual and systemic level (Luber et al., 2009; Putchala & Putchala, 2024; Ying et al., 2024).

8.6 Ethical Considerations

8.6.1 Accessibility and Equity

Ensuring accessibility and equity in the development and application of neuroplasticity-enhancing technologies is a pressing ethical challenge. To fully realize the social value of these technologies, research populations must reflect the diversity of society. This imperative stems from the recognition that convenience samples and secondary data often skew toward younger, wealthier, and healthier individuals, thereby excluding those from underrepresented groups. Such imbalances disproportionately affect individuals from racial minorities, lower-income groups, and those in underserved geographic locations, further exacerbating existing inequalities in access to healthcare and technological advancements. Digital health technologies, such as telemedicine, wearables, and neuroplasticity-enhancing tools, often face established racial and socioeconomic adoption gaps. These gaps highlight the need for researchers to actively engage participants from diverse backgrounds. By creating representative research populations, researchers can ensure that the benefits and burdens of these technologies are equitably distributed. This approach might involve tailored recruitment strategies to capture the experiences of marginalized groups and designing interventions that address specific barriers to access. A fairer distribution of these technologies ensures they do not inadvertently widen existing disparities in healthcare and rehabilitation (Levin et al., 2014; Putchala & Putchala, 2024).

8.6.2 Ethical Use in Forensic Contexts

The use of neuroplasticity-enhancing technologies in forensic contexts presents unique ethical considerations. While these technologies hold significant promise for offender rehabilitation—such as fostering emotional regulation and impulse control—there are legitimate concerns about coercion and misuse. The integration of these tools in correctional settings raises questions about informed consent. Individuals in such settings may not feel entirely free to decline participation, especially if refusing could lead to perceived disadvantages or missed opportunities for early parole (Levin et al., 2014; Ying et al., 2024).

Additionally, the privacy implications of collecting and utilizing neurological data in forensic contexts cannot be ignored. The potential misuse of such data—whether by manipulating behavior or infringing on individual rights—poses a significant ethical challenge. These considerations call for strict oversight and clear ethical frameworks to ensure that neuroplasticity-enhancing interventions in forensic settings are used responsibly and primarily for rehabilitation, not punishment or coercion. Policymakers and practitioners must approach the use of these technologies with caution, balancing the potential benefits with the risks of exploitation. Prioritizing transparency, ensuring informed consent, and safeguarding against the misuse of data are crucial steps in mitigating the ethical challenges associated with these interventions (Ying et al., 2024; Zotey et al., 2023).

8.6.3 Long-Term Effects

The long-term effects of sustained use of neuroplasticity-enhancing technologies remain an area of significant ethical concern. While these tools offer promising short-term benefits, the potential risks associated with prolonged use on brain function and structure are not yet fully understood. Research indicates that frequent stimulation or training targeting specific neural networks may inadvertently affect brain development, cognitive flexibility, or emotional regulation over time. Ongoing research is critical to understanding the broader implications of these technologies. Longitudinal studies with rigorous follow-ups can help identify any adverse effects that might emerge after extended use. Ethical research design must include robust safety protocols and ensure that participants are fully informed of potential long-term risks. This includes obtaining informed consent that explicitly addresses these uncertainties, ensuring participants understand the potential benefits and drawbacks of engaging with these interventions (Levin et al., 2014; Putchala & Putchala, 2024; Zotey et al., 2023). Beyond research, policymakers must remain vigilant in regulating these technologies as they become more widely adopted. Regular monitoring and updates to clinical guidelines can help mitigate risks while maximizing benefits. For instance, establishing ethical standards for the use of neuroplasticity-enhancing tools in clinical and non-clinical settings can ensure that safety and equity remain at the forefront of their application (Luber et al., 2009; Zilverstand et al., 2016).

8.7 Future Directions

8.7.1 Integration of AI and Machine Learning

The integration of artificial intelligence (AI) and machine learning (ML) into neuroplasticity research is opening new frontiers in personalized and precise interventions. AI-powered platforms are revolutionizing neuroplasticity training by tailoring interventions to the unique needs of each individual. Using sophisticated algorithms, these platforms can adapt in real time, adjusting the difficulty and type of exercises based on a person's brain characteristics, learning patterns, and performance. This personalized approach not only enhances learning but also optimizes the stimulation of neuroplasticity, making interventions more effective and efficient (Brahma & Vimal, 2024; Oluwafemidiakhoa, 2024).

Predictive modeling is another groundbreaking application of AI in neuroplasticity. By analyzing large datasets, including neuroimaging results, behavioral assessments, and genetic information, AI is helping researchers predict treatment outcomes with remarkable accuracy. These models enable clinicians to identify individuals who are most likely to benefit from specific neuroplasticity-based interventions, thus improving the precision and efficacy of treatments. Deep learning models have shown exceptional capabilities in analyzing complex neuroimaging data, aiding in the early diagnosis of neurodegenerative diseases and informing personalized treatment strategies. Additionally, AI is being used to automate tasks like neuroimaging analysis, significantly reducing the time required for diagnosis and enabling faster implementation of treatment plans. This automation not only improves diagnostic accuracy but also frees up valuable time for clinicians, allowing them to focus on patient care and intervention planning (Khaneja & Arora, 2024; Putchala & Putchala, 2024).

8.7.2 Advancements in Neuroimaging

Advancements in neuroimaging technologies are transforming our ability to track and understand neuroplastic changes in real time. Techniques like functional MRI (fMRI) and diffusion MRI (dMRI) have become essential tools for precision psychiatry, allowing researchers to monitor brain activity and structural changes with unprecedented detail. High-resolution imaging technologies now offer insights across multiple scales—from macroscopic brain regions to microscopic neural pathways—helping us understand the intricate relationships between brain structure and behavior (Chen et al., 2018).

One of the most promising developments in neuroimaging is its ability to enable real-time tracking of neuroplasticity. This capability allows researchers and clinicians to observe how the brain responds to specific interventions, providing immediate feedback and facilitating more dynamic and adaptive treatment plans. In addition to fMRI and dMRI, other advanced methods like electroencephalography (EEG), magnetoencephalography (MEG), and functional near-infrared spectroscopy (fNIRS) are further enhancing our understanding of brain function. The discovery of novel biomarkers through neuroimaging is another exciting frontier. By combining imaging techniques with AI, researchers are identifying biomarkers that signal the presence of psychiatric or neurological disorders. These biomarkers not only aid in diagnosis but also provide critical information for tailoring interventions, thereby improving outcomes for individuals with conditions like depression, anxiety, or neurodegenerative diseases (Chen et al., 2022; Hain et al., 2023).

8.7.3 Cross-Disciplinary Collaboration

The rapid advancements in neuroplasticity research would not be possible without collaboration across multiple disciplines. Neuroscience, psychology, engineering, and computer science are converging to create innovative applications and deepen our understanding of brain function and behavior. This interdisciplinary approach is enabling researchers to tackle complex questions about the brain in ways that were previously unimaginable. For instance, computational models from engineering and computer science are being used to simulate and predict brain dynamics. These models allow researchers to explore how different interventions might influence neuroplasticity, paving the way for more targeted and effective therapies. By integrating insights from diverse fields, scientists are developing sophisticated models that not only enhance our understanding of neuroplasticity but also improve the design and implementation of interventions (Brahma & Vimal, 2024; Khaneja & Arora, 2024; Oluwafemidiakhoa, 2024). Ethical considerations are an integral part of this cross-disciplinary collaboration. As neuroplasticity research advances, it is essential to address emerging ethical issues, such as the implications of using AI to influence brain function and the potential for misuse of neuroimaging data.

Researchers, clinicians, and ethicists must work together to establish guidelines that ensure the responsible development and application of these technologies. This collaboration is particularly important as neuroplasticity-enhancing tools become more accessible, raising questions about equity, privacy, and consent (Brahma & Vimal, 2024; Putchala & Putchala, 2024).

8.8 Conclusion

As we reach the end of this chapter, it becomes clear that neuroplasticity—the brain's incredible ability to reorganize and adapt—holds immense promise for transforming lives. From recovering motor functions after a stroke to enhancing cognitive abilities in conditions like ADHD and Alzheimer's, neuroplasticity is at the heart of many groundbreaking advancements in neuroscience and rehabilitation. The journey through various neuroplasticity-enhancing technologies, such as Brain-Computer Interfaces (BCIs), Virtual Reality (VR), Augmented Reality (AR), transcranial Direct Current Stimulation (tDCS), and neurofeedback, showcases how these tools are revolutionizing therapeutic practices. BCIs bridge the gap between neural activity and external devices, offering stroke survivors new ways to regain control over their movements. Meanwhile, VR and AR create immersive environments that make rehabilitation engaging and effective, allowing patients to practice real-life scenarios safely and repeatedly.

tDCS and neurofeedback offer non-invasive methods to modulate brain activity, supporting emotional regulation and improving attention and memory. These technologies are particularly promising for treating depression, anxiety, and other mood disorders by targeting specific brain regions involved in emotional control. The ability to personalize these interventions ensures that each individual receives the most effective treatment tailored to their unique brain patterns and needs. These exciting advancements come with important ethical considerations. Ensuring accessibility and equity is crucial to prevent widening existing disparities in healthcare. It's essential that diverse populations are included in research to ensure that neuroplasticity-based interventions benefit everyone, regardless of race, income, or geographic location. Additionally, the use of these technologies in forensic settings raises significant ethical questions about consent, privacy, and the potential for misuse. Balancing the benefits of rehabilitation with the need to protect individual rights is a delicate but necessary task. Long-term effects of these technologies also warrant careful attention. While the short-term benefits are promising, understanding how sustained use impacts brain function and structure is vital. Ongoing research and ethical oversight are necessary to ensure that these interventions promote healthy, adaptive changes without unintended consequences. This includes rigorous testing, transparent reporting, and the development of guidelines to safeguard participant well-being.

AI-powered platforms are making personalized training more precise, while predictive models help tailor interventions to maximize their effectiveness.

Neuroimaging technologies, such as real-time fMRI and advanced EEG, provide deeper insights into how the brain adapts, enabling more targeted and effective therapies. Collaboration across disciplines—bringing together neuroscience, psychology, engineering, and computer science—further accelerates innovation. These interdisciplinary teams are essential for developing sophisticated models that enhance our understanding of brain dynamics and improve intervention strategies. Additionally, addressing ethical issues through collaborative efforts ensures that the development and application of these technologies remain responsible and equitable.

Neuroplasticity-enhancing technologies represent a transformative shift in how we approach brain health and rehabilitation. They offer hope for individuals recovering from neurological injuries, managing cognitive impairments, and seeking to optimize their cognitive and emotional well-being. By embracing these advancements with a commitment to ethical responsibility and inclusivity, we can unlock the full potential of the brain's adaptability. As we continue to explore and expand the boundaries of neuroplasticity, it is crucial to maintain a balanced approach that prioritizes both scientific progress and ethical integrity. This ensures that the benefits of these technologies are accessible to all, fostering a future where enhanced brain health and cognitive resilience are within everyone's reach. The ongoing evolution of neuroplasticity research not only deepens our understanding of the brain's remarkable capabilities but also paves the way for innovative interventions that can significantly improve the quality of life for countless individuals. In embracing the promise of neuroplasticity, we stand on the brink of a new era in neuroscience—one where the brain's capacity for change becomes a foundation for healing, growth, and human flourishing.

References

Aderinto, N., Olatunji, G., Abdulbasit, M. O., Edun, M., Aboderin, G., & Egbunu, E. (2023). Exploring the efficacy of virtual reality-based rehabilitation in stroke: A narrative review of current evidence. *Annals of Medicine, 55*(2), 2285907. https://doi.org/10.1080/07853890.2023.2285907

Alloush, T., Fayez, E., Al-Azab, I. A.-A., Bouls, M., & Hamdy, M. (2020). Role of transcranial direct current stimulation in the management of post-stroke depression. *Neuroscience and Medicine, 11*(1), 1. https://doi.org/10.4236/nm.2020.111001

Appelbaum, L. G., Shenasa, M. A., Stolz, L., & Daskalakis, Z. (2023). Synaptic plasticity and mental health: Methods, challenges and opportunities. *Neuropsychopharmacology, 48*(1), 113–120. https://doi.org/10.1038/s41386-022-01370-w

Badoiu, A., Mitran, S. I., Catalin, B., Balseanu, T. A., Popa-Wagner, A., Gherghina, F. L., Albu, C. V., & Sandu, R. E. (2023). From molecule to patient rehabilitation: The impact of transcranial direct current stimulation and magnetic stimulation on stroke—A narrative review. *Neural Plasticity, 2023*, 5044065. https://doi.org/10.1155/2023/5044065

Bielas, J., & Michalczyk, Ł. (2021). Beta neurofeedback training improves attentional control in the elderly. *Psychological Reports, 124*(1), 54–69. https://doi.org/10.1177/0033294119900348

Brahma, N., & Vimal, S. (2024). Artificial intelligence in neuroimaging: Opportunities and ethical challenges. *Brain & Spine, 4*, 102919. https://doi.org/10.1016/j.bas.2024.102919

Caeyenberghs, K., Clemente, A., Imms, P., Egan, G., Hocking, D. R., Leemans, A., Metzler-Baddeley, C., Jones, D. K., & Wilson, P. H. (2018). Evidence for training-dependent structural neuroplasticity in brain-injured patients: A critical review. *Neurorehabilitation and Neural Repair, 32*(2), 99–114. https://doi.org/10.1177/1545968317753076

Calderone, A., Latella, D., Bonanno, M., Quartarone, A., Mojdehdehbaher, S., Celesti, A., & Calabrò, R. S. (2024). Towards transforming neurorehabilitation: The impact of artificial intelligence on diagnosis and treatment of neurological disorders. *Biomedicine, 12*(10), 10. https://doi.org/10.3390/biomedicines12102415

Catania, V., Rundo, F., Panerai, S., & Ferri, R. (2023). Virtual reality for the rehabilitation of acquired cognitive disorders: A narrative review. *Bioengineering, 11*(1), 35. https://doi.org/10.3390/bioengineering11010035

Chen, H., Epstein, J., & Stern, E. (2010). Neural plasticity after acquired brain injury: Evidence from functional neuroimaging. *PM&R, 2*(12, Supplement), S306–S312. https://doi.org/10.1016/j.pmrj.2010.10.006

Chen, Z. S., Kulkarni, P., Galatzer-Levy, I. R., Bigio, B., Nasca, C., & Zhang, Y. (2022). Modern views of machine learning for precision psychiatry. *Patterns, 3*(11), 100602. https://doi.org/10.1016/j.patter.2022.100602

Chen, K.-L., Lin, C.-H., Yu, T.-Y., Huang, C.-Y., & Chen, Y.-D. (2018). Differences between the childhood autism rating scale and the social responsiveness scale in assessing symptoms of children with autistic spectrum disorder. *Journal of Autism & Developmental Disorders, 48*(9), 3191–3198. https://doi.org/10.1007/s10803-018-3585-y

Cherry, K. (2024). How brain neurons change over time from life experience. *Verywell Mind.* https://www.verywellmind.com/what-is-brain-plasticity-2794886

Ciccarelli, G., Federico, G., Mele, G., Di Cecca, A., Migliaccio, M., Ilardi, C. R., Alfano, V., Salvatore, M., & Cavaliere, C. (2023). Simultaneous real-time EEG-fMRI neurofeedback: A systematic review. *Frontiers in Human Neuroscience, 17*, 1123014. https://doi.org/10.3389/fnhum.2023.1123014

Cognitive Training Research – BrainTrain, Inc. (n.d.). Retrieved January 2, 2025, from https://www.braintrain.com/cognitive-training-research/

Cramer, S. C., Sur, M., Dobkin, B. H., O'Brien, C., Sanger, T. D., Trojanowski, J. Q., Rumsey, J. M., Hicks, R., Cameron, J., Chen, D., Chen, W. G., Cohen, L. G., deCharms, C., Duffy, C. J., Eden, G. F., Fetz, E. E., Filart, R., Freund, M., Grant, S. J., et al. (2011). Harnessing neuroplasticity for clinical applications. *Brain, 134*(6), 1591–1609. https://doi.org/10.1093/brain/awr039

Cramer, S. C., Wolf, S. L., Adams, H. P., Chen, D., Dromerick, A. W., Dunning, K., Ellerbe, C., Grande, A., Janis, S., Lansberg, M. G., Lazar, R. M., Palesch, Y. Y., Richards, L., Roth, E., Savitz, S. I., Wechsler, L. R., Wintermark, M., & Broderick, J. P. (2017). Stroke recovery and rehabilitation research. *Stroke, 48*(3), 813–819. https://doi.org/10.1161/STROKEAHA.116.015501

deCharms, R. C. (2007). Reading and controlling human brain activation using real-time functional magnetic resonance imaging. *Trends in Cognitive Sciences, 11*(11), 473–481. https://doi.org/10.1016/j.tics.2007.08.014

Drigas, A., & Sideraki, A. (2024). Brain neuroplasticity leveraging virtual reality and brain–computer interface technologies. *Sensors, 24*(17), 17. https://doi.org/10.3390/s24175725

Fotuhi, M., Khorrami, N. D., & Raji, C. A. (2023). Benefits of a 12-week non-drug "brain fitness program" for patients with attention-deficit/hyperactive disorder, post-concussion syndrome, or memory loss. *Journal of Alzheimer's Disease Reports, 7*(1), 675–697. https://doi.org/10.3233/ADR-220091

Galetto, V., & Sacco, K. (2017). Neuroplastic changes induced by cognitive rehabilitation in traumatic brain injury: A review. *Neurorehabilitation and Neural Repair, 31*(9), 800–813. https://doi.org/10.1177/1545968317723748

Gangemi, A., De Luca, R., Fabio, R. A., Lauria, P., Rifici, C., Pollicino, P., Marra, A., Olivo, A., Quartarone, A., & Calabrò, R. S. (2023). Effects of virtual reality cognitive training on neuroplasticity: A quasi-randomized clinical trial in patients with stroke. *Biomedicine, 11*(12), 3225. https://doi.org/10.3390/biomedicines11123225

Georgiev, D. D., Georgieva, I., Gong, Z., Nanjappan, V., & Georgiev, G. V. (2021). Virtual reality for neurorehabilitation and cognitive enhancement. *Brain Sciences, 11*(2), 221. https://doi.org/10.3390/brainsci11020221

Gomez Palacio Schjetnan, A., Faraji, J., Metz, G. A., Tatsuno, M., & Luczak, A. (2013). Transcranial direct current stimulation in stroke rehabilitation: A review of recent advancements. *Stroke Research and Treatment, 2013*(1), 170256. https://doi.org/10.1155/2013/170256

Gunawan, H., Gunawan, I., Hambarsari, Y., Danuaji, R., Hamidi, B. L., & Benedictus, B. (2024). Virtual reality intervention for improving cognitive function in post-stroke patient: A systematic review and meta-analysis. *Brain Disorders, 15*, 100152. https://doi.org/10.1016/j.dscb.2024.100152

Hain, D., Jurowetzki, R., Squicciarini, M., & Lihui, X. (2023). *Unveiling the neurotechnology landscape: Scientific advancements innovations and major trends*. UNESCO. https://unesdoc.unesco.org/ark:/48223/pf0000386137

INVITED SYMPOSIUM. (n.d.). https://doi.org/10.1002/ijop.13169.

Israsena, P., Jirayucharoensak, S., Hemrungrojn, S., & Pan-Ngum, S. (2021). Brain exercising games with consumer-grade single-channel electroencephalogram neurofeedback: Pre-post intervention study. *JMIR Serious Games, 9*(2), e26872. https://doi.org/10.2196/26872

Jiang, Y., Abiri, R., & Zhao, X. (2017). Tuning up the old brain with new tricks: Attention training via neurofeedback. *Frontiers in Aging Neuroscience, 9*, 52. https://doi.org/10.3389/fnagi.2017.00052

Jones, T. A. (2011). Experience-dependent changes in nonhumans. In *Neuroplasticity and rehabilitation* (pp. 38–64). The Guilford Press.

Kania, B. F., Wrońska, D., & Zięba, D. (2017). Introduction to neural plasticity mechanism. *Journal of Behavioral and Brain Science, 7*(2), 41–49. https://doi.org/10.4236/jbbs.2017.72005

Kerr, A. L., Cheng, S.-Y., & Jones, T. A. (2011). Experience-dependent neural plasticity in the adult damaged brain. *Journal of Communication Disorders, 44*(5), 538–548. https://doi.org/10.1016/j.jcomdis.2011.04.011

Khaneja, S., & Arora, T. (2024). The potential of neuroscience in transforming business: A meta-analysis. *Future Business Journal, 10*(1), 77. https://doi.org/10.1186/s43093-024-00369-7

Kricheldorff, J., Göke, K., Kiebs, M., Kasten, F. H., Herrmann, C. S., Witt, K., & Hurlemann, R. (2022). Evidence of neuroplastic changes after transcranial magnetic, electric, and deep brain stimulation. *Brain Sciences, 12*(7), 929. https://doi.org/10.3390/brainsci12070929

Ku, J., & Kang, Y. J. (2018). Novel virtual reality application in field of neurorehabilitation. *Brain & Neurorehabilitation, 11*(1), e5. https://doi.org/10.12786/bn.2018.11.e5

Lavy, Y., Dwolatzky, T., Kaplan, Z., Guez, J., & Todder, D. (2021). Mild cognitive impairment and neurofeedback: A randomized controlled trial. *Frontiers in Aging Neuroscience, 13*, 657646. https://doi.org/10.3389/fnagi.2021.657646

Levin, M. E., MacLane, C., Daflos, S., Seeley, J., Hayes, S. C., Biglan, A., & Pistorello, J. (2014). Examining psychological inflexibility as a transdiagnostic process across psychological disorders. *Journal of Contextual Behavioral Science, 3*(3), 155–163. https://doi.org/10.1016/j.jcbs.2014.06.003

Li, M. (2022). Associations of body image with depressive symptoms and PTG among breast cancer patients: The mediating role of social support. *Frontiers in Psychology, 13*, 953306. https://www.frontiersin.org/articles/10.3389/fpsyg.2022.953306

Luber, B., Fisher, C., Appelbaum, P. S., Ploesser, M., & Lisanby, S. H. (2009). Non-invasive brain stimulation in the detection of deception: Scientific challenges and ethical consequences. *Behavioral Sciences & the Law, 27*(2), 191–208. https://doi.org/10.1002/bsl.860

Marlats, F., Bao, G., Chevallier, S., Boubaya, M., Djabelkhir-Jemmi, L., Wu, Y.-H., Lenoir, H., Rigaud, A.-S., & Azabou, E. (2020). SMR/theta neurofeedback training improves cognitive performance and EEG activity in elderly with mild cognitive impairment: A pilot study. *Frontiers in Aging Neuroscience, 12*, 147. https://doi.org/10.3389/fnagi.2020.00147

Marzola, P., Melzer, T., Pavesi, E., Gil-Mohapel, J., & Brocardo, P. S. (2023). Exploring the role of neuroplasticity in development, aging, and neurodegeneration. *Brain Sciences, 13*(12), 12. https://doi.org/10.3390/brainsci13121610

Mehler, D. M. A., Sokunbi, M. O., Habes, I., Barawi, K., Subramanian, L., Range, M., Evans, J., Hood, K., Lührs, M., Keedwell, P., Goebel, R., & Linden, D. E. J. (2018). Targeting the affective brain-a randomized controlled trial of real-time fMRI neurofeedback in patients with depression. *Neuropsychopharmacology: Official Publication of the American College of Neuropsychopharmacology, 43*(13), 2578–2585. https://doi.org/10.1038/s41386-018-0126-5

Morya, E., Monte-Silva, K., Bikson, M., Esmaeilpour, Z., Biazoli, C. E., Fonseca, A., Bocci, T., Farzan, F., Chatterjee, R., Hausdorff, J. M., da Silva Machado, D. G., Brunoni, A. R., Mezger, E., Moscaleski, L. A., Pegado, R., Sato, J. R., Caetano, M. S., Sá, K. N., Tanaka, C., et al. (2019). Beyond the target area: An integrative view of tDCS-induced motor cortex modulation in patients and athletes. *Journal of Neuroengineering and Rehabilitation, 16*(1), 141. https://doi.org/10.1186/s12984-019-0581-1

Neurofeedback For Cognitive Decline: Preserving Brain Health As We Age. (2023). *FamilyTime*. https://www.familytimecenters.com/post/neurofeedback-for-cognitive-decline-preserving-brain-health-as-we-age

Neuroplasticity. (n.d.). Physiopedia. Retrieved December 29, 2024, from https://www.physio--pedia.com/Neuroplasticity

Neuroplasticity | Different Types, Facts, & Research | Britannica. (n.d.). Retrieved December 29, 2024, from https://www.britannica.com/science/neuroplasticity

Neuroplasticity in stroke and brain injury: Shaping modern rehabilitation… (n.d.). *MedLink neurology*. Retrieved December 29, 2024, from https://www.medlink.com/news/neuroplasticity-in-stroke-and-brain-injury-shaping-modern-rehabilitation-practices

Neuroplasticity—An overview | ScienceDirect Topics. (n.d.-a). Retrieved December 29, 2024, from https://www.sciencedirect.com/topics/agricultural-and-biological-sciences/neuroplasticity

Neuroplasticity—An overview | ScienceDirect Topics. (n.d.-b). Retrieved January 1, 2025, from https://www.sciencedirect.com/topics/agricultural-and-biological-sciences/neuroplasticity

Oluwafemidiakhoa. (2024, April 6). From silicon to neurons: Bridging AI and neuroscience to decode the human mind. *Kinomoto.Mag AI*. https://medium.com/kinomoto-mag/from-silicon-to-neurons-bridging-ai-and-neuroscience-to-decode-the-human-mind-a7bd60ec6009

Prigatano, G. P., Braga, L. W., Johnson, S. F., & Souza, L. M. N. (n.d.). Neuropsychological rehabilitation, neuroimaging and neuroplasticity: A clinical commentary. *NeuroRehabilitation, 49*(2), 255–265. https://doi.org/10.3233/NRE-218024

Pub, O. A. (n.d.). Neuroplasticity and cognitive rehabilitation. *Journal of Human Psychology*. Open Access Pub. Retrieved December 29, 2024, from https://openaccesspub.org

Puderbaugh, M., & Emmady, P. D. (2024). Neuroplasticity. In *StatPearls*. StatPearls Publishing. http://www.ncbi.nlm.nih.gov/books/NBK557811/

Putchala, G., & Putchala, C. S. (2024). The role of artificial intelligence in neurosciences: An approach to neuroplasticity in the era of AI for personalized rehabilitation. *International Journal of Communication Networks and Information Security, 16*(1).

Rogers, J. M., Duckworth, J., Middleton, S., Steenbergen, B., & Wilson, P. H. (2019). Elements virtual rehabilitation improves motor, cognitive, and functional outcomes in adult stroke: Evidence from a randomized controlled pilot study. *Journal of Neuroengineering and Rehabilitation, 16*(1), 56. https://doi.org/10.1186/s12984-019-0531-y

Role of Neuroplasticity in Neuro-rehabilitation. (n.d.). Physiopedia. Retrieved December 14, 2024, from https://www.physio-pedia.com/Role_of_Neuroplasticity_in_Neuro-rehabilitation

Saj, A., Pierce, J. E., Ronchi, R., Ros, T., Thomasson, M., Bernati, T., Van De Ville, D., Serino, A., & Vuilleumier, P. (2021). Real-time fMRI and EEG neurofeedback: A perspective on applications for the rehabilitation of spatial neglect. *Annals of Physical and Rehabilitation Medicine, 64*(5), 101561. https://doi.org/10.1016/j.rehab.2021.101561

Schlaug, G., Renga, V., & Nair, D. (2008). Transcranial direct current stimulation in stroke recovery. *Archives of Neurology, 65*(12), 1571–1576. https://doi.org/10.1001/archneur.65.12.1571

Shen, J., Gu, X., Fu, J., Yao, Y., Li, Y., Zeng, M., Liu, Z., & Lu, C. (2023). Virtual reality-induced motor function of the upper extremity and brain activation in stroke: Study protocol for a randomized controlled trial. *Frontiers in Neurology, 14*, 1094617. https://doi.org/10.3389/fneur.2023.1094617

Sloane, K. L., & Hamilton, R. H. (2024). Transcranial direct current stimulation to ameliorate post-stroke cognitive impairment. *Brain Sciences, 14*(6), 6. https://doi.org/10.3390/brainsci14060614

Stagg, C. J., & Johansen-Berg, H. (2013). Studying the effects of transcranial direct-current stimulation in stroke recovery using magnetic resonance imaging. *Frontiers in Human Neuroscience, 7*, 857. https://doi.org/10.3389/fnhum.2013.00857

Stein, J., Rief, K., & Hull, D. (2017, March 14). Virtual reality and robotic applications in rehabilitation. *PM&RKnowledgeNOW*. https://now.aapmr.org/virtual-reality-robotic-applications-in-rehabilitation/

Stoeckel, L. E., Garrison, K. A., Ghosh, S., Wighton, P., Hanlon, C. A., Gilman, J. M., Greer, S., Turk-Browne, N. B., deBettencourt, M. T., Scheinost, D., Craddock, C., Thompson, T., Calderon, V., Bauer, C. C., George, M., Breiter, H. C., Whitfield-Gabrieli, S., Gabrieli, J. D., LaConte, S. M., et al. (2014). Optimizing real time fMRI neurofeedback for therapeutic discovery and development. *NeuroImage: Clinical, 5*, 245–255. https://doi.org/10.1016/j.nicl.2014.07.002

Tonti, E., Budini, M., & Vingolo, E. M. (2021). Visuo-acoustic stimulation's role in synaptic plasticity: A review of the literature. *International Journal of Molecular Sciences, 22*(19), 10783. https://doi.org/10.3390/ijms221910783

Torrisi, M., Maggio, M. G., De Cola, M. C., Zichittella, C., Carmela, C., Porcari, B., la Rosa, G., De Luca, R., Naro, A., & Calabrò, R. S. (2021). Beyond motor recovery after stroke: The role of hand robotic rehabilitation plus virtual reality in improving cognitive function. *Journal of Clinical Neuroscience, 92*, 11–16. https://doi.org/10.1016/j.jocn.2021.07.053

Treves, I. N., Greene, K. D., Bajwa, Z., Wool, E., Kim, N., Bauer, C. C. C., Bloom, P. A., Pagliaccio, D., Zhang, J., Whitfield-Gabrieli, S., & Auerbach, R. P. (2024). Mindfulness-based neurofeedback: A systematic review of EEG and fMRI studies. *Imaging Neuroscience, 2*, 1–25. https://doi.org/10.1162/imag_a_00396

Tymofiyeva, O., & Gaschler, R. (2021). Training-induced neural plasticity in youth: A systematic review of structural and functional MRI studies. *Frontiers in Human Neuroscience, 14*, 497245. https://doi.org/10.3389/fnhum.2020.497245

Vilou, I., Varka, A., Parisis, D., Afrantou, T., & Ioannidis, P. (2023). EEG-neurofeedback as a potential therapeutic approach for cognitive deficits in patients with dementia, multiple sclerosis, stroke and traumatic brain injury. *Life, 13*(2), 365. https://doi.org/10.3390/life13020365

Virtual Reality for Individuals Affected by Stroke. (n.d.). Physiopedia. Retrieved December 28, 2024, from https://www.physio-pedia.com/Virtual_Reality_for_Individuals_Affected_by_Stroke

Wairimu, A. (2024). Neuroplasticity and augmented reality rehabilitation. *Research Output Journal of Engineering and Scientific Research, 3*(1), 105–109.

Ying, H., Pranolo, A., Nuryana, Z., & Syafitri, A. I. (2024). Emerging trends in the evolution of neuropsychology and artificial intelligence: A comprehensive analysis. *Telematics and Informatics Reports, 16*, 100171. https://doi.org/10.1016/j.teler.2024.100171

Zhao, H., Cao, J., Xie, J., Liao, W.-H., Lei, Y., Cao, H., Qu, Q., & Bowen, C. (2023). Wearable sensors and features for diagnosis of neurodegenerative diseases: A systematic review. *Digital Health, 9*, 20552076231173569. https://doi.org/10.1177/20552076231173569

Zilverstand, A., Parvaz, M. A., Moeller, S. J., & Goldstein, R. Z. (2016). Chapter 13—Cognitive interventions for addiction medicine: Understanding the underlying neurobiological mechanisms. In H. Ekhtiari & M. P. Paulus (Eds.), *Progress in brain research* (Vol. 224, pp. 285–304). Elsevier. https://doi.org/10.1016/bs.pbr.2015.07.019

Zotey, V., Andhale, A., Shegekar, T., & Juganavar, A. (2023). Adaptive neuroplasticity in brain injury recovery: Strategies and insights. *Cureus, 15*(9), e45873. https://doi.org/10.7759/cureus.45873

Chapter 9
Cognitive Training Programs

9.1 Introduction

Cognitive training has emerged as an intervention designed to maintain or enhance cognitive functions, particularly in response to challenges posed by aging, neurological disorders, and cognitive impairments. At its core, cognitive training leverages the concept of neuroplasticity, the brain's inherent ability to reorganize and form new neural connections through repeated practice and exposure to targeted tasks. Rooted in theories of cognitive reserve and learning transfer, these programs aim not only to strengthen specific cognitive domains, such as memory, attention, and executive functions, but also to foster generalization of these skills to everyday tasks. While the promise of cognitive training is widely acknowledged, its efficacy and long-term impact remain topics of ongoing investigation and debate. Key meta-analyses have demonstrated small but statistically significant improvements in global cognitive functioning among cognitively healthy older adults (Lampit et al., 2014). These findings support the idea that systematic, computerized interventions can enhance performance in targeted areas. Similarly, the ACTIVE Study (Advanced Cognitive Training for Independent and Vital Elderly) provided strong evidence for training-induced benefits in executive functions and specific daily tasks, though the extent and durability of these gains varied (Mendes et al., 2022; Rebok et al., 2014).

Theories of task-specific gains and learning transfer play a central role in understanding cognitive training outcomes. Task-specific gains refer to improvements in tasks directly practiced during training, while learning transfer involves the application of these skills to untrained areas or real-world scenarios. Evidence for transfer effects, however, remains inconclusive. Tetlow and Edwards highlighted the limited generalization of training gains to everyday life, emphasizing that improvements in attention, processing speed, and visuospatial memory often do not extend to functional activities. These findings align with critiques suggesting that training-induced

K. J. Reddy, *Innovations in Neurocognitive Rehabilitation*,
https://doi.org/10.1007/978-3-031-88117-6_9

changes may be confined to laboratory contexts, necessitating further exploration of mechanisms that facilitate real-world impact (de Jager Loots et al., 2024).

The variability in cognitive training outcomes is influenced by factors such as age, baseline cognitive status, and individual differences in motivation and adherence. Theories like self-determination theory emphasize the importance of intrinsic motivation in sustaining engagement with training programs, while practical barriers such as technological literacy and accessibility further complicate implementation. Compliance remains a significant challenge, with some studies reporting dropout rates as high as 60% (de Jager Loots et al., 2024). Addressing these barriers through gamification, adaptive algorithms, and user-centered design has shown potential in enhancing user engagement and improving adherence. Advancements in technology have revolutionized cognitive training, incorporating tools such as artificial intelligence (AI), virtual reality (VR), and wearable devices. AI-driven platforms dynamically adapt task difficulty based on performance, optimizing the training experience for individual users (Gómez-Soria et al., 2023). VR offers immersive environments for training spatial memory and multitasking, while wearable devices provide real-time feedback using metrics like EEG signals. These technologies align with adaptive learning theories, which posit that personalized feedback and task adjustment maximize the efficacy of training.

The integration of cognitive training with complementary interventions further underscores its potential. The biopsychosocial model suggests that combining cognitive exercises with physical activity, social engagement, and stress management may yield synergistic effects. Physical activity, for instance, promotes neurogenesis and vascular health, while social interactions stimulate emotional and cognitive processes that reinforce learning. González-Palau et al. demonstrated how combined interventions improved verbal memory, attention, and mood in older adults with mild cognitive impairment (Ge et al., 2018; Gómez-Soria et al., 2023).

Despite these advancements, ethical considerations and methodological limitations remain significant. Overhyped claims about the effectiveness of cognitive training have raised concerns about transparency and credibility, necessitating rigorous evaluation of program efficacy. Data privacy and equitable access to technology-enhanced tools are also pressing issues, particularly in low-resource settings (de Jager Loots et al., 2024; Ge et al., 2018). Looking forward, the field is moving toward more comprehensive frameworks for evaluating cognitive training. Longitudinal research is essential to determine the durability of training-induced benefits and their applicability to real-world functioning. Theories of ecological validity emphasize the importance of designing interventions that reflect everyday challenges, ensuring that cognitive gains translate into meaningful improvements in daily living. Innovations such as neuroadaptive training, which uses neural data to dynamically adjust tasks, and group-based cognitive interventions incorporating social dynamics are promising avenues for future development. Cognitive training programs represent a valuable tool for enhancing cognitive functioning across diverse populations. While existing evidence highlights the potential for improvements in memory, attention, and executive functions, challenges related to transfer effects, adherence, and equity remain. By integrating theories of neuroplasticity, task specificity, and learning transfer with emerging technologies and holistic approaches, the next generation of cognitive training programs can achieve greater efficacy and accessibility. Through rigorous research and ethical implementation, these interventions have the potential to significantly improve cognitive health, quality of life, and functional independence for individuals worldwide.

9.2 Foundations of Cognitive Training

Cognitive training programs are rooted in the principles of neuroplasticity, the brain's ability to reorganize itself by forming new neural connections in response to learning and experience. By engaging in repetitive and structured cognitive exercises, these programs aim to enhance the efficiency and adaptability of neural networks. Neuroplasticity underpins the concept that cognitive functions, even in aging or impaired brains, can be strengthened or restored through systematic training. These programs are designed to target specific cognitive domains, including memory, attention, and executive functions, with the goal of improving overall cognitive performance. The foundation of cognitive training lies in structured, evidence-based interventions that employ various techniques to optimize performance in these areas. These techniques are informed by well-established theories of learning, neuroplasticity, and cognitive reserve.

9.2.1 Core Domains of Cognitive Training

9.2.1.1 Memory Training

Memory training techniques have long been a cornerstone of cognitive enhancement, offering powerful tools to improve retention, recall, and the overall efficiency of cognitive processes. Central to these techniques is the principle of spaced repetition, a strategy rooted in cognitive science that systematically spaced review sessions over time to optimize memory retention. The concept, first systematically studied by Hermann Ebbinghaus, demonstrated that information is more effectively retained when reviewed at increasing intervals, counteracting the exponential decline in memory retention illustrated by the forgetting curve ("Boost Learning With a Simple Cognitive Trick," 2024; Smolen et al., 2016). Unlike massed repetition, where learners repeatedly review material in a single session, spaced repetition strengthens neural pathways by gradually reactivating memory traces, leading to more durable long-term memory formation.

Modern cognitive science has not only confirmed the efficacy of spaced repetition but has also advanced its application through computational models and technology. Research suggests that irregular spacing intervals, determined by algorithms that adjust based on individual performance, can further enhance learning outcomes("Boost Learning With a Simple Cognitive Trick," 2024; Smolen et al., 2016). These findings have been translated into practical tools, such as adaptive learning platforms, which automate the process of spacing and personalization. Such platforms have made spaced repetition more accessible, offering learners a convenient way to improve retention, whether they are mastering academic content, acquiring professional skills, or pursuing personal development.

Memory training techniques like spaced repetition exemplify the intersection of cognitive science, education, and technology. Grounded in decades of research, these methods offer a practical, evidence-based approach to improving memory and learning efficiency across diverse populations and contexts. As new tools and insights continue to emerge, the potential of spaced repetition to enrich cognitive training programs and enhance lifelong learning becomes increasingly clear, marking it as a cornerstone of modern educational and cognitive interventions (Johnson & Miles, 2019).

- *The role of neuroplasticity in memory training:* The effectiveness of these tasks in developmental training, aging, and brain injury recovery is underpinned by the principles of neuroplasticity. By engaging in repetitive, targeted exercises, individuals can strengthen neural connections, create new synaptic pathways, and even compensate for damaged regions of the brain. Across the lifespan, memory training tasks serve as vital tools for cognitive enhancement, promoting mental agility in children, preserving cognitive function in older adults, and supporting recovery in individuals with brain injuries. By tailoring interventions to meet the specific needs of each population, memory training tasks can help maximize cognitive potential, improve quality of life, and foster lifelong brain health. As research continues to refine these techniques and incorporate emerging technologies, such as gamified applications and adaptive algorithms, the scope and impact of memory training will undoubtedly expand, making it an increasingly valuable component of cognitive intervention programs (Abdul Razak et al., 2024; Frankenmolen et al., 2018).
- *Developmental training:* In children, memory training often focuses on enhancing foundational cognitive skills required for learning and academic success. Phonological awareness tasks, such as identifying rhymes or breaking down word families, play a crucial role in improving early literacy by strengthening the brain's ability to process and manipulate language sounds (Upham, 2024). Similarly, number sequencing activities—such as counting backward or completing numerical patterns—build mathematical fluency and working memory. Working memory exercises, including n-back tasks and digit span tests, challenge children to hold and manipulate information actively, a critical skill for problem-solving and decision-making (Barman et al., 2016). Attention training activities, like the Stroop test or Simon Says, help young learners develop focus and filter out distractions in an increasingly stimuli-rich environment (Abdul Razak et al., 2024). Finally, executive function tasks such as the Tower of London puzzle cultivate planning, goal-setting, and problem-solving abilities, essential

for academic and social success (Upham, 2024). These activities, when practiced consistently, nurture cognitive growth during critical developmental periods.

- *Aging:* In older adults, memory training focuses on preserving and enhancing cognitive abilities to mitigate age-related cognitive decline and promote mental sharpness. Recall exercises, such as remembering lists of items, family birthdays, or past events, stimulate episodic memory and help maintain neural connections involved in everyday memory tasks (*Understanding Aging Brains, How to Improve Memory and When to Seek Help*, n.d.). Episodic memory games, like storytelling or maintaining a memory journal, foster autobiographical recall, which has been shown to improve emotional well-being and cognitive resilience. Cognitive stimulation activities, including crosswords and Sudoku puzzles, engage multiple cognitive domains simultaneously, promoting sustained mental engagement and cognitive reserve. Attention and focus drills, such as following detailed instructions or ignoring distractions, are particularly beneficial in addressing declines in sustained attention. Additionally, tasks aimed at improving processing speed, like dot cancellation tests, help maintain quick cognitive reactions essential for daily activities. These memory training tasks not only strengthen neural pathways but also help older adults remain active and engaged, enhancing their quality of life (Barman et al., 2016; Frankenmolen et al., 2018; Upham, 2024).
- *Traumatic brain injury*: In individuals recovering from brain injuries, memory training tasks are specifically designed to restore cognitive function and compensate for lost abilities. Sensory integration exercises, such as walking on balance beams or proprioceptive activities, help patients process and integrate sensory input, improving overall sensory-motor coordination (Attout et al., 2020). Motor rehabilitation tasks, like fine motor puzzles or hand manipulation exercises, support the recovery of dexterity and coordination critical for daily tasks. Attention restoration therapy, which may involve nature walks or focused breathing exercises, helps reduce mental fatigue and improve sustained attention in individuals experiencing cognitive overload (Johnson & Miles, 2019). Cognitive compensation strategies, such as breaking complex tasks into manageable steps or utilizing assistive technology, enable individuals to adapt and perform daily activities despite residual cognitive challenges. Emotional regulation exercises, including mindfulness meditation and emotional labeling, help patients manage stress and rebuild emotional resilience (Johnson & Miles, 2019).

9.2.1.2 Attention Training

Attention is a fundamental cognitive skill that allows individuals to focus on relevant stimuli while filtering out distractions. The ability to allocate attention effectively is crucial for learning, decision-making, and emotional regulation. Attention training, particularly techniques like the Attention Training Technique (ATT), has garnered significant attention for its potential to enhance mental health, cognitive development, and neuroplasticity. Grounded in both theoretical and empirical research, attention training has shown promise in addressing a range of cognitive and psychological challenges across diverse populations. Recent findings highlight the efficacy of ATT, a technique developed by Adrian Wells in 1990, in treating mental health conditions such as anxiety, depression, and schizophrenia. ATT focuses on redirecting attention away from distressing internal stimuli and promoting flexible and adaptive attention control. Systematic reviews and meta-analyses have demonstrated significant benefits, with effect sizes ranging from moderate to large (pooled ES: 0.74–1.00) for individuals experiencing anxiety and depressive disorders. For instance, randomized controlled trials (RCTs) have shown that ATT outperforms reference groups, with Cohen's d values ranging from 0.40 to 1.23 in improving emotional and attentional outcomes (Soule et al., 2024).

Beyond clinical applications, attention training plays a pivotal role in cognitive development and neuroplasticity. Neuroimaging studies reveal that attention training induces changes in brain regions such as the prefrontal cortex and parietal lobes, which are critical for attentional control (Ponsford et al., 2023). These findings align with Hebb's theory further suggesting that repeated activation of attentional networks strengthens neural connections and enhances cognitive efficiency. Practices like mindfulness meditation, which involves sustained attention to the present moment, have been linked to increased gray matter density in attention-related brain areas, further supporting the neuroplastic potential of attention training Kaurani et al., 2024).

Attention training also leverages theoretical frameworks to guide its design and application. The Attentional Control Theory posits that attentional resources are finite and must be strategically allocated to maintain goal-directed behavior. Techniques like ATT embody this theory by helping individuals shift their focus from distressing stimuli to goal-relevant tasks (Kaurani et al., 2024). Similarly, Load Theory explains how working memory capacity limits attentional processing, emphasizing the importance of reducing cognitive load to optimize attentional performance (Kaurani et al., 2024; Ponsford et al., 2023).

Evidence-based practices like task-switching exercises or Stroop tests are designed to enhance selective and divided attention, aligning with these theoretical principles. Repetition plays a critical role in the success of attention training. Regular practice strengthens neural pathways associated with attentional control, leading to lasting improvements in cognitive performance. Spaced repetition, a method where tasks are revisited at increasing intervals, has proven particularly effective in consolidating attentional gains (Kaurani et al., 2024; Soule et al., 2024). Participants in studies employing spaced repetition techniques for attentional tasks, such as identifying patterns or sustaining focus on stimuli over time, showed improved performance and reduced cognitive fatigue (Kaurani et al., 2024). Applications of attention training extend across clinical and non-clinical settings. In clinical contexts, ATT has proven effective for anxiety and depressive disorders, with systematic reviews affirming its utility in reducing symptoms and improving emotional regulation. In educational settings, attention training has been integrated into interventions for children with attention-deficit/hyperactivity disorder (ADHD), helping to improve sustained attention and classroom behavior. Furthermore, in aging populations, attention training has been employed to mitigate cognitive decline, with tasks like dual-task training and selective attention drills improving focus and processing speed (Kaurani et al., 2024; Ponsford et al., 2023; Soule et al., 2024).

Technology has further enhanced the accessibility and efficacy of attention training. Interactive platforms utilizing gamification and real-time feedback make training more engaging and effective. For instance, adaptive attention-training apps dynamically adjust task difficulty based on user performance, optimizing the learning experience and ensuring sustained engagement (Ponsford et al., 2023). Virtual reality (VR) environments, which immerse users in complex, dynamic settings, have also been used to train selective and sustained attention, showing promise for populations ranging from students to individuals recovering from traumatic brain injuries (Kaurani et al., 2024; Soule et al., 2024). While attention training holds significant promise, challenges remain. Many studies suffer from small sample sizes, limiting the generalizability of findings. There is also a need for more longitudinal research to examine the durability of attention-training effects and their impact on real-world outcomes (Kaurani et al., 2024; Soule et al., 2024). Additionally, individual differences in baseline attention, motivation, and adherence to training protocols influence outcomes, emphasizing the need for personalized approaches.

- *Attention training in neurodevelopmental disorders:* Attention training is critical for addressing the unique cognitive challenges faced by children with neurodevelopmental disorders, such as ADHD and autism spectrum disorders. One of the most widely recognized approaches is Attention Process Training (APT), which systematically targets five components of attention: focused, sustained, selective, alternating, and divided attention. This evidence-based approach begins with simple tasks, such as identifying a target amidst minimal distractions, and gradually progresses to more complex activities, like alternating attention between tasks or managing multiple stimuli simultaneously. Incorporating cognitive remediation strategies alongside APT has been shown to enhance its effectiveness. Using memory notebooks to track tasks and combining direct attention training with cognitive behavioral therapy (CBT) can significantly improve divided attention, emotional regulation, and task-switching abilities. These integrated approaches are particularly effective for children with neurodevelopmental conditions, where emotional and cognitive challenges are often intertwined (*Pediatric Traumatic Brain Injury*, n.d.). Evidence-based practices also emphasize the use of engaging and developmentally appropriate tasks, such as phonological awareness activities, number sequencing, and executive function exercises. Phonological awareness tasks, like identifying rhyming words, strengthen early literacy skills, while working memory activities, such as n-back tasks and digit span tests, enhance cognitive flexibility (Soule et al., 2024). Cultural considerations play a crucial role in the success of these interventions. Integrating culturally relevant games, stories, or symbols into attention training programs can improve engagement and comprehension. Additionally, involving family members in training enhances adherence and ensures that skills learned in structured settings are generalized to the child's daily life, fostering sustained improvement (Ponsford et al., 2023).

- *Attention training in traumatic brain injury:* Traumatic brain injury (TBI) often results in significant deficits in attention, impacting an individual's ability to focus, manage tasks, and sustain attention. Attention Process Training (APT) remains the gold standard for rehabilitating attentional deficits in TBI survivors. APT involves structured exercises tailored to the individual's needs, beginning with tasks like identifying simple visual targets and progressing to more complex activities such as alternating or dividing attention between stimuli. Technology-based interventions are also gaining traction in TBI rehabilitation. Virtual reality (VR) environments offer immersive scenarios, such as navigating crowded spaces or reacting to traffic signals, to train selective and sustained attention. These real-world simulations allow patients to practice attentional control in safe yet challenging settings, with studies showing measurable improvements in attentional flexibility and task accuracy. Mobile apps with gamified elements and real-time feedback have further increased accessibility, enabling patients to engage in training from home, thereby maintaining consistency and adherence (Abdul Razak et al., 2024).
- Another evidence-based approach is Attention Restoration Therapy (ART), which emphasizes the use of restorative environments, such as nature-based activities, to reduce cognitive fatigue and enhance sustained attention. Activities like guided nature walks or focused breathing in outdoor settings have shown significant benefits for TBI patients, especially those experiencing mental overload (Barman et al., 2016). Cultural factors are crucial when designing attention training programs for TBI survivors. Tailoring training activities to reflect the patient's cultural and linguistic background improves engagement and comprehension. Additionally, involving caregivers and family members in therapy sessions not only supports skill application but also ensures emotional and social support during the recovery process (Buccilli et al., 2024).
- *Attention training in aging:* As cognitive processes slow with age, attention training becomes essential for preserving attentional capacity and overall cognitive health. Evidence-based approaches include activities like crosswords, Sudoku puzzles, and memory recall games, which engage multiple cognitive domains and help improve selective and sustained attention. These tasks are particularly effective because they simultaneously stimulate mental flexibility and problem-solving skills (Soule et al., 2024). Computerized cognitive training programs are also widely used for older adults. These programs incorporate interactive exercises like visual search tasks, auditory discrimination drills, and dot cancellation tests, which adapt dynamically to the user's performance. Studies show that such programs not only enhance attention but also improve processing speed and reaction times, critical for maintaining functional independence (Soule et al., 2024).
- Combining attention training with other interventions, such as physical activity or social engagement, further amplifies its benefits. Group exercise classes that

integrate dual-task activities, such as following verbal instructions while coordinating movements, have been shown to improve attentional control and foster social connections. Similarly, storytelling sessions and group discussions promote attention by encouraging active listening and cognitive engagement (Soule et al., 2024). Culturally relevant strategies are essential for older populations. Incorporating elements such as traditional music, regional storytelling practices, or culturally significant games into training activities can enhance emotional resonance and motivation. Additionally, addressing technological barriers by providing training on how to use digital tools ensures that older adults from diverse socioeconomic backgrounds can benefit from computerized attention training programs (Pediatric Traumatic Brain Injury, n.d.).

9.2.1.3 Evidence-Based Approaches Across Populations

Across all populations, the effectiveness of attention training is maximized through evidence-based practices. Programs should be tailored to individual needs, taking into account factors such as age, cognitive capacity, injury severity, and cultural context. Progressive difficulty is a critical component, starting with simple tasks and gradually increasing complexity to challenge attentional capabilities while avoiding frustration. Regular practice is essential for consolidating gains, as consistent engagement strengthens neural pathways associated with attention. Additionally, integrating attention training with other therapies, such as CBT for emotional regulation or physical activity for cognitive stimulation, enhances overall outcomes. Emphasizing ecological validity—by incorporating real-world tasks into training—ensures that improvements translate into daily life (Cicerone et al., 2019; Power & D'Amato, 2021).

9.2.1.4 Executive Function Training

Executive functions—skills like decision-making, problem-solving, and inhibitory control—are essential for navigating the complexities of everyday life. They allow us to weigh options, regulate impulses, and tackle challenges effectively. However, when these functions are impaired, whether due to developmental disorders, brain injuries, or aging, they can create significant barriers to independence and well-being. Executive function training aims to strengthen these skills through evidence-based interventions, empowering individuals to lead more adaptive and fulfilling lives.

- *Theoretical frameworks:* Executive function training is guided by several theoretical models. The Executive Attention Theory highlights the interdependence of attention, memory, and executive processes, emphasizing the need for integrated approaches in training. Meanwhile, the Working Memory Capacity (WMC) Theory posits that working memory and executive attention are closely linked, suggesting that improvements in one domain can support the other Pasqualotto et al., 2021. The distinction between hot and cold executive functions—emotionally charged versus neutral cognitive processes—further informs the design of tasks, ensuring they align with the specific challenges of different populations (Pasqualotto et al., 2021). Executive function training offers a powerful tool for improving decision-making, problem-solving, and inhibitory control across diverse populations. By incorporating evidence-based practices like regular, adaptive training and embedding tasks into real-life contexts, these programs can create meaningful and lasting improvements. Cultural sensitivity enhances engagement and ensures interventions resonate with participants' experiences, while theoretical frameworks provide a foundation for targeted and effective design. Although challenges like achieving far transfer effects persist, ongoing innovation and research hold promise for the continued evolution of executive function training, empowering individuals to navigate life's complexities with greater confidence and competence.

- *Decision-making and problem-solving:* Decision-making and problem-solving are at the heart of executive functioning. These skills enable individuals to plan, evaluate options, and execute solutions effectively. Research has shown that training programs tailored to these processes can have meaningful impacts. Computerized cognitive training programs use dynamic, interactive tasks to enhance decision-making skills. These tasks often simulate real-world scenarios, such as managing a budget or planning a daily schedule, allowing participants to practice strategic thinking in a controlled setting. For individuals with ADHD, such interventions have proven particularly effective, reducing impulsivity and fostering more thoughtful decision-making (Bombonato et al., 2024; Pasqualotto et al., 2021).
- Older adults have also benefited from such training. Programs like the ACTIVE Study demonstrated how targeted exercises in problem-solving can help participants better manage daily tasks, such as coordinating medical appointments or handling unexpected events. The ability to embed these tasks into real-life contexts ensures that the training directly translates to practical improvements (Pasqualotto et al., 2021). Cultural sensitivity enhances the effectiveness of training programs. For instance, incorporating scenarios relevant to an individual's cultural background—like organizing a community gathering or navigating traditional marketplaces—can make the training more relatable and engaging. Providing materials in participants' native languages further ensures accessibility and meaningful participation (Pasqualotto et al., 2021).
- *Inhibitory control:* Inhibitory control is the ability to resist distractions, suppress impulses, and stay focused on tasks. For individuals with ADHD, autism spectrum disorders, or brain injuries, inhibitory control challenges can significantly impact daily life. Training programs targeting this skill often involve tasks like go/no-go exercises, where participants practice suppressing automatic responses, or stop-signal tasks that require quick decision-making under controlled conditions. These tasks have been shown to improve impulse regulation and attention, particularly in individuals with ADHD (Bombonato et al., 2024). Computerized

working memory training offers another avenue for developing inhibitory control. For example, participants may be asked to remember and manipulate information while ignoring irrelevant stimuli. In one study, a 10-week computerized intervention led to significant reductions in impulsivity and better emotional regulation in children with ADHD (Bombonato et al., 2024; Iversen & Lewis, 2021). Cultural context plays a critical role in designing inhibitory control tasks. In cultures that emphasize collective behavior, exercises might incorporate scenarios like waiting for group consensus before acting. This approach not only respects cultural norms but also makes the training more meaningful and applicable (Pasqualotto et al., 2021).

- *Evidence-based practices in executive function training:* Several practices have emerged as cornerstones of effective executive function training:

1. *Intensity and frequency:* Regular, short sessions (20–40 minutes) conducted multiple times a week (5 times) have shown to produce stronger outcomes. This consistent repetition reinforces skills and promotes long-term gains (Bombonato et al., 2024).
2. *Real-life integration:* Embedding training in daily activities increases its relevance and effectiveness. For instance, problem-solving exercises that involve planning meals or managing errands allow participants to apply skills immediately in familiar contexts (Bombonato et al., 2024).
3. *Home-based interventions:* Training programs delivered at home or in hybrid formats have been particularly effective. They allow participants to practice in their natural environments, ensuring a deeper connection between training and daily life. A study by Weiss et al. found that a home/school-based program significantly improved social skills and emotional regulation in children with autism spectrum disorder (Bombonato et al., 2024).
4. *Progressive complexity:* Starting with simple tasks and gradually increasing difficulty ensures participants build confidence and competence. This scaffolding approach avoids overwhelming participants while maintaining a sense of achievement (Bombonato et al., 2024).
5. *Adaptive technology:* Personalized training, where difficulty levels adjust based on performance, keeps participants engaged and challenged at an appropriate level. This adaptability has proven effective in computerized interventions for ADHD and other conditions (Bombonato et al., 2024).
6. *Overcoming challenges in transfer effects*: One of the biggest challenges in executive function training is achieving far transfer effects—the generalization of skills learned in training to untrained tasks and real-life scenarios. While many programs show strong near transfer effects (improvements in tasks similar to those practiced), far transfer remains less consistent. To address this, training programs are increasingly incorporating dual-task exercises that overlap multiple domains, such as combining memory and decision-making tasks. Additionally, integrating mindfulness and physical activity into training has shown promise in engaging broader cognitive and emotional systems, potentially enhancing far transfer effects (Bombonato et al., 2024; Iversen & Lewis, 2021).

9.2.2 Mechanisms of Change

9.2.2.1 Task-Specific Gains

Cognitive training has gained prominence as a tool to enhance mental abilities such as memory, attention, and problem-solving. However, the mechanisms driving improvements in trained tasks remain a subject of considerable research. Emerging evidence suggests that cognitive training primarily enhances the efficiency of using existing cognitive resources rather than expanding cognitive capacity itself. This distinction sheds light on why significant improvements are often observed in trained tasks, but generalization to untrained tasks, or "far transfer effects," remains elusive.

Recent studies have clarified the nuances of training-induced changes. For instance, research on visual working memory (WM) training has shown that while training improves the quality of visual WM representations—enhancing how effectively information is stored and retrieved—it does not increase the quantity or overall capacity of these representations (Jiang et al., 2017). This finding underscores the importance of efficiency in cognitive performance, where individuals become better at allocating attention, reducing distractions, and optimizing task-specific strategies. A key framework for understanding these outcomes is the capacity-efficiency model, which explains cognitive training effects through two pathways: either expanding capacity or improving efficiency in using existing capacity. Evidence suggests that most training-induced gains are attributable to the latter—enhanced efficiency. In a recent study, participants in WM training programs often demonstrate improvements through compression (learning to identify and use patterns and redundancies in stimuli) and optimization (allocating attentional resources more effectively). These strategies allow individuals to achieve better results within the limits of their inherent cognitive capacity (Jiang et al., 2017).

Meta-analyses provide further insights into task-specific gains. Consistently, they report large improvements in trained tasks, such as n-back or visual search tasks, but limited evidence of transfer to untrained tasks or broader cognitive domains. A systematic review highlighted that while participants became adept at specific exercises, such as remembering sequences of shapes or digits, these skills did not generalize to unrelated tasks like problem-solving or decision-making (Jiang et al., 2017; von Bastian et al., 2022). This lack of robust transfer effects suggests that cognitive training primarily refines task-specific processes rather than inducing widespread cognitive changes. Theoretical models also provide valuable perspectives on the mechanisms of task-specific gains. Executive Attention Theory, for instance, emphasizes the interaction between attention, working memory, and executive processes. According to this theory, cognitive training strengthens these interconnected systems, improving how individuals manage attention and memory during demanding tasks (Jiang et al., 2017). Similarly, the Working Memory Capacity (WMC) Theory posits that training enhances executive attention capacity, enabling individuals to better filter irrelevant information and focus on task-relevant details (Jiang et al., 2017; von Bastian et al., 2022). These theories align with the observed improvements in efficiency rather than capacity expansion.

The distinction between hot and cold executive functions further elucidates why task-specific gains may not generalize. Training in emotionally neutral (cold) contexts, such as abstract problem-solving tasks, does not necessarily transfer to emotionally charged (hot) scenarios like decision-making in high-stress environments. This division highlights the importance of designing training programs that reflect the diversity of real-world cognitive demands (Jiang et al., 2017).

Repetition is another cornerstone of effective cognitive training. Frequent and intensive interventions—such as training sessions conducted five times a week for 5–7 weeks—have been shown to produce greater task-specific gains compared to less frequent sessions. Shorter sessions (20–40 min) repeatedly appear to be particularly effective, allowing participants to refine their strategies without becoming overwhelmed (Jiang et al., 2017; von Bastian et al., 2022). These findings emphasize the importance of sustained engagement in achieving meaningful improvements. Despite the clear benefits of task-specific training, the field faces challenges in achieving broader generalization. The variability in study designs and the lack of standardized definitions for "far transfer effects" complicate comparisons across studies. While some researchers report improvements in related cognitive domains, others find no evidence of transfer beyond the trained tasks (Jiang et al., 2017; von Bastian et al., 2022).

This inconsistency underscores the need for clearer methodologies and outcome measures to advance the field. Task-specific gains are rooted in mechanisms such as compression, optimization, and improved executive attention, as supported by theoretical frameworks like the capacity-efficiency model and Working Memory Capacity Theory. However, the limited transfer effects observed highlight the need for innovative approaches to bridge the gap between task-specific improvements and broader cognitive benefits. Resolving these challenges through rigorous research and methodological consistency, cognitive training can unlock its full potential to enhance both task performance and real-world functioning.

9.2.2.2 Transfer Effects

Cognitive training has been widely researched as a means of enhancing cognitive abilities, with particular interest in its potential to promote transfer effects. Transfer effects refer to the extent to which improvements from training generalize beyond the specific tasks practiced during intervention. While some studies show promising outcomes, others highlight the challenges in achieving robust, consistent transfer effects. This has sparked ongoing debate and calls for further exploration to better understand the conditions that facilitate transfer. One of the more encouraging findings comes from research involving older adults with mild cognitive impairment. In these populations, cognitive training targeting working memory has shown significant benefits. For instance, participants in a working memory training program demonstrated not only improvements in executive functions but also gains in their ability to manage daily living activities, such as organizing tasks or remembering appointments (Weng et al., 2019).

These results suggest that targeted training can influence both cognitive domains and practical, real-world abilities. However, other studies have painted a more complex picture. An 8-week working memory training program with middle-aged adults failed to produce significant transfer effects to untrained cognitive tasks or neural changes. This highlights a common issue in the field: while near transfer effects (improvements in tasks similar to those trained) are often observed, far transfer effects (gains in unrelated cognitive domains or daily functioning) remain elusive (Ripp et al., 2022). This discrepancy underscores the need to better understand the mechanisms that facilitate—or inhibit—transfer effects.

The inconsistency in findings is further reflected in meta-analyses and systematic reviews, which consistently report mixed results. While some studies show meaningful improvements in untrained areas, others reveal limited evidence for far transfer effects, suggesting that cognitive training benefits are often confined to the tasks practiced (Kelly et al., 2014; Zelinski, 2009). Literature deduces that while participants showed clear gains in tasks like memory recall or pattern recognition, these improvements did not extend to more complex domains like problem-solving or decision-making.

Understanding these transfer effects requires grounding in theoretical frameworks. The capacity-efficiency model provides one explanation, suggesting that cognitive training can either expand the brain's overall capacity or improve efficiency in how existing resources are used. However, evidence indicates that many observed improvements stem from enhanced efficiency—better organization, attention allocation, or strategy use—rather than an actual increase in capacity. This aligns with findings that training often produces task-specific gains without widespread generalization.

Another important framework is the transfer taxonomy, which distinguishes between near and far transfer based on the similarity between trained and tested tasks (Zelinski, 2009). Near transfer, such as improved performance on tasks closely related to the training (e.g., from one type of memory recall task to another), is more commonly observed because the skills required for both tasks overlap. Far transfer, on the other hand, demands that the cognitive processes learned in training extend to entirely different contexts—a much harder outcome to achieve.

The role of repetition and intensity is also critical in facilitating transfer effects. Studies show that frequent, intensive interventions—such as training five times a week for 5–7 weeks—are more likely to yield significant results. Shorter sessions (20–40 min) repeatedly appear to be more effective than fewer, longer sessions, as they allow participants to regularly reinforce newly acquired skills without overwhelming cognitive resources (Stojanoski et al., 2018; Weng et al., 2019). Additionally, programs that integrate training into real-life contexts, such as engaging participants in tasks resembling everyday challenges, hold greater promise for achieving meaningful transfer (Greenwood, 2003).

One area of optimism is multidomain cognitive training, which incorporates exercises targeting a variety of cognitive functions—such as memory, attention, and problem-solving—within a single program. By engaging multiple cognitive systems simultaneously, multidomain approaches may increase the likelihood of transfer effects. For example, a study combining working memory training with mindfulness exercises reported modest improvements in both cognitive performance and emotional regulation, suggesting that interventions engaging multiple domains can lead to more holistic benefits (Ben Izhak & Lavidor, 2023; Weng et al., 2019). While some studies demonstrate clear benefits in untrained domains, others suggest that gains are largely confined to the tasks practiced. This variability highlights the importance of repetition, theoretical grounding, and innovative program design to improve generalization. By addressing methodological challenges and exploring multidomain approaches, future research can pave the way for more effective, transferable cognitive training interventions that enhance both cognitive abilities and real-world functioning.

9.2.2.3 Adaptive Algorithms

Cognitive training has evolved significantly with the advent of computerized systems, particularly through the integration of adaptive algorithms. These systems adjust training tasks in real time based on an individual's performance, offering a personalized and more engaging approach to cognitive enhancement. Adaptive algorithms are designed to dynamically modify the difficulty of tasks, ensuring that participants remain challenged but not overwhelmed. This balance is critical for maximizing the effectiveness of training while maintaining motivation and adherence.

Recent advances in adaptive algorithms highlight their potential to revolutionize cognitive training. One example is the MAKSCog application, which incorporates a learning system that selects exercises tailored to the user's skill level. This adaptive approach ensures that training begins at an appropriate baseline, gradually increasing in complexity as the participant's performance improves. In studies using MAKSCog, individuals engaged in tasks that targeted multiple cognitive domains, with difficulty levels adjusted based on real-time feedback (Book et al., 2022). Such systems allow for a highly individualized training experience, making interventions more relevant and effective.

A key study involving adaptive computerized cognitive training (CCT) provided participants with a tablet-based application that used algorithms to tailor tasks across cognitive domains. The training started with simple exercises, progressing to more challenging tasks only when participants consistently met predefined accuracy thresholds. This progression ensures that individuals are neither under- nor overstimulated, a limitation often seen in traditional, non-adaptive cognitive training programs (Kong et al., 2024). Another comparative study demonstrated the superiority of adaptive CCT over a non-adaptive control condition, showing that personalized difficulty adjustments led to better engagement and improved cognitive outcomes (Book et al., 2022). These findings align with theories of cognitive plasticity, which emphasize the brain's ability to reorganize and adapt in response to training. Adaptive algorithms optimize this process by tailoring interventions to an individual's capabilities, ensuring that cognitive exercises remain challenging yet achievable. Theories of learning suggest that real-time feedback enhances self-awareness, helping participants understand their progress and encouraging sustained effort (Kong et al., 2024).

Adaptive systems also address a critical limitation of traditional cognitive training: the one-size-fits-all approach. By personalizing task difficulty, these algorithms make training more accessible to a diverse range of participants, including those with mild cognitive impairment (MCI). For instance, systematic reviews have shown that CCT can provide moderate cognitive benefits for individuals with MCI, but outcomes are often limited by a lack of personalization in task design. Adaptive algorithms mitigate this issue by adjusting the training to meet the unique needs of each participant (Book et al., 2022).

The role of repetition and intensity in adaptive systems is also noteworthy. Studies have found that frequent, short-duration sessions (e.g., 20–40 min, five times per week) are more effective than longer, less frequent sessions. Adaptive algorithms take advantage of these findings by optimizing task scheduling and progression, ensuring consistent cognitive stimulation without overloading the participant (Book et al., 2022; Kong et al., 2024). This approach not only improves cognitive outcomes but also enhances the overall user experience, encouraging long-term adherence to training programs. One of the most exciting applications of adaptive algorithms lies in their potential to promote real-world functionality. By targeting specific cognitive deficits and gradually expanding to broader domains, these systems can help participants translate training benefits into daily life. For example, the MAKSCog application has been shown to enhance attention, memory, and executive functions, which are critical for managing daily tasks such as remembering appointments or making financial decisions (Book et al., 2022; Kong et al., 2024).

Despite these promising developments, challenges remain in implementing adaptive algorithms on a broader scale. Designing algorithms that effectively balance complexity and accessibility requires a deep understanding of cognitive neuroscience and individual variability. Additionally, ensuring that these systems are user-friendly and accessible to individuals with varying levels of technological literacy is essential for maximizing their impact (Book et al., 2022; Interventions and Diagnostics – Neuroscape, 2022). By leveraging real-time feedback and dynamic task adjustments, these systems address the limitations of traditional programs and align with contemporary theories of learning and neuroplasticity. While further research is needed to refine these systems and explore their full potential, adaptive algorithms hold great promise for improving cognitive training outcomes and enhancing quality of life for diverse populations.

9.3 Technology-Enhanced Cognitive Training

Recent advancements in technology have transformed cognitive training, making it more effective and accessible. Key innovations include:

9.3.1 Gamification and Digital Platforms

The integration of gamification and digital platforms into cognitive training has transformed the way individuals engage with mental exercises. By incorporating game-like elements such as scoring systems, rewards, and challenges, these platforms enhance user engagement, motivation, and adherence to training programs. Applications like Elevate and Peak exemplify this trend, offering cognitive exercises with user-friendly interfaces and gamified features that make mental training more enjoyable and accessible (Bang et al., 2023; Chanda, 2024). At the heart of these platforms is the goal of creating a balance between challenge and enjoyment. The Mechanics, Dynamics, and Aesthetics (MDA) framework, often used in game design, has been adopted to develop engaging cognitive training tasks. By focusing on mechanics (rules and actions), dynamics (interactions and responses), and aesthetics (emotional experiences), developers aim to create games that not only entertain but also deliver measurable cognitive benefits. Tasks that reward users for completing puzzles or achieving streaks help sustain engagement while reinforcing skills like memory and attention (Khaleghi et al., 2021). Another foundational theory driving the success of gamification in cognitive training is Self-Determination Theory (SDT), which explains how intrinsic motivation is fostered through autonomy, competence, and relatedness. Gamified platforms like Peak capitalize on these principles by allowing users to choose their training focus, track their progress, and share achievements with friends. This sense of autonomy and accomplishment motivates users to stay consistent with their training, ultimately improving adherence and outcomes (Khaleghi et al., 2021; "Unconventional Practice Tools: Gamification and Brain Training Apps," 2024).

The Mobile Application Rating Scale (MARS) has become a common tool for assessing the quality of cognitive training apps, particularly in their ability to deliver meaningful cognitive engagement. Studies show that applications targeting multiple cognitive domains, such as attention, memory, and problem-solving, tend to achieve higher quality scores. However, simply addressing multiple domains does not guarantee effectiveness. The design and delivery of the tasks, as well as the user experience, play critical roles in determining their success. Real-world applications of gamified cognitive training have demonstrated promising results. Applications such as Constant Therapy, designed for individuals recovering from traumatic brain injury (TBI), uses personalized, gamified exercises to support rehabilitation. Studies have found that patients using Constant Therapy show improvements in cognitive functions such as attention and memory, highlighting the potential of gamified platforms in clinical settings. Similarly, Peak Brain Training has been shown to improve cognitive performance in healthy adults, further showcasing the potential of gamified platforms for general cognitive enhancement (Bang et al., 2023).

Despite these successes, the evidence for the overall effectiveness of gamified cognitive training remains mixed. While many users report subjective improvements and enjoy the gamified experience, meta-analyses reveal variability in measurable outcomes. For instance, near transfer effects, where improvements are observed in tasks similar to those practiced, are more common than far transfer effects, which involve gains in unrelated cognitive domains or daily functioning (Bang et al., 2023; Sardi et al., 2017). This suggests that while gamification can boost engagement, its ability to drive broad cognitive changes requires further investigation. As gamified platforms grow in popularity, ethical considerations also come to the forefront. Many apps track user performance to provide personalized feedback and adaptive task difficulty, but this data collection raises concerns about privacy and consent. Developers must ensure that users are fully informed about how their data will be used and implement robust safeguards to protect sensitive information (Khaleghi et al., 2021).

The accessibility of gamified cognitive training is another area of promise. Mobile platforms allow users to train anytime and anywhere, democratizing access to cognitive enhancement tools. This is particularly beneficial for populations with limited access to traditional therapeutic interventions, such as individuals in remote areas or those with mobility challenges. Moreover, as digital therapeutics continue to evolve, there is growing interest in incorporating these platforms into formal clinical protocols for managing cognitive impairments (Bang et al., 2023; Shah, 2023). Gamification and digital platforms have transformed cognitive training by making it more engaging, accessible, and user-centered. Through frameworks like MDA and theories such as SDT, developers have created tools that motivate users to stick with training while delivering measurable cognitive benefits. However, the field faces challenges in demonstrating consistent effectiveness and addressing ethical concerns. As research continues and technologies advance, gamified cognitive training has the potential to become an integral part of cognitive enhancement and rehabilitation strategies, empowering users to achieve their mental fitness goals in an enjoyable and meaningful way.

9.3.2 Artificial Intelligence (AI) and Personalization

Artificial intelligence (AI) is transforming the landscape of cognitive training for Attention Deficit Hyperactivity Disorder (ADHD), offering innovative and personalized approaches to treatment. By leveraging advanced technologies like augmented reality (AR), virtual reality (VR), and machine learning, AI-driven cognitive training programs are addressing key challenges in ADHD management, such as improving attention, impulse control, and working memory. These technologies provide real-time feedback and adapt to individual performance, creating dynamic and engaging training environments that are tailored to the unique needs of each participant. Recent advancements highlight the promise of AI in ADHD interventions. Applications like CogniFit, BrainBeat, NeuroPlus, and Lumosity integrate real-time performance monitoring and feedback, enhancing user motivation and treatment effectiveness. These apps use adaptive algorithms to adjust task difficulty, ensuring that exercises remain challenging yet achievable. NeuroPlus incorporates VR tasks that train attention and self-regulation, while BrainBeat uses rhythm-based exercises to improve focus and cognitive control (Frolli et al., 2024). Such tools not only provide targeted cognitive training but also foster engagement through interactive and gamified experiences.

AI is also revolutionizing ADHD diagnosis. Traditional diagnostic methods often rely on subjective assessments, but AI-powered tools offer more precise and data-driven approaches. For instance, systems like QbTest, Adeo, NeuroElectrics, and Mindstrong Health analyze data from neuropsychological, behavioral, and neurophysiological tests to identify complex patterns associated with ADHD. These tools leverage machine learning algorithms to detect subtle markers of the disorder, facilitating early and accurate diagnoses (Frolli et al., 2024; Stevens et al., 2016). This precision allows clinicians to tailor interventions more effectively, addressing the specific needs of each individual. Among the most notable advancements is EndeavorRx, a cognitive training game developed by Akili Interactive that became the first FDA-approved treatment for ADHD in children. EndeavorRx combines attention and multitasking tasks within an engaging gaming environment, providing a non-pharmaceutical intervention to improve cognitive abilities. Studies on EndeavorRx have shown that children who engage with the game demonstrate significant improvements in attention and impulse control, underscoring the potential of AI-driven gamification in therapeutic contexts (Frolli et al., 2024; Somma et al., 2019).

The integration of AI in cognitive training aligns with theories of cognitive plasticity, which emphasize the brain's ability to reorganize and adapt in response to targeted interventions. By tailoring exercises to individual capabilities and providing real-time feedback, AI systems optimize learning experiences and promote meaningful cognitive improvements. For instance, VR environments that simulate real-world scenarios allow users to practice managing distractions and maintaining focus, translating these skills into daily life (Cao et al., 2023; Frolli et al., 2024). This personalized approach is particularly valuable for children with ADHD, who often face unique challenges that require individualized solutions. However, the adoption of AI-driven cognitive training is not without challenges. Accessibility remains a significant concern, as not all families have access to the technology required for these interventions. Ensuring equitable distribution of AI tools is critical to avoid exacerbating existing disparities in ADHD care.

Data privacy and security are paramount, as AI systems often collect and analyze sensitive personal information. Developers must implement robust safeguards and obtain explicit consent from users to address these ethical concerns (Frolli et al., 2024; Medina et al., 2021). The integration of AI into ADHD treatment represents a paradigm shift, combining the precision of machine learning with the adaptability of cognitive training. By personalizing interventions and incorporating engaging technologies, AI has the potential to significantly improve the quality of life for individuals with ADHD. While further research is needed to explore the long-term effects and refine implementation strategies, the initial findings are promising. As the field continues to evolve, AI-driven cognitive training is poised to become a cornerstone of ADHD management, offering hope and innovation to families and clinicians alike.

9.3.3 Virtual Reality (VR) and Immersive Environments

Virtual reality (VR) has emerged as a powerful tool in cognitive training, offering dynamic, realistic, and immersive environments that can replicate real-world scenarios with high precision. In particular, VR-based interventions have shown significant promise in enhancing spatial memory and navigation abilities, providing innovative solutions for individuals with cognitive impairments due to conditions such as Alzheimer's disease, mild cognitive impairment, or traumatic brain injuries. By leveraging VR's unique capabilities, cognitive training programs can engage users in interactive and controlled environments, fostering both learning and rehabilitation. Research demonstrates that VR training significantly improves spatial memory and navigation skills. Studies have shown that active exploration of virtual environments—where participants can navigate freely rather than passively observing—plays a crucial role in acquiring spatial knowledge. This learning process involves multiple components, such as motor orders, proprioceptive and vestibular feedback, attention allocation, cognitive decision-making, and mental manipulation of spatial information. These elements work together to help users develop an internal representation of the virtual space, which can translate to better navigation and orientation abilities in real-world settings (Calderone et al., 2024; Montana et al., 2019).

Key design features in VR systems, such as visual cueing, further enhance spatial learning. For instance, the use of visual aids like arrows or maps within the virtual environment has been shown to significantly improve navigational performance, particularly for individuals with spatial memory deficits. A large visual arrow that synchronizes allocentric (map-based) and egocentric (self-centered) perspectives can help users maintain cognitive alignment between their location and the broader spatial layout. Additionally, interactive aerial views within VR environments facilitate the retrieval of stored spatial layouts, offering comprehensive information about the user's heading and surroundings. These features are especially beneficial for patients with Alzheimer's disease or mild cognitive impairment, helping them navigate complex environments more effectively (Belger et al., 2019; Montana et al., 2019). One of the most significant advantages of VR is its ability to promote neuroplasticity—the brain's ability to reorganize and adapt. Through carefully designed navigational tasks, VR training stimulates brain plasticity mechanisms by engaging neurotransmitter systems and neural pathways associated with spatial memory. This is particularly relevant in neurorehabilitation, where repeated exposure to immersive VR environments can support recovery in individuals with traumatic brain injuries. Evidence suggests that shorter intervals between the onset of brain injury and the initiation of VR-based rehabilitation are associated with better outcomes, emphasizing the importance of timely intervention (De Luca et al., 2023; Montana et al., 2019).

However, the effectiveness of VR training depends on several factors, including the mode of exploration and the structure of the training protocol. Active exploration—where participants interact with the environment and make decisions—has been found to be more effective than passive observation in facilitating spatial learning. Additionally, the duration of training plays a critical role; short training sessions (less than 4 h) may be insufficient to induce meaningful cognitive changes, highlighting the need for well-structured, long-term interventions. These findings underscore the importance of tailoring VR-based training protocols to the specific needs of participants (Belger et al., 2019). Despite the growing evidence supporting VR's effectiveness, there are challenges and limitations that must be addressed.

Comparing exploration modes between immersive and non-immersive VR systems can affect training outcomes. While immersive systems offer greater engagement and realism, they may not always be accessible or feasible for all users. Non-immersive systems, on the other hand, are more widely available but may lack the depth and interactivity of fully immersive experiences. Researchers must continue to explore ways to balance accessibility with the richness of the training environment to maximize the benefits of VR-based cognitive training (Belger et al., 2019).

VR's potential for cognitive rehabilitation extends beyond individual benefits. It provides researchers and clinicians with high experimental control, allowing them to manipulate variables and assess outcomes with precision. This makes VR a valuable tool not only for training but also for understanding the underlying mechanisms of spatial memory and navigation. By integrating these insights into practice, VR can help advance the field of neurorehabilitation and cognitive enhancement (Catania et al., 2023; Montana et al., 2019). VR-based cognitive training offers a promising approach for enhancing spatial memory and navigation abilities. Its ability to create realistic, interactive, and controlled environments enables users to develop critical cognitive skills in a safe and engaging manner. While challenges remain in optimizing protocols and addressing accessibility, the potential for VR to facilitate neuroplasticity and support recovery is undeniable. As the technology continues to evolve, VR is poised to play an increasingly central role in cognitive rehabilitation and enhancement strategies, particularly for individuals with spatial memory impairments or traumatic brain injuries.

9.3.4 Wearable Technology

Wearable technologies, particularly electroencephalogram (EEG) headsets, have gained significant traction as tools for cognitive training and remote health monitoring. These devices offer the ability to measure brain activity in real-time, enabling interventions that are both precise and accessible outside traditional clinical settings. By harnessing EEG technology, cognitive training programs can provide personalized neurofeedback, promote mindfulness, and address symptoms of various neurological and mental health disorders.

EEG headsets have been particularly effective in neurofeedback interventions, which train individuals to regulate their brain activity by providing real-time feedback on neural signals. For instance, these devices have been used to manage conditions such as ADHD and PTSD. In these cases, neurofeedback sessions involve monitoring specific EEG frequency bands associated with attention or emotional regulation, guiding participants to alter their brain activity patterns toward desired states. This targeted approach allows for tailored training, which has been shown to improve focus and reduce symptoms of hyperactivity or stress in individuals with these disorders (Emish & Young, 2024). The potential for wearable EEG devices extends beyond clinical interventions. They are increasingly being used for cognitive training in areas such as mindfulness, sleep quality, and general cognitive performance. EEG neurofeedback training has demonstrated benefits in enhancing mindfulness by helping participants achieve states of calm and focus. Similarly, wearable EEG devices have shown promise in improving sleep quality by tracking brainwave activity during different sleep stages and providing tailored feedback to encourage healthier sleep patterns (du Bois et al., 2021; Emish & Young, 2024).

A notable advantage of wearable EEG technology is its ability to make brain-based therapies accessible outside traditional clinical environments. Devices such as the DSI-24 headset and Emotiv EEG systems allow users to engage in cognitive training at home or on the go. This accessibility is particularly important for individuals in remote or underserved areas, enabling them to benefit from advanced neurofeedback interventions without requiring frequent visits to specialized clinics (Emish & Young, 2024; Sugden et al., 2023). While the promise of EEG-based cognitive training is clear, there are challenges that must be addressed to maximize its effectiveness. Signal quality remains one of the most significant hurdles. EEG signals are highly sensitive to noise from physical movements, external electrical interference, and other environmental factors. These issues are particularly pronounced in real-world settings, where conditions are less controlled than in clinical environments. To address this, rigorous signal processing techniques and validation strategies are essential to ensure the reliability of the data collected by wearable EEG devices (*DSI-24*, n.d.; Emish & Young, 2024).

Combining EEG technology with other neuroimaging techniques, such as functional Near-Infrared Spectroscopy (fNIRS), presents an exciting opportunity to overcome some of these challenges. While EEG offers high temporal resolution, allowing for precise tracking of brain activity over time, fNIRS provides high spatial resolution, capturing detailed information about brain regions involved in specific tasks. Together, these complementary modalities can create a more comprehensive picture of cognitive processes, enhancing the effectiveness of cognitive training programs (Emish & Young, 2024). Ethical considerations are another important aspect of wearable EEG technology. Given that these devices collect sensitive neural data, ensuring data privacy and security is paramount. Users must be fully informed about how their data will be used, and developers must implement robust safeguards to protect this information. Additionally, making these devices affordable and user-friendly is essential for ensuring equitable access, particularly for populations with limited resources (Cunnington, 2019; Flanagan & Saikia, 2023).

Despite these challenges, wearable EEG technology has demonstrated significant potential in enhancing cognitive training and supporting remote monitoring of health conditions. Remote neurofeedback programs that leverage wearable EEG devices have shown promise in managing chronic pain, tracking mental health conditions, and improving emotional regulation. These interventions rely on the real-time feedback provided by EEG devices to guide participants toward healthier neural activity patterns, demonstrating the versatility of this technology across various applications (du Bois et al., 2021; Emish & Young, 2024). EEG technology represents a groundbreaking advancement in cognitive training, offering accessible and personalized interventions for a wide range of applications. By addressing challenges related to signal quality, data privacy, and usability, these devices have the potential to transform how cognitive training and neurofeedback therapies are delivered. As the field continues to evolve, wearable EEG systems are poised to play a central role in promoting cognitive health and well-being, empowering individuals to take control of their mental fitness in innovative and impactful ways.

9.4 Applications of Cognitive Training Programs

9.4.1 Cognitive Aging

Cognitive aging is a natural process that can lead to declines in memory, reasoning, and processing speed, impacting the ability of older adults to maintain independence. However, research has shown that cognitive training can play a pivotal role in mitigating these effects, enhancing cognitive reserve, and even delaying the onset of dementia. One of the most influential studies in this field is the Advanced Cognitive Training for Independent and Vital Elderly (ACTIVE) study, which has provided compelling evidence of the long-term benefits of cognitive interventions for older adults. The ACTIVE study, funded by the National Institute on Aging and the National Institute for Nursing Research, was designed in response to growing concerns about the healthcare burden associated with aging populations. Initiated in 1996 across six field centers in the United States, the study targeted abilities that tend to decline with age, starting in the mid-sixties. The primary goal was to evaluate the effectiveness of three cognitive training interventions—memory, reasoning, and visual speed of processing—in preserving cognitive health and functional independence in older adults (Rebok et al., 2014; Tennstedt & Unverzagt, 2013).

The study's conceptual model was grounded in the hypothesis that domain-specific cognitive training could lead to both targeted and generalized improvements. For instance, memory training was expected to enhance recall, reasoning training to improve problem-solving, and visual speed-of-processing training to support quicker decision-making in everyday contexts. These improvements were anticipated to transfer to real-world functional outcomes, such as managing finances, driving safely, and maintaining personal health. By linking cognitive training to practical abilities, the ACTIVE study sought to address a key gap in previous research, which often failed to demonstrate such real-world applications (*Researchers Identify First Brain Training Exercise Positively Linked to Dementia Prevention*, 2017; Tennstedt & Unverzagt, 2013).

The findings from the ACTIVE study were groundbreaking. Participants who underwent cognitive training demonstrated significant improvements in the targeted cognitive domains, with some benefits persisting for years after the intervention. The speed-of-processing training not only improved reaction times but also reduced the risk of driving accidents—a critical factor for maintaining independence in later life. Similarly, reasoning training led to better problem-solving in everyday tasks, while memory training improved the ability to recall information critical for daily functioning. One of the study's most compelling contributions was its long-term follow-up, which revealed that cognitive training could delay cognitive decline and reduce dependency in older adults. For instance, participants who received speed-of-processing training were found to have a reduced risk of developing dementia years later compared to those in the control group. This aligns with the broader understanding that enhancing cognitive reserve—an individual's capacity to withstand age-related brain changes—can delay the clinical onset of cognitive impairments. These findings highlight the potential of targeted cognitive interventions to promote healthier cognitive aging (Budson, 2021; Hara, 2017; Tennstedt & Unverzagt, 2013).

The ACTIVE study also underscored the importance of tailoring cognitive training programs to individual needs. Each intervention focused on specific cognitive domains, but the training was flexible enough to accommodate variations in participants' baseline abilities and learning paces. This individualized approach is crucial, as cognitive aging manifests differently across individuals. By addressing specific areas of decline while also encouraging general cognitive engagement, the study demonstrated that cognitive training could be both effective and accessible. While the ACTIVE study provided valuable insights, it also pointed to areas for future research. While the interventions were effective in slowing cognitive decline, the extent to which these gains could be sustained or enhanced through ongoing training remains an open question. Additionally, integrating cognitive training with other interventions, such as physical activity or social engagement, may yield even greater benefits. Research has shown that combining cognitive and physical activities can have synergistic effects, further enhancing cognitive reserve and overall quality of life (Miller, 2017; Tennstedt & Unverzagt, 2013).

The implications of the ACTIVE study extend beyond its research findings. By demonstrating the feasibility and benefits of cognitive training for older adults, the study has influenced the development of public health interventions aimed at promoting cognitive health in aging populations. Programs inspired by the ACTIVE study are now being implemented in community centers and healthcare facilities, providing older adults with tools to maintain their cognitive health and independence. Furthermore, the study has highlighted the importance of early intervention, suggesting that engaging in cognitive training before significant declines occur can maximize benefits (Guzick, 2016; Tennstedt & Unverzagt, 2013). The on-going insights from the study can be applied by researchers and clinicians to develop even more effective strategies for supporting cognitive health in aging populations.

9.4.2 Neurological Rehabilitation

Cognitive impairments are a common consequence of strokes, affecting domains like attention, memory, and executive functions. These deficits often hinder recovery and reduce quality of life for stroke survivors. To address these challenges, computer-based cognitive training has emerged as a promising tool, offering targeted interventions that can complement traditional therapies. By leveraging principles of neuroplasticity—the brain's ability to reorganize and adapt—computer-based attention retraining aims to support recovery and promote functional independence in post-stroke patients. Studies have consistently highlighted the effectiveness of computer-based interventions in improving cognitive functions for stroke survivors. For instance, platforms like BrainHQ and RehaCom offer structured exercises designed to enhance attention, memory, and other cognitive abilities. These programs often include tasks that challenge sustained, selective, divided, and shifting attention, enabling patients to gradually rebuild their cognitive capacities. Exercises on these platforms are scientifically tailored to adapt to individual performance, ensuring that participants are neither overwhelmed nor under-challenged (Fava-Felix et al., 2022; Maggio et al., 2023).

Beyond traditional software, innovative tools like the Reh@Task platform have incorporated virtual reality (VR) for dual training of upper limb mobility and cognitive functions. These immersive environments allow patients to practice real-world tasks in a controlled setting, enhancing both motor and cognitive recovery. Similarly, brain–computer interfaces (BCIs) and multimodal stimulation approaches have shown promise in delivering comprehensive rehabilitation that integrates physical and cognitive therapy. BCIs use neural feedback to guide patients in retraining their cognitive abilities, fostering a deeper connection between mental effort and tangible progress (De Luca et al., 2023; Fava-Felix et al., 2022).

Computer-based attention retraining also addresses barriers to care, such as geographical or socioeconomic constraints. Home-based platforms allow patients to engage in cognitive rehabilitation without needing frequent visits to clinical facilities. This accessibility not only reduces healthcare disparities but also enables continuity of care, which is crucial for long-term recovery. A patient recovering in a rural area can access programs like BrainHQ from home, maintaining their cognitive rehabilitation regimen without the need for costly or logistically challenging travel (Jung et al., 2021; Maggio et al., 2023).

The benefits of these interventions are supported by theories of neuroplasticity, which suggest that targeted cognitive training can stimulate recovery in damaged brain areas. Structured and repetitive tasks help strengthen neural pathways associated with attention and other cognitive domains, creating opportunities for compensatory mechanisms to emerge. Combining cognitive training with other therapies, such as occupational therapy and aerobic exercise, has been shown to enhance overall outcomes. By integrating physical activity into cognitive rehabilitation not only supports general health but also promotes better neural connectivity, amplifying the benefits of attention retraining (Fava-Felix et al., 2022; Maggio et al., 2023).

Despite its promise, computer-based cognitive training is not without limitations. While many studies report improvements in trained cognitive domains, the transfer of these gains to daily functioning remains inconsistent. A patient might show significant improvement in selective attention tasks but struggle to apply this progress to complex, real-world activities like managing finances or returning to work. Furthermore, the optimal duration and frequency of training remain areas of ongoing research. Current evidence suggests that programs lasting 4–12 weeks with multiple sessions per week yield the most significant benefits, but more work is needed to refine these protocols (Maggio et al., 2023; van de Ven et al., 2015).

The accessibility and cost-effectiveness of computer-based interventions also raise important considerations for widespread adoption. While home-based platforms reduce logistical barriers, ensuring these technologies are affordable and user-friendly is essential. Programs must be designed with the needs of diverse populations in mind, accounting for variations in digital literacy, socioeconomic status, and physical or cognitive abilities. Addressing these factors will be crucial for maximizing the impact of computer-based cognitive training on a global scale (Fava-Felix et al., 2022; Maggio et al., 2023).

9.4.3 Pediatric and Educational Contexts

Cognitive training has emerged as a valuable tool for supporting children with learning disabilities and Attention Deficit Hyperactivity Disorder (ADHD). By targeting specific cognitive domains such as attention, working memory, and executive functions, these interventions aim to enhance academic performance, reduce cognitive challenges, and foster greater independence in children with diverse learning needs. Research highlights the potential of integrating cognitive training into both clinical and educational settings, offering a complementary approach to traditional teaching methods. Numerous studies have demonstrated the efficacy of cognitive training for children with learning disabilities or ADHD. For instance, interventions focused on improving working memory have led to notable gains in cognitive flexibility, processing speed, and overall academic performance. Children who completed working memory training protocols showed improvements in attention and enhanced capabilities in reading and mathematics. These findings underscore the importance of targeted cognitive exercises in addressing specific learning deficits (Looney et al., 2024; Wiest et al., 2022).

Programs like BrainHQ and RehaCom have gained attention for their effectiveness in enhancing cognitive abilities among children with learning differences. These computer-based platforms provide structured, interactive tasks that adapt to the learner's progress, ensuring that training remains both engaging and challenging. Additionally, custom-designed interventions often target specific aspects of attention—such as sustained, selective, divided, and shifting attention—helping children develop better focus and self-regulation (Looney et al., 2024; Veloso et al., 2020).

Classroom-based implementations of cognitive training have shown significant promise. A study conducted with students in grades three through six integrated cognitive training into the school day, resulting in improved working memory, cognitive flexibility, and processing speed. By embedding these interventions into the academic schedule, the study demonstrated that cognitive training could be seamlessly incorporated into educational settings without disrupting regular curricula. Such models serve as a blueprint for how cognitive training can be scaled to benefit larger student populations (Looney et al., 2024; Moore & Ledbetter, 2019). The effectiveness of cognitive training aligns with theories of neuroplasticity, which emphasize the brain's ability to adapt and reorganize in response to targeted interventions. These programs stimulate neural pathways associated with attention and memory, promoting both cognitive development and recovery of impaired brain areas. Combining cognitive training with other educational strategies, such as personalized learning plans and teacher-led interventions, can further enhance outcomes. A study obtained that integrating cognitive exercises with reading-focused activities significantly boosted reading comprehension and retention in children with dyslexia (Studer-Luethi et al., 2022; Wiest et al., 2022).

While evidence supports the effectiveness of cognitive training, researchers continue to explore optimal protocols, including the duration, frequency, and intensity of interventions. Programs lasting between 4 and 12 weeks with multiple weekly sessions have shown the most promising results. However, the variability in outcomes suggests that individualized approaches may be necessary to address the diverse needs of children with learning disabilities or ADHD. Understanding how specific task features and participant characteristics influence training outcomes remains a critical area of investigation (Rapport et al., 2013; Studer-Luethi et al., 2022). Accessibility and cost-effectiveness are key considerations for the long-term success of cognitive training programs in educational contexts. While classroom-integrated programs reduce logistical barriers, the availability of resources such as computers, software, and trained facilitators can impact implementation. Developing affordable, user-friendly tools that align with diverse educational systems is essential for ensuring widespread adoption. Moreover, ensuring that interventions are inclusive and tailored to children from varying socioeconomic backgrounds can help address disparities in educational access (Looney et al., 2024; Roording-Ragetlie et al., 2016). The potential of cognitive training extends beyond the classroom. Home-based programs provide an opportunity for parents to support their children's cognitive development outside of school. Apps and online platforms that offer guided exercises in attention and memory can serve as a supplemental tool for children with ADHD or learning disabilities. These tools empower families to take an active role in their children's cognitive development, reinforcing the skills learned through school-based or clinical interventions (Looney et al., 2024; Rao, 2018).

9.5 Challenges

Cognitive training holds immense potential to improve mental capabilities across various populations, but its practical application faces significant challenges. One of the primary obstacles is the variability in outcomes across individuals. Studies reveal that factors such as self-reported brain health, cognitive failures, grit, ambition, and even socioeconomic status have limited predictive value for determining participants' commitment to or persistence with cognitive training. While education level shows some association with commitment, its influence remains inconsistent across diverse groups, suggesting that adherence to cognitive training may depend more on external factors than intrinsic participant characteristics. This variability underscores the need for a more nuanced understanding of what drives successful engagement and outcomes in cognitive training programs (Chan et al., 2024; Tullo et al., 2023).

Access to high-quality cognitive training programs presents another significant hurdle, particularly in low-resource settings. While not explicitly covered in all studies, the unequal distribution of cognitive training resources creates barriers for underserved communities, limiting their ability to benefit from these interventions. This disparity highlights the importance of developing cost-effective and scalable training solutions that can reach diverse populations. Bridging this gap is critical for ensuring equitable access to the cognitive and functional benefits these programs offer (Chan et al., 2024).

Ensuring long-term adherence to cognitive training protocols is perhaps one of the most complex challenges. Motivation plays a pivotal role in determining whether participants stick with training regimens, and various strategies have been employed to enhance it. Gamification, as seen in platforms like Lumosity and Cogmed, has shown promise in engaging users by making training enjoyable and rewarding. Offering participants a variety of training tasks and allowing them to choose their preferred paradigms has also been found to increase engagement. Personalized coaching, another approach used in programs like Cogmed, provides tailored support and encouragement, further boosting adherence. However, studies indicate that simply adhering to a schedule, such as training at regular intervals, does not necessarily correlate with improved performance or sustained engagement. This finding suggests that motivation requires more than routine; it needs to be intrinsic, supported by strategies that resonate with individual participants (He et al., 2023; Hoorelbeke et al., 2022).

The inconsistency in performance trajectories across participants also reflects a broader challenge in cognitive training: the disconnect between perceived capabilities and actual needs. For instance, participants may disengage if they feel the training tasks are too easy or too difficult, highlighting the importance of designing adaptive training programs that dynamically adjust to individual skill levels. Lessons from other fields, such as behavioral therapy and online cognitive behavioral therapy, emphasize the significance of addressing optimal levels of challenge and motivation to enhance adherence. Translating these insights into cognitive training could pave the way for more effective interventions (Singh et al., 2024; Tullo et al., 2023).

Another layer of complexity lies in understanding why cognitive training yields inconsistent benefits. Meta-analyses and systematic reviews have revealed that baseline cognitive abilities often interact with training outcomes, influencing the degree of improvement observed. This interplay suggests that a one-size-fits-all approach is unlikely to work. Instead, interventions need to be tailored based on participants' initial cognitive profiles, focusing on targeted improvements that align with their specific needs. Such an approach not only maximizes the potential for meaningful gains but also addresses the heterogeneity in training effects across populations (Trenorden et al., 2022). The cost and scalability of cognitive training programs remain critical concerns. While digital platforms have made these interventions more accessible, there is still a need to ensure that they are affordable and usable across diverse socioeconomic groups. Programs designed with inclusivity in mind—both in terms of cost and cultural relevance—are essential for reaching broader audiences. Advancing the science of cognitive training requires robust methodologies to assess not only the immediate outcomes of these programs but also their long-term impacts on real-world functioning (Ball et al., 2002).

9.6 Ethical Considerations

As cognitive training and AI-driven interventions continue to evolve, ethical considerations are becoming increasingly critical to ensure responsible development and implementation. These considerations encompass the integrity of claims about program effectiveness, the safeguarding of user data, and the equitable distribution of these transformative technologies. A primary concern in the field of cognitive training is avoiding overhyped claims about program effectiveness. Although not directly addressed in the search results, this issue has long plagued both academic research and commercial applications. To uphold scientific integrity, researchers and developers must adhere to rigorous standards, including conducting peer-reviewed studies, transparently disclosing limitations and potential biases, and providing detailed information about the demographics, sample sizes, and outcome measures of their research. Overstating the capabilities of cognitive training programs can lead to public mistrust, misplaced investments, and ethical questions about accountability.

Data privacy and security are critical in digital tools, particularly those powered by artificial intelligence (AI) and machine learning (ML). Cognitive training platforms often collect sensitive user data to provide personalized interventions, but this data must be handled responsibly. Studies emphasize the importance of safeguarding patient confidentiality, implementing robust cybersecurity measures to prevent unauthorized access, and obtaining informed consent for data collection and usage. This is especially crucial as AI technologies in cognitive training increasingly integrate user data to refine algorithms and optimize outcomes. Another ethical imperative is ensuring equitable access to cognitive training and AI-driven interventions. Digital tools hold the promise of democratizing cognitive healthcare, but they also risk exacerbating existing inequalities. To address this, developers must prioritize affordability and accessibility, particularly for underserved populations. For instance, offering low-cost or subsidized programs can help bridge the digital divide. Moreover, ensuring these tools are user-friendly and accessible to individuals with varying levels of technological literacy is critical. Without these considerations, cognitive training programs risk excluding vulnerable populations, undermining their broader social benefits.

Transparency in algorithm development and decision-making processes is another ethical consideration. Users and stakeholders must be able to trust the systems they interact with, and this trust is built on clear explanations of how algorithms work and how decisions are made. For example, developers should openly communicate how data is used to personalize interventions or how biases in datasets are mitigated. This transparency extends to addressing potential algorithmic biases that could inadvertently disadvantage certain groups. By proactively identifying and correcting these biases, developers can create more inclusive and fair cognitive training systems. Cultural sensitivity in training content is another crucial factor. Cognitive training programs should be designed with an awareness of the diverse cultural contexts in which they will be used. This includes considering language differences, cultural norms, and region-specific needs to ensure that interventions are relevant and effective for all users. Ignoring these factors could lead to tools that are less effective or even culturally inappropriate, limiting their impact and uptake.

Another ethical dimension is the potential for dependency on cognitive training technologies. While these tools offer significant benefits, they could inadvertently lead to over-reliance, with users prioritizing digital interventions over traditional cognitive activities or social engagement. Developers and clinicians must strike a balance, ensuring that cognitive training complements, rather than replaces, other meaningful activities and supports broader cognitive development goals. The use of cognitive training data also raises questions about balancing individual rights with public health interests. For example, data collected from large-scale cognitive training programs could be invaluable for understanding cognitive health trends and informing public health strategies. However, such uses must be weighed against the need to protect individual privacy and autonomy. Robust frameworks for data anonymization and ethical use must be established to ensure that public health benefits do not come at the expense of individual rights.

9.7 Future Directions

The field of cognitive training is entering an era of rapid innovation, driven by technological advances, emerging neuroscientific insights, and a growing understanding of personalized learning. These developments promise to revolutionize how cognitive abilities are assessed, enhanced, and sustained, setting the stage for interventions that are not only more effective but also more accessible and engaging. One of the most promising advancements lies in the integration of AI-powered feedback systems. Artificial intelligence (AI) can process vast amounts of cognitive data in real time, offering unprecedented levels of personalization. For example, AI can dynamically adjust the difficulty of tasks based on an individual's performance, ensuring that each session remains optimally challenging and engaging. By leveraging predictive analytics, AI can also forecast potential cognitive challenges, proactively tailoring training protocols to meet the user's evolving needs. These capabilities are already being explored in applications such as dynamic task adaptation for working memory and attention training, demonstrating the potential to significantly improve outcomes (Elendu et al., 2023; Harishbhai Tilala et al., 2024).

Equally transformative is the emergence of neuroadaptive training systems, which integrate neural data directly into the training process. By using real-time brain signals—captured through technologies like functional Near-Infrared Spectroscopy (fNIRS)—these systems create a closed-loop learning environment. The system continuously learns from the user's brain activity, adjusting the training parameters in real time to optimize engagement and efficacy. Neuroadaptive methods are particularly promising for interventions requiring precise targeting, such as those for attention disorders or post-stroke rehabilitation. For example, in flight simulators, fNIRS-based neuroadaptive training has shown remarkable success in enhancing cognitive performance by adapting tasks to the user's cognitive state (*Artificial Intelligence and Privacy – Issues and Challenges*, n.d.; Nguyen et al., 2023).

Social dimensions of cognitive training are also gaining attention, with group-based cognitive training emerging as an innovative approach. These interventions leverage the benefits of social interaction to enhance cognitive outcomes. In a group setting, participants can engage in collaborative problem-solving, share strategies, and provide mutual support, fostering a sense of community and motivation. Shared cognitive goals and peer-to-peer learning create a dynamic environment that enhances the training experience, particularly for older adults and individuals recovering from neurological injuries. Research suggests that social engagement itself has cognitive benefits, making group-based training a valuable complement to individual-focused interventions (Al-Zahrani, 2024; Harishbhai Tilala et al., 2024).

Looking ahead, longitudinal research will play a crucial role in understanding the long-term impacts of cognitive training. Extended follow-up studies are needed to assess whether the improvements achieved during training persist over time and translate into meaningful real-world outcomes. For instance, tracking participants over years can provide valuable insights into how cognitive training affects daily functioning, independence, and quality of life. Such research could also help refine training protocols by identifying which methods yield the most durable benefits. Comparative analyses across various training paradigms will enable researchers to determine the relative effectiveness of different approaches, offering evidence-based guidance for practitioners and policymakers (*Artificial Intelligence and Privacy – Issues and Challenges,* n.d.; Harishbhai Tilala et al., 2024).

Another promising direction involves the integration of multiple modalities to create holistic cognitive training programs. Combining cognitive exercises with physical activity, nutrition counseling, and stress management has the potential to amplify the benefits of training by addressing multiple aspects of brain health simultaneously. For example, studies have shown that physical exercise can enhance neuroplasticity, suggesting that combining it with cognitive training could lead to synergistic effects. Similarly, personalized biomarker profiles, which take into account genetic and biological markers, could help tailor interventions to the unique needs of each individual, further enhancing their effectiveness.

Technological advancements such as virtual and augmented reality (VR/AR) are also reshaping the landscape of cognitive training. Immersive VR environments allow users to engage in highly interactive and realistic scenarios, making training both engaging and effective. For instance, VR has been successfully used for spatial memory and navigation training, particularly in individuals with cognitive impairments. As VR and AR technologies continue to evolve, their applications in cognitive training are likely to expand, offering new ways to simulate complex, real-world tasks in safe and controlled settings (Nguyen et al., 2023). Ethical considerations will remain central as cognitive training technologies advance. Ensuring equitable access to AI-driven tools is essential to prevent widening disparities in cognitive healthcare. Transparent algorithm development and efforts to mitigate biases are critical to maintaining public trust. Furthermore, as cognitive training becomes increasingly reliant on data-driven systems, safeguarding user privacy and security will be paramount. Emerging technologies like blockchain and decentralized AI offer promising solutions for creating secure, transparent, and accessible platforms.

9.8 Conclusion

Cognitive training programs have transformed from basic exercises into highly sophisticated interventions that improve lives across diverse populations. From children with ADHD to older adults experiencing cognitive decline, these programs leverage the brain's incredible ability to adapt and grow through targeted, repetitive exercises. While challenges remain in ensuring equitable access and consistent outcomes, advances in technology—such as AI, virtual reality, and wearable devices—have made cognitive training more personalized, engaging, and effective. The applications of cognitive training are vast, from helping students improve focus and memory in classrooms to aiding stroke survivors in their recovery. Programs like the ACTIVE study show the long-term benefits of cognitive interventions for aging adults, while school and home-based programs demonstrate how training can be seamlessly integrated into daily life. However, ethical considerations like data privacy, transparency, and accessibility must guide future development to ensure these innovations benefit everyone. The future of cognitive training is promising. As AI and neuroscience continue to evolve, interventions will become even more adaptive and impactful, offering tailored solutions for individuals' unique needs. While much remains to be explored, cognitive training already stands as a powerful tool for improving mental abilities, fostering independence, and enhancing quality of life at any stage of life. With thoughtful innovation and a focus on inclusivity, its potential is boundless. Ethical considerations will remain central as cognitive training technologies advance. Ensuring equitable access to AI-driven tools is essential to prevent widening disparities in cognitive healthcare. Transparent algorithm development and efforts to mitigate biases are critical to maintaining public trust. Furthermore, as cognitive training becomes increasingly reliant on data-driven systems, safeguarding user privacy and security will be paramount. Emerging technologies like blockchain and decentralized AI offer promising solutions for creating secure, transparent, and accessible platforms.

References

Abdul Razak, L. H., Denis, T., Murugiah, Y., Yoong, W. K., Idris, Z., & Senik, M. H. (2024). The effect of traumatic brain injury on memory. *The Malaysian Journal of Medical Sciences: MJMS, 31*(3), 52–74. https://doi.org/10.21315/mjms2024.31.3.4

Al-Zahrani, A. M. (2024). Unveiling the shadows: Beyond the hype of AI in education. *Heliyon, 10*(9), e30696. https://doi.org/10.1016/j.heliyon.2024.e30696

Artificial Intelligence and Privacy – Issues and Challenges – Office of the Victorian Information Commissioner. (n.d.). Retrieved January 31, 2025, from https://ovic.vic.gov.au/privacy/resources-for-organisations/artificial-intelligence-and-privacy-issues-and-challenges/

Attout, L., Ordonez Magro, L., Szmalec, A., & Majerus, S. (2020). The developmental neural substrates of Hebb repetition learning and their link with reading ability. *Human Brain Mapping, 41*(14), 3956–3969. https://doi.org/10.1002/hbm.25099

Ball, K., Berch, D. B., Helmers, K. F., Jobe, J. B., Leveck, M. D., Marsiske, M., Morris, J. N., Rebok, G. W., Smith, D. M., Tennstedt, S. L., Unverzagt, F. W., Willis, S. L., & For the ACTIVE Study Group. (2002). Effects of cognitive training interventions with older AdultsA randomized controlled trial. *JAMA, 288*(18), 2271–2281. https://doi.org/10.1001/jama.288.18.2271

Bang, M., Jang, C. W., Kim, H. S., Park, J. H., & Cho, H. E. (2023). Mobile applications for cognitive training: Content analysis and quality review. *Internet Interventions, 33*, 100632. https://doi.org/10.1016/j.invent.2023.100632

Barman, A., Chatterjee, A., & Bhide, R. (2016). Cognitive impairment and rehabilitation strategies after traumatic brain injury. *Indian Journal of Psychological Medicine, 38*(3), 172–181. https://doi.org/10.4103/0253-7176.183086

Belger, J., Krohn, S., Finke, C., Tromp, J., Klotzsche, F., Villringer, A., Gaebler, M., Chojecki, P., Quinque, E., & Thöne-Otto, A. (2019). Immersive virtual reality for the assessment and training of spatial memory: Feasibility in individuals with brain injury. In *2019 international conference on virtual rehabilitation (ICVR)* (pp. 1–2) 10.1109/ICVR46560.2019.8994342.

Ben Izhak, S., & Lavidor, M. (2023). Strategy and Core cognitive training effects on working memory performance: A systematic review and meta-analysis. *Journal of Cognition and Development, 24*(4), 486–513. https://doi.org/10.1080/15248372.2023.2172413

Bombonato, C., Del Lucchese, B., Ruffini, C., Di Lieto, M. C., Brovedani, P., Sgandurra, G., Cioni, G., & Pecini, C. (2024). Far transfer effects of trainings on executive functions in neurodevelopmental disorders: A systematic review and Metanalysis. *Neuropsychology Review, 34*(1), 98–133. https://doi.org/10.1007/s11065-022-09574-z

Book, S., Jank, M., Pendergrass, A., & Graessel, E. (2022). Individualised computerised cognitive training for community-dwelling people with mild cognitive impairment: Study protocol of a completely virtual, randomised, controlled trial. *Trials, 23*, 371. https://doi.org/10.1186/s13063-022-06152-9

Boost Learning With a Simple Cognitive Trick: Spaced Repetition. (2024, May 30). *Training Industry*. https://trainingindustry.com/articles/strategy-alignment-and-planning/boost-learning-with-a-simple-cognitive-trick-spaced-repetition/

Buccilli, B., Alan, A., Baha, A., Shahzad, A., Almealawy, Y. F., Chisvo, N. S., Ennabe, M., & Weinand, M. (2024). The importance of behavioral interventions in traumatic brain injury. *Surgical Neurology International, 15*, 22. https://doi.org/10.25259/SNI_776_2023

Budson, A. E. (2021, September 16). *Can physical or cognitive activity prevent dementia?* Harvard Health. https://www.health.harvard.edu/blog/can-physical-or-cognitive-activity-prevent-dementia-202109162595

Calderone, A., Latella, D., Bonanno, M., Quartarone, A., Mojdehdehbaher, S., Celesti, A., & Calabrò, R. S. (2024). Towards transforming Neurorehabilitation: The impact of artificial intelligence on diagnosis and treatment of neurological disorders. *Biomedicine, 12*(10), Article 10. https://doi.org/10.3390/biomedicines12102415

Cao, M., Martin, E., & Li, X. (2023). Machine learning in attention-deficit/hyperactivity disorder: New approaches toward understanding the neural mechanisms. *Translational Psychiatry, 13*(1), 1–12. https://doi.org/10.1038/s41398-023-02536-w

Catania, V., Rundo, F., Panerai, S., & Ferri, R. (2023). Virtual reality for the rehabilitation of acquired cognitive disorders: A narrative review. *Bioengineering, 11*(1), 35. https://doi.org/10.3390/bioengineering11010035

Chan, A. T. C., Ip, R. T. F., Tran, J. Y. S., Chan, J. Y. C., & Tsoi, K. K. F. (2024). Computerized cognitive training for memory functions in mild cognitive impairment or dementia: A systematic review and meta-analysis. *npj Digital Medicine, 7*(1), 1–11. https://doi.org/10.1038/s41746-023-00987-5

Chanda, D. (2024, July 22). Develop a brain training app like peak—cost and features. *IdeaUsher*. https://ideausher.com/blog/develop-a-brain-training-app-like-peak/

Cicerone, K. D., Goldin, Y., Ganci, K., Rosenbaum, A., Wethe, J. V., Langenbahn, D. M., Malec, J. F., Bergquist, T. F., Kingsley, K., Nagele, D., Trexler, L., Fraas, M., Bogdanova, Y., & Harley, J. P. (2019). Evidence-based cognitive rehabilitation: Systematic review of the literature from 2009 through 2014. *Archives of Physical Medicine and Rehabilitation, 100*(8), 1515–1533. https://doi.org/10.1016/j.apmr.2019.02.011

Cunnington, R. (2019). *Wearable technology for neuroscience in classrooms*. IBE — Science of Learning Portal. https://solportal.ibe-unesco.org/articles/wearable-technology-for-neuroscience-in-classrooms/

de Jager Loots, C. A., Price, G., Barbera, M., Neely, A. S., Gavelin, H. M., Lehtisalo, J., Ngandu, T., Solomon, A., Mangialasche, F., & Kivipelto, M. (2024). Development of a cognitive training support Programme for prevention of dementia and cognitive decline in at-risk older adults. *Frontiers in Dementia, 3*, 1331741. https://doi.org/10.3389/frdem.2024.1331741

De Luca, R., Bonanno, M., Marra, A., Rifici, C., Pollicino, P., Caminiti, A., Castorina, M. V., Santamato, A., Quartarone, A., & Calabrò, R. S. (2023). Can virtual reality cognitive rehabilitation improve executive functioning and coping strategies in traumatic brain injury? A pilot study. *Brain Sciences, 13*(4), *Article 4*. https://doi.org/10.3390/brainsci13040578

DSI-24. (n.d.). *Wearable Sensing|Dry EEG*. Retrieved January 10, 2025, from https://wearable-sensing.com/dsi-24/

du Bois, N., Bigirimana, A. D., Korik, A., Kéthina, L. G., Rutembesa, E., Mutabaruka, J., Mutesa, L., Prasad, G., Jansen, S., & Coyle, D. H. (2021). Neurofeedback with low-cost, wearable electroencephalography (EEG) reduces symptoms in chronic post-traumatic stress disorder. *Journal of Affective Disorders, 295*, 1319–1334. https://doi.org/10.1016/j.jad.2021.08.071

Elendu, C., Amaechi, D. C., Elendu, T. C., Jingwa, K. A., Okoye, O. K., John Okah, M., Ladele, J. A., Farah, A. H., & Alimi, H. A. (2023). Ethical implications of AI and robotics in healthcare: A review. *Medicine, 102*(50), e36671. https://doi.org/10.1097/MD.0000000000036671

Emish, M., & Young, S. D. (2024). Remote wearable neuroimaging devices for health monitoring and Neurophenotyping: A scoping review. *Biomimetics, 9*(4), 237. https://doi.org/10.3390/biomimetics9040237

Fava-Felix, P. E., Bonome-Vanzelli, S. R. C., Ribeiro, F. S., & Santos, F. H. (2022). Systematic review on post-stroke computerized cognitive training: Unveiling the impact of confounding factors. *Frontiers in Psychology, 13*. https://doi.org/10.3389/fpsyg.2022.985438

Flanagan, K., & Saikia, M. J. (2023). Consumer-grade electroencephalogram and functional near-infrared spectroscopy Neurofeedback Technologies for Mental Health and Wellbeing. *Sensors, 23*(20), Article 20. https://doi.org/10.3390/s23208482

Frankenmolen, N. L., Overdorp, E. J., Fasotti, L., Claassen, J. A. H. R., Kessels, R. P. C., & Oosterman, J. M. (2018). Memory strategy training in older adults with subjective memory complaints: A randomized controlled trial. *Journal of the International Neuropsychological Society, 24*(10), 1110–1120. https://doi.org/10.1017/S1355617718000619

Frolli, A., Saviano, E., Ricci, M., Sica, S. L., & Romano, M. (2024). Artificial intelligence in ADHD treatment: A brief narrative review. *Digital Innovations for Learning and Neurodevelopmental Disorders, 3751*.

Ge, S., Zhu, Z., Wu, B., & McConnell, E. S. (2018). Technology-based cognitive training and rehabilitation interventions for individuals with mild cognitive impairment: A systematic review. *BMC Geriatrics, 18*(1), 213. https://doi.org/10.1186/s12877-018-0893-1

Gómez-Soria, I., Iguacel, I., Aguilar-Latorre, A., Peralta-Marrupe, P., Latorre, E., Zaldívar, J. N. C., & Calatayud, E. (2023). Cognitive stimulation and cognitive results in older adults: A systematic review and meta-analysis. *Archives of Gerontology and Geriatrics, 104*, 104807. https://doi.org/10.1016/j.archger.2022.104807

Greenwood, R. (Ed.). (2003). *Handbook of neurological rehabilitation* (2nd ed.). Psychology Press.

Guzick, D. S. (2016). *Augmenting cognitive training in older adults: The "ACT Study."* UFHealth. https://ufhealth.org/news/2016/augmenting-cognitive-training-older-adults-act-study

Hara, Y. (2017). *Brain training can reduce dementia risk*. Cognitive vitality. https://www.alzdiscovery.org/cognitive-vitality/blog/brain-training-can-reduce-dementia-risk

Harishbhai Tilala, M., Kumar Chenchala, P., Choppadandi, A., Kaur, J., Naguri, S., Saoji, R., & Devaguptapu, B. (2024). Ethical considerations in the use of artificial intelligence and machine learning in health care: A comprehensive review. *Cureus, 16*(6), e62443. https://doi.org/10.7759/cureus.62443

He, Z., Dieciuc, M., Carr, D., Chakraborty, S., Singh, A., Fowe, I. E., Zhang, S., Lustria, M. L. A., Terracciano, A., Charness, N., & Boot, W. R. (2023). New opportunities for the early detection and treatment of cognitive decline: Adherence challenges and the promise of smart and person-centered technologies. *BMC Digital Health, 1*(1), 7. https://doi.org/10.1186/s44247-023-00008-1

Hoorelbeke, K., Vervaeke, J., Siegle, G. J., Baeken, C., & Koster, E. H. W. (2022). Individual differences associated with treatment adherence and transfer effects following gamified web-based

cognitive control training for repetitive negative thinking. *Internet Interventions, 27*, 100507. https://doi.org/10.1016/j.invent.2022.100507
Interventions and Diagnostics—Neuroscape. (2022, April 26). https://neuroscape.ucsf.edu/technology/interventions-and-diagnostics/
Iversen, R. K., & Lewis, C. (2021). Executive function skills are linked to restricted and repetitive behaviors: Three correlational meta analyses. *Autism Research, 14*(6), 1163–1185. https://doi.org/10.1002/aur.2468
Jiang, Y., Abiri, R., & Zhao, X. (2017). Tuning up the old brain with new tricks: Attention training via Neurofeedback. *Frontiers in Aging Neuroscience, 9*, 52. https://doi.org/10.3389/fnagi.2017.00052
Johnson, A. J., & Miles, C. (2019). Visual Hebb repetition effects: The role of psychological distinctiveness revisited. *Frontiers in Psychology, 10*. https://doi.org/10.3389/fpsyg.2019.00017
Jung, H., Jeong, J.-G., Cheong, Y.-S., Nam, T.-W., Kim, J.-H., Park, C.-H., Park, E., & Jung, T.-D. (2021). The effectiveness of computer-assisted cognitive rehabilitation and the degree of recovery in patients with traumatic brain injury and stroke. *Journal of Clinical Medicine, 10*(24), Article 24. https://doi.org/10.3390/jcm10245728
Kaurani, P., de Marchi, M., Apolaro, A. V., Kunchala, K., Maini, S., Rges, H. A. F., Isaac, A., Lakkimsetti, M., Raake, M., & Nazir, Z. (2024). Advances in Neurorehabilitation: Strategies and outcomes for traumatic brain injury recovery. *Cureus, 16*(6), e62242. https://doi.org/10.7759/cureus.62242
Kelly, M. E., Loughrey, D., Lawlor, B. A., Robertson, I. H., Walsh, C., & Brennan, S. (2014). The impact of cognitive training and mental stimulation on cognitive and everyday functioning of healthy older adults: A systematic review and meta-analysis. *Ageing Research Reviews, 15*, 28–43. https://doi.org/10.1016/j.arr.2014.02.004
Khaleghi, A., Aghaei, Z., & Mahdavi, M. A. (2021). A Gamification framework for cognitive assessment and cognitive training: Qualitative study. *JMIR Serious Games, 9*(2), e21900. https://doi.org/10.2196/21900
Kong, Y., Guo, Q. H., Zhou, L., He, L., Zeng, Y., Du, X., Dong, J. Z., Jiang, C., Wang, J. G., & Ma, C. S. (2024). Digital computerised cognitive training for preventing cognitive decline among hypertensive patients: A study protocol for a multicentre randomised controlled trial (DELIGHT trial). *BMJ Open, 14*(2), e079305. https://doi.org/10.1136/bmjopen-2023-079305
Lampit, A., Hallock, H., & Valenzuela, M. (2014). Computerized cognitive training in cognitively healthy older adults: A systematic review and meta-analysis of effect modifiers. *PLoS Medicine, 11*(11), e1001756. https://doi.org/10.1371/journal.pmed.1001756
Looney, L., Wong, E. H., Rosales, K. P., Bacon, J. M., & Wiest, D. J. (2024). Supporting learning differences: Effects of cognitive training on cognitive abilities in a school-based sample. *Education Sciences, 14*(1), Article 1. https://doi.org/10.3390/educsci14010089
Maggio, M. G., De Bartolo, D., Calabrò, R. S., Ciancarelli, I., Cerasa, A., Tonin, P., Di Iulio, F., Paolucci, S., Antonucci, G., Morone, G., & Iosa, M. (2023). Computer-assisted cognitive rehabilitation in neurological patients: State-of-art and future perspectives. *Frontiers in Neurology, 14*. https://doi.org/10.3389/fneur.2023.1255319
Medina, R., Bouhaben, J., de Ramón, I., Cuesta, P., Antón-Toro, L., Pacios, J., Quintero, J., Ramos-Quiroga, J. A., & Maestú, F. (2021). Electrophysiological brain changes associated with cognitive improvement in a pediatric attention deficit hyperactivity disorder digital artificial intelligence-driven intervention: Randomized controlled trial. *Journal of Medical Internet Research, 23*(11), e25466. https://doi.org/10.2196/25466
Mendes, L., Oliveira, J., Barbosa, F., & Castelo-Branco, M. (2022). A conceptual view of cognitive intervention in older adults with and without cognitive decline—A systemic review. *Frontiers in Aging, 3*. https://doi.org/10.3389/fragi.2022.844725
Miller, M. S. (2017). *Digital training program reduces risk of dementia among older adults*. PennState. https://www.psu.edu/news/research/story/digital-training-program-reduces-risk-dementia-among-older-adults

Montana, J. I., Tuena, C., Serino, S., Cipresso, P., & Riva, G. (2019). Neurorehabilitation of spatial memory using virtual environments: A systematic review. *Journal of Clinical Medicine, 8*(10), Article 10. https://doi.org/10.3390/jcm8101516

Moore, A. L., & Ledbetter, C. (2019). The promise of clinician-delivered cognitive training for children diagnosed with ADHD. *Journal of Mental Health & Clinical Psychology, 3*(3) https://www.mentalhealthjournal.org/articles/the-promise-of-cliniciandelivered-cognitive-training-for-children-diagnosed-with-adhd.html

Nguyen, A., Ngo, H. N., Hong, Y., Dang, B., & Nguyen, B.-P. T. (2023). Ethical principles for artificial intelligence in education. *Education and Information Technologies, 28*(4), 4221–4241. https://doi.org/10.1007/s10639-022-11316-w

Pasqualotto, A., Mazzoni, N., Bentenuto, A., Mulè, A., Benso, F., & Venuti, P. (2021). Effects of cognitive training programs on executive function in children and adolescents with autism Spectrum disorder: A systematic review. *Brain Sciences, 11*(10), 1280. https://doi.org/10.3390/brainsci11101280

Pediatric Traumatic Brain Injury. (n.d.). *American Speech-Language-Hearing Association; American Speech-Language-Hearing Association.* Retrieved January 10, 2025, from https://www.asha.org/practice-portal/clinical-topics/pediatric-traumatic-brain-injury/?srsltid=AfmBOorHEvCsnLTy0ca9HApXBfXEM-Xpood5mD55SwJlbuu8jFbMO4vX

Ponsford, J., Velikonja, D., Janzen, S., Harnett, A., McIntyre, A., Wiseman-Hakes, C., Togher, L., Teasell, R., Kua, A., Patsakos, E., Welch-West, P., & Bayley, M. T. (2023). INCOG 2.0 guidelines for cognitive rehabilitation following traumatic brain injury, part II: Attention and information processing speed. *The Journal of Head Trauma Rehabilitation, 38*(1), 38. https://doi.org/10.1097/HTR.0000000000000839

Power, E. M., & D'Amato, R. C. (2021). Using a standardized neuropsychological model to guide a qualitative and quantitative assessment for evidence-based interventions. In R. C. D'Amato, A. S. Davis, E. M. Power, & E. C. Eusebio (Eds.), *Understanding the biological basis of behavior: Developing evidence-based interventions for clinical, counseling and school psychologists* (pp. 213–256). Springer International Publishing. https://doi.org/10.1007/978-3-030-59162-5_9

Rao, V. C. (2018). *The use of English language in research* (Vol. 2, p. 8).

Rapport, M. D., Orban, S. A., Kofler, M. J., & Friedman, L. M. (2013). Do programs designed to train working memory, other executive functions, and attention benefit children with ADHD? A meta-analytic review of cognitive, academic, and behavioral outcomes. *Clinical Psychology Review, 33*(8), 1237–1252. https://doi.org/10.1016/j.cpr.2013.08.005

Rebok, G. W., Ball, K., Guey, L. T., Jones, R. N., Kim, H.-Y., King, J. W., Marsiske, M., Morris, J. N., Tennstedt, S. L., Unverzagt, F. W., & Willis, S. L. (2014). Ten-year effects of the advanced cognitive training for independent and vital elderly cognitive training trial on cognition and everyday functioning in older adults. *Journal of the American Geriatrics Society, 62*(1), 16–24. https://doi.org/10.1111/jgs.12607

Researchers identify first brain training exercise positively linked to dementia prevention. (2017, November 16). News. https://medicine.iu.edu/news/2017/11/brain-exercise-dementia-prevention

Ripp, I., Emch, M., Wu, Q., Lizarraga, A., Udale, R., von Bastian, C. C., Koch, K., & Yakushev, I. (2022). Adaptive working memory training does not produce transfer effects in cognition and neuroimaging. *Translational Psychiatry, 12*(1), 1–13. https://doi.org/10.1038/s41398-022-02272-7

Roording-Ragetlie, S., Klip, H., Buitelaar, J., & Slaats-Willemse, D. (2016). Working memory training in children with neurodevelopmental disorders. *Psychology, 7*(3), Article 3. https://doi.org/10.4236/psych.2016.73034

Sardi, L., Idri, A., & Fernández-Alemán, J. L. (2017). A systematic review of gamification in e-health. *Journal of Biomedical Informatics, 71*, 31–48. https://doi.org/10.1016/j.jbi.2017.05.011

Shah, M. (2023, May 31). *A comprehensive guide of brain training apps market: Unlocking the potential of cognitive enhancement.* EIN Presswire. https://www.einpresswire.com/

article/636850764/a-comprehensive-guide-of-brain-training-apps-market-unlocking-the-potential-of-cognitive-enhancement.
Singh, A., Chakraborty, S., He, Z., Pang, Y., Zhang, S., Subedi, R., Lustria, M. L., Charness, N., & Boot, W. (2024). Predicting adherence to computer-based cognitive training programs among older adults: Study of domain adaptation and deep learning. *JMIR Aging, 7*(1), e53793. https://doi.org/10.2196/53793
Smolen, P., Zhang, Y., & Byrne, J. H. (2016). The right time to learn: Mechanisms and optimization of spaced learning. *Nature Reviews. Neuroscience, 17*(2), 77–88. https://doi.org/10.1038/nrn.2015.18
Somma, F., Rega, A., & Gigliotta, O. (2019). Artificial intelligence-powered cognitive training applications for children with attention deficit hyperactivity disorder: A brief review. *Proceedings of the First Symposium on Psychology-Based Technologies*, 2524. https://www.semanticscholar.org/paper/Artificial-intelligence-powered-cognitive-training-Somma-Rega/daebe86d68d89f6b58d2f3dcb302337c181fc4c4
Soule, A. C., Fish, T., Winegardner, J., & Schrieff-Brown, L. (2024). Implementing neuropsychological rehabilitation following severe traumatic brain injury in a low-to-middle income country: A case report. *Frontiers in Rehabilitation Sciences, 5*. https://doi.org/10.3389/fresc.2024.1393302
Stevens, M. C., Gaynor, A., Bessette, K. L., & Pearlson, G. D. (2016). A preliminary study of the effects of working memory training on brain function in attention-deficit/hyperactivity disorder. *Brain Imaging and Behavior, 10*(2), 387–407. https://doi.org/10.1007/s11682-015-9416-2
Stojanoski, B., Lyons, K. M., Pearce, A. A. A., & Owen, A. M. (2018). Targeted training: Converging evidence against the transferable benefits of online brain training on cognitive function. *Neuropsychologia, 117*, 541–550. https://doi.org/10.1016/j.neuropsychologia.2018.07.013
Studer-Luethi, B., Toermaenen, M., Margelisch, K., Hogrefe, A. B., & Perrig, W. J. (2022). Effects of working memory training on Children's memory and academic performance: The role of training task features and Trainee's characteristics. *Journal of Cognitive Enhancement, 6*(3), 340–357. https://doi.org/10.1007/s41465-022-00242-x
Sugden, R. J., Pham-Kim-Nghiem-Phu, V.-L. L., Campbell, I., Leon, A., & Diamandis, P. (2023). Remote collection of electrophysiological data with brain wearables: Opportunities and challenges. *Bioelectronic Medicine, 9*(1), 12. https://doi.org/10.1186/s42234-023-00114-5
Tennstedt, S. L., & Unverzagt, F. W. (2013). The ACTIVE study: Study overview and major findings. *Journal of Aging and Health, 25*(8), 3S–20S. https://doi.org/10.1177/0898264313518133
Trenorden, K. I., Hull, M. J., Lampit, A., Greaves, D., & Keage, H. A. D. (2022). Older adults' experiences of a computerised cognitive training intervention: A mixed methods study. *Australian Journal of Psychology, 74*(1), 2036581. https://doi.org/10.1080/00049530.2022.2036581
Tullo, D., Feng, Y., Pahor, A., Cote, J. M., Seitz, A. R., & Jaeggi, S. M. (2023). Investigating the role of individual differences in adherence to cognitive training. *Journal of Cognition, 6*(1), 48. https://doi.org/10.5334/joc.315
Unconventional Practice Tools: Gamification and Brain Training Apps. (2024). *Psicosmart*. https://psicosmart.net/blogs/blog-unconventional-practice-tools-gamification-and-brain-training-apps-190276
Understanding aging brains, how to improve memory and when to seek help. (n.d.). https://www.apa.org. Retrieved January 6, 2025, from https://www.apa.org/topics/aging-older-adults/memory-brain-changes
Upham, B. (2024, December 17). *10 memory-boosting games and activities to keep your brain sharp. EverydayHealth.Com*. https://www.everydayhealth.com/longevity/mental-fitness/brain-exercises-for-memory.aspx
van de Ven, R. M., Schmand, B., Groet, E., Veltman, D. J., & Murre, J. M. J. (2015). The effect of computer-based cognitive flexibility training on recovery of executive function after stroke: Rationale, design and methods of the TAPASS study. *BMC Neurology, 15*(1), 144. https://doi.org/10.1186/s12883-015-0397-y

Veloso, A., Vicente, S. G., & Filipe, M. G. (2020). Effectiveness of cognitive training for school-aged children and adolescents with attention deficit/hyperactivity disorder: A systematic review. *Frontiers in Psychology, 10*. https://doi.org/10.3389/fpsyg.2019.02983

von Bastian, C. C., Belleville, S., Udale, R. C., Reinhartz, A., Essounni, M., & Strobach, T. (2022). Mechanisms underlying training-induced cognitive change. *Nature Reviews Psychology, 1*(1), 30–41. https://doi.org/10.1038/s44159-021-00001-3

Weng, W., Liang, J., Xue, J., Zhu, T., Jiang, Y., Wang, J., & Chen, S. (2019). The transfer effects of cognitive training on working memory among Chinese older adults with mild cognitive impairment: A randomized controlled trial. *Frontiers in Aging Neuroscience, 11*, 212. https://doi.org/10.3389/fnagi.2019.00212

Wiest, G. M., Rosales, K. P., Looney, L., Wong, E. H., & Wiest, D. J. (2022). Utilizing cognitive training to improve working memory, attention, and impulsivity in school-aged children with ADHD and SLD. *Brain Sciences, 12*(2), 141. https://doi.org/10.3390/brainsci12020141

Zelinski, E. M. (2009). Far transfer in cognitive training of older adults. *Restorative Neurology and Neuroscience, 27*(5), 455–471. https://doi.org/10.3233/RNN-2009-0495

Chapter 10
Speech and Language Rehabilitation Technologies

10.1 Introduction

Communication is a cornerstone of human connection, shaping our relationships, interactions, and sense of identity. For individuals with speech and language disorders, the loss or impairment of communication can be a profound barrier to social, professional, and personal fulfillment. Disorders such as aphasia, dysarthria, stuttering, and developmental delays not only impact the ability to speak or understand language but often influence self-esteem and overall quality of life. Fortunately, the field of speech and language rehabilitation offers hope through interventions aimed at restoring or compensating for these abilities (Des Roches & Kiran, 2017). Traditionally, speech and language rehabilitation relied on clinician-led therapies focused on improving specific aspects of communication, such as articulation, fluency, or comprehension (What Is the Role of Speech Therapy in Brain Injury Recovery?, 2023). These approaches, while effective, often faced challenges in meeting the diverse needs of individuals. With the advent of modern technology, however, the field has witnessed a transformative shift. Innovations such as brain–computer interfaces (BCIs), artificial intelligence (AI), and augmented reality (AR) have enhanced the capacity to deliver personalized and dynamic interventions, broadening the horizons of what is achievable in rehabilitation (Des Roches & Kiran, 2017; Telepractice, n.d.).

The Scope of Speech and Language Disorders Speech and language disorders encompass a wide spectrum of conditions, each with unique challenges and rehabilitation needs. For example, aphasia, commonly caused by strokes or brain injuries, impairs the ability to express or understand language, often affecting reading and writing as well. Dysarthria, on the other hand, involves difficulty controlling the muscles used for speech, leading to slurred or slowed speech. Apraxia of speech adds another layer of complexity, as individuals struggle to coordinate the movements required for articulate speech (What Is the Role of Speech Therapy in Brain

K. J. Reddy, *Innovations in Neurocognitive Rehabilitation*,
https://doi.org/10.1007/978-3-031-88117-6_10

Injury Recovery?, 2023). Developmental disorders, such as autism spectrum disorder (ASD), may also include communication difficulties, often requiring early and sustained interventions. Other conditions, such as dysphagia—impairment in swallowing—and cognitive-linguistic disorders that affect attention, memory, reasoning, and problem-solving, highlight the interconnected nature of speech, language, and broader cognitive functions. For instance, individuals recovering from traumatic brain injuries may face difficulties in both communication and cognitive processing, which require holistic therapeutic strategies to address these overlapping challenges (Megan, 2019; Speech Disorder - an Overview, n.d.; "What Is the Role of Speech Therapy in Brain Injury Recovery?", 2023). In all these cases, the role of a speech-language pathologist (SLP) is to identify the underlying issues and design a rehabilitation program tailored to the individual's needs and goals.

Principles of Speech and Language Rehabilitation At the heart of speech and language rehabilitation lies a blend of restorative and compensatory strategies. Restorative approaches focus on rebuilding the skills and connections needed for speech and language, leveraging neuroplasticity—the brain's remarkable ability to adapt and reorganize itself in response to therapy. Techniques such as repetitive practice of speech movements or targeted language exercises aim to strengthen neural pathways, ultimately improving communication abilities. One compelling example of restorative therapy is melodic intonation therapy (MIT), often used for individuals with aphasia. This method harnesses the brain's capacity to use musical elements like rhythm and melody to bypass damaged language areas and facilitate verbal expression (Speech Disorder - an Overview, n.d.). Another technique, constraint-induced language therapy (CILT), encourages individuals to use verbal communication by restricting alternative methods, effectively enhancing neural plasticity. Compensatory strategies, on the other hand, focus on helping individuals navigate daily communication challenges using alternative tools and techniques. These may include teaching patients to use augmentative and alternative communication (AAC) devices or implementing strategies to enhance comprehension and expression (Megan, 2019; What Is the Role of Speech Therapy in Brain Injury Recovery?, 2023). For example, picture-based communication systems are often used for nonverbal children with ASD, allowing them to express needs and preferences more effectively. Additionally, speech-generating devices that convert text or symbols into spoken words have been transformative for individuals with severe speech impairments. The integration of these approaches ensures that therapy addresses both immediate needs and long-term recovery goals (Megan, 2019; Telepractice, n.d.).

The Transformative Role of Technology While traditional therapies remain foundational, recent advancements in technology have revolutionized the field of speech and language rehabilitation. For example, brain–computer interfaces enable individuals with severe paralysis or aphasia to communicate by decoding neural signals into speech or text. These systems translate thoughts into words, offering a breakthrough for patients with locked-in syndrome who were previously unable to express themselves. Speech recognition systems, often powered by AI, analyze speech patterns in real-time to provide corrective feedback, making interventions more engaging and precise (Des Roches & Kiran, 2017; Lundgren et al., 2010). For instance, stuttering therapy apps now incorporate AI algorithms to detect and correct disfluencies, allowing users to practice fluency exercises independently. Augmented and virtual reality technologies create immersive environments that simulate conversational scenarios, allowing individuals to practice social communication in a safe and controlled setting. For example, VR-based therapies can replicate real-world situations, such as ordering food at a restaurant or engaging in a job interview, providing a practical and engaging way to build confidence (Des Roches & Kiran, 2017; Lundgren et al., 2010). These innovations are particularly beneficial for individuals with ASD, who may struggle with social interactions in real-world environments. Wearable devices equipped with smart sensors further extend the reach of therapy by enabling real-time monitoring of vocal fold vibrations or speech parameters in natural settings. For instance, a wearable throat sensor can track vocal strain in individuals recovering from vocal cord surgery, providing immediate feedback to prevent overuse (Speech Disorder - an Overview, n.d.; Telepractice, n.d.; "What Is the Role of Speech Therapy in Brain Injury Recovery?", 2023). These devices not only support ongoing therapy but also empower individuals to take an active role in managing their rehabilitation.

10.2 Foundations of Speech and Language Rehabilitation

Effective speech and language rehabilitation is grounded in foundational principles that harness the brain's capacity for change, leverage motor learning, and utilize feedback mechanisms. These principles provide a framework for designing interventions that address diverse speech and language disorders. By integrating insights from neuroplasticity, motor learning theory, and feedback strategies, clinicians can develop comprehensive and evidence-based approaches to improve communication outcomes for individuals.

10.2.1 Principles of Speech Rehabilitation

- *Neuroplasticity*: At the heart of speech rehabilitation lies neuroplasticity, the brain's ability to reorganize itself in response to learning and targeted therapy. This remarkable capacity allows neurons to form new connections, enabling recovery of function even after significant injury. For individuals recovering from stroke or brain injury, experience-dependent neuronal plasticity can play a critical role in regaining lost communication abilities (Maier et al., 2019). Research highlights the importance of structured, repetitive training in driving neuroplastic changes. Studies with animal models have demonstrated that envi-

ronmental enrichment and task-specific training can stimulate neural activation and synaptogenesis, laying the foundation for functional recovery. In humans, these findings underscore the importance of designing therapy sessions that are engaging, varied, and tailored to individual needs. Clinical applications of neuroplasticity include techniques such as melodic intonation therapy (MIT), which uses rhythm and melody to bypass damaged language networks, and constraint-induced language therapy (CILT), which encourages verbal communication by limiting the use of alternative modalities (Ludlow et al., 2008). These approaches not only activate specific brain regions but also foster new neural pathways that support language recovery.
- *Motor learning*: Motor learning principles are central to speech rehabilitation, particularly for disorders that involve impaired control of speech movements, such as dysarthria or apraxia of speech. Effective motor learning relies on structured practice and feedback strategies that promote skill acquisition and retention (Leech et al., 2022; Maier et al., 2019).

- Key principles of motor learning include (Leech et al., 2022; Maier et al., 2019):

1. *Massed vs. spaced practice*: While massed practice involves concentrated, repetitive drills, spaced practice spreads sessions over time, often yielding better retention and transfer of skills.
2. *Task-specific practice*: Therapy activities should closely mimic real-world tasks to enhance functional outcomes. For example, practicing conversational speech may be more beneficial than isolated sound drills for individuals with dysarthria.
3. *Goal-oriented practice*: Setting meaningful goals motivates individuals and aligns therapy with their communication priorities.
4. *Variable practice*: Introducing variability, such as practicing speech in different contexts or with varying complexity, improves adaptability and generalization.
5. *Increasing difficulty*: Gradually challenging the individual's skills fosters growth and prevents plateauing.
6. *Multisensory stimulation*: Combining visual, auditory, and tactile cues enhances engagement and facilitates learning.

Delayed and reduced-frequency feedback also plays a pivotal role in motor learning. Studies suggest that delaying feedback allows individuals to evaluate their own performance internally, promoting deeper learning and retention. For instance, a patient practicing speech articulation may benefit from receiving feedback after completing several trials rather than immediately after each attempt. Reduced feedback frequency, where feedback is provided less consistently, encourages patients to rely on intrinsic rather than extrinsic cues, further solidifying motor patterns.

- *Feedback mechanisms:* Feedback is a cornerstone of effective rehabilitation, helping individuals refine their speech and language skills through real-time input and adjustments. There are two primary types of feedback in speech rehabilitation (Maier et al., 2019):

1. *Explicit feedback*: Also known as knowledge of results, this involves providing clear information about the outcome of a task. For example, a clinician might inform a patient whether their pronunciation of a word was correct.
2. *Implicit feedback*: Also known as knowledge of performance, this focuses on the process rather than the outcome. For instance, a patient might be guided to adjust their tongue placement during articulation exercises.

   Auditory and visual feedback are particularly powerful in refining speech production. Visual tools, such as spectrograms or waveform displays, allow patients to see their speech patterns in real time, providing an additional layer of understanding. For example, individuals with stuttering can use software that visualizes fluency patterns, helping them identify moments of disfluency and work toward smoother speech. Innovative technologies, such as biofeedback devices, are increasingly being integrated into therapy to provide multisensory input. For instance, wearable throat sensors can track vocal fold vibrations, offering tactile feedback that complements auditory and visual cues (Neuroplasticity in Clinical Practice, n.d.). These tools not only enhance learning but also empower individuals to take an active role in their rehabilitation.

- *Integrating principles in clinical practice:* The principles of neuroplasticity, motor learning, and feedback mechanisms interact in complex ways, requiring clinicians to adopt a nuanced approach to therapy. One key consideration is the distinction between performance during practice and long-term retention and transfer of skills. While a patient may demonstrate immediate improvements during therapy sessions, true success lies in their ability to apply these skills in everyday communication contexts. For example, a patient recovering from a stroke might show significant progress in pronouncing specific sounds during therapy but struggle to use those sounds fluently in conversation. By incorporating task-specific practice, variable practice, and reduced feedback frequency, clinicians can bridge this gap and facilitate real-world application of learned skills.

10.2.2 Language Rehabilitation Approaches

Language is an intricate tapestry of meaning, structure, and social interaction. For individuals with language impairments, challenges in expressing thoughts, understanding others, or navigating social contexts can be profound. Language rehabilitation seeks to address these challenges by targeting specific domains of language function, including semantics, syntax, and pragmatics. By integrating neuropsychological rehabilitation principles with emerging technologies, clinicians can deliver more effective and personalized interventions.

- *Semantic therapy:* Semantic therapy focuses on enhancing word retrieval and comprehension, which are often compromised in conditions such as aphasia or traumatic brain injury (TBI). Techniques like semantic feature analysis and semantic mapping are frequently employed to strengthen the connections between words and their meanings. For example, a clinician might guide a patient to generate attributes of a target word (e.g., for "apple": red, fruit, grows on trees), thereby reinforcing neural associations and improving retrieval. Neuropsychological rehabilitation emphasizes the role of structured, repetitive practice in rebuilding semantic networks. Semantic therapy can also incorporate technology, such as tablet-based apps that use interactive exercises to expand vocabulary and improve conceptual understanding. These tools provide real-time feedback, which is crucial for reinforcing correct associations and maintaining engagement (Liégeois et al., 2013; Social Communication Disorder, n.d.).

- *Syntactic training:* Syntactic training addresses sentence formation and grammatical accuracy, both of which are essential for effective communication. Individuals with syntactic deficits may struggle with constructing sentences, using proper verb tenses, or maintaining subject–verb agreement. Rehabilitation strategies often include structured elicitation tasks, such as sentence completion or rearranging scrambled words into grammatically correct sentences. The integration of technology into syntactic training has shown promising results. For instance, AI-powered platforms can analyze a patient's spoken or written output to identify grammatical errors and provide tailored exercises. Virtual reality (VR) environments also offer immersive settings for practicing syntactic skills in real-world scenarios, such as composing sentences during a simulated interview or casual conversation. These innovations align with the principles of neuroplasticity, enabling patients to practice syntax in a contextually relevant and engaging manner (Alduais et al., 2023; Liégeois et al., 2013; *Social Communication Disorder*, n.d.).
- *Pragmatic skills:* Pragmatic skills—the ability to use language effectively in social contexts—are a critical but often overlooked aspect of language rehabilitation. Pragmatic impairments can manifest as difficulties with turn-taking, understanding implied meanings, or appropriately initiating and concluding conversations. These challenges are common in individuals with social communication disorders, TBI, or autism spectrum disorder (ASD). Rehabilitation approaches for pragmatics emphasize real-world applicability. Role-playing activities allow patients to practice conversational scenarios, such as greeting someone or resolving a misunderstanding. Social stories, which use narratives to model appropriate social behaviors, are another effective tool, particularly for children with ASD. Pragmatic therapy often incorporates elements of cognitive rehabilitation, such as teaching theory of mind (the ability to understand others' perspectives) and executive function skills like self-monitoring and impulse control. Technology is increasingly being leveraged to enhance pragmatic skills. For example, virtual avatars in VR environments can simulate diverse social interactions, enabling patients to practice turn-taking or interpreting nonverbal cues in a controlled setting. Mobile apps with video-based exercises also allow individuals to rehearse and refine their social communication skills in a structured and accessible way (Alduais et al., 2023; Spaccavento et al., 2024).

10.3 Technological Advancements in Speech and Language Rehabilitation

Technological advancements have transformed the field of speech and language rehabilitation, offering innovative solutions that combine personalized care with state-of-the-art tools. These technologies leverage neuropsychological principles, ensuring that interventions are both scientifically grounded and patient-centric. By addressing the unique challenges faced by individuals with communication impairments, these advancements pave the way for more effective and engaging therapy options.

10.3.1 Speech Recognition Systems

Speech recognition systems have emerged as a cornerstone in modern speech therapy, enabling real-time analysis of speech patterns and providing corrective feedback. These systems are particularly beneficial for individuals with conditions like stuttering, where immediate feedback can help improve fluency. Automatic speech recognition (ASR) tools analyze speech in real time, identifying disfluencies and guiding users toward smoother speech production. For example, a person practicing fluency exercises can receive immediate insights into their speech rate, intonation, and rhythm. This instant feedback allows users to adjust their speaking patterns during practice, reinforcing positive behaviors and discouraging disfluent tendencies. These tools empower individuals to engage in self-guided practice, extending the benefits of therapy beyond clinical settings. The integration of artificial intelligence (AI) with ASR systems has further enhanced their capabilities. AI-driven speech recognition tools can detect subtle errors, analyze user progress, and adapt exercises based on individual needs. This personalization ensures that therapy remains challenging yet achievable, fostering consistent improvement over time (Arntz et al., 2023). Additionally, ASR technologies have been incorporated into telerehabilitation platforms, making high-quality therapy accessible to individuals in remote or underserved areas (Arntz et al., 2023). Speech recognition systems are not limited to stuttering therapy. They are also used in accent modification, articulation training, and language learning. By analyzing pronunciation accuracy and providing tailored feedback, these systems help users refine their communication skills across various contexts. The adaptability and scalability of speech recognition technologies make them an invaluable asset in modern rehabilitation.

10.3.2 Brain–Computer Interfaces (BCIs)

Brain–computer interfaces (BCIs) represent one of the most groundbreaking innovations in speech and language rehabilitation. These systems enable communication by decoding neural signals and translating them into speech or text. For individuals with severe paralysis or locked-in syndrome, BCIs provide a lifeline, restoring their ability to communicate (Arntz et al., 2023). The core principle of BCIs is to harness the brain's neural activity and convert it into actionable outputs. For instance, an individual with amyotrophic lateral sclerosis (ALS) can use a BCI to compose messages by selecting letters on a virtual keyboard, guided by their neural signals. This process not only fosters independence but also enhances quality of life, enabling users to express their thoughts and emotions effectively (Lazcano-Herrera et al., 2022).

Recent advancements in machine learning and signal processing have significantly improved the accuracy and usability of BCIs. These systems can adapt to users' unique neural patterns, ensuring that communication remains intuitive and efficient. By reinforcing neural pathways associated with communication, BCIs also leverage neuroplasticity to enhance long-term outcomes. Beyond individual applications, BCIs are being explored for collaborative use in group settings, such as classrooms or rehabilitation centers. This expansion underscores the versatility of BCIs, demonstrating their potential to address a wide range of communication challenges. Additionally, researchers are investigating the integration of BCIs with other technologies, such as augmented reality (AR), to create more immersive and interactive communication experiences (Lazcano-Herrera et al., 2022).

10.3.3 Augmented and Virtual Reality (AR/VR)

Augmented reality (AR) and virtual reality (VR) technologies are revolutionizing rehabilitation by creating immersive environments that enhance engagement and facilitate skill development. VR offers fully immersive experiences, allowing users to practice communication skills in simulated real-world scenarios. For example, VR-based therapy helps individuals with autism spectrum disorder (ASD) practice social interactions in safe and controlled settings, building confidence and reducing anxiety. AR complements VR by overlaying digital elements onto real-world environments. For instance, AR can be used in speech motor rehabilitation by pairing digital visualizations with robotic gloves or motion-capturing devices. These tools provide real-time feedback on users' movements, helping them refine their techniques and achieve their therapeutic goals. One of the most significant advantages of AR and VR technologies is their ability to replicate specific scenarios for practice. For example, a VR program might simulate ordering food at a restaurant or participating in a job interview, enabling users to rehearse these situations in a risk-free environment. These targeted interventions not only improve communication skills but also promote generalization, ensuring that users can apply their learning in real-life contexts. Combining AR and VR with gamification elements has proven particularly effective in engaging younger patients. By integrating therapy tasks into interactive games, these technologies make rehabilitation enjoyable and motivating, fostering greater participation and commitment. Furthermore, advancements in hardware and software design have made AR and VR systems more accessible, ensuring that their benefits reach a broader audience (Arntz et al., 2023).

10.3.4 Artificial Intelligence (AI) in Rehabilitation

Artificial intelligence (AI) is at the forefront of innovation in speech and language therapy, offering personalized, data-driven solutions that enhance traditional rehabilitation methods. AI-powered platforms analyze speech patterns, predict therapy outcomes, and adapt exercises based on individual progress. These tools provide clinicians with valuable insights into patient performance, enabling more precise and effective interventions (Arntz et al., 2023; Rasa, 2024). One of the most impactful applications of AI in rehabilitation is early diagnosis. Machine learning algorithms can analyze subtle changes in speech patterns to identify early signs of neurodegenerative diseases, such as Parkinson's or Alzheimer's. This early intervention allows clinicians to initiate therapy at a stage when it is most effective, improving long-term outcomes. AI also plays a critical role in patient monitoring. By processing data from wearable devices, AI systems track key metrics like vocal fold vibrations, phonation quality, and speech clarity. This information enables clinicians to evaluate progress objectively and make data-driven adjustments to treatment plans. Moreover, AI-powered virtual assistants can guide patients through therapy exercises, providing real-time feedback and encouragement. In addition to enhancing individual care, AI is being integrated into tele-rehabilitation platforms to expand access to therapy. These platforms use AI to personalize therapy sessions, ensuring that remote users receive the same quality of care as those in clinical settings (Arntz et al., 2023; Pearce et al., 2024; Rasa, 2024). The scalability and adaptability of AI-driven tools make them a cornerstone of modern rehabilitation.

10.3.5 Wearable Devices

Wearable technology has emerged as a vital component of speech and language rehabilitation, offering continuous monitoring and feedback in real-world settings. Devices such as smart sensors, activity trackers, and biometric monitors allow clinicians to gather comprehensive data on patients' speech and voice parameters. For example, wearable sensors can track vocal fold vibrations, providing insights into phonation and vocal strain during daily activities (Arntz et al., 2023). Beyond data collection, wearable devices empower users by delivering actionable feedback. For instance, a wearable throat sensor might alert a patient to overuse or incorrect vocal techniques, enabling immediate corrections. These devices also contribute to a holistic approach to rehabilitation by tracking physical and emotional well-being, which often influence communication abilities. The portability and accessibility of wearable devices make them invaluable for extending therapy beyond the clinic. For example, smartwatches equipped with speech analysis capabilities can prompt users to perform specific exercises throughout the day, reinforcing learning and promoting consistency (Rasa, 2024). Additionally, wearable devices can integrate with tele-rehabilitation platforms, providing therapists with real-time data to guide remote sessions effectively (Arntz et al., 2023). Wearable technology is also being explored for its potential in gamified therapy applications. By combining biofeedback with interactive games, these devices make rehabilitation engaging and enjoyable, particularly for younger patients. As advancements in wearable technology continue, their applications in speech and language therapy are expected to expand, further enhancing their impact.

10.4 Applications of Speech and Language Rehabilitation Technologies

Speech and language rehabilitation technologies have opened new avenues for addressing communication impairments across a variety of conditions. These innovations are tailored to meet the unique needs of individuals with neurological disorders, developmental disabilities, degenerative diseases, and hearing impairments. By integrating neuropsychological rehabilitation principles and the latest technologies, these interventions aim to enhance communication abilities, independence, and quality of life.

10.4.1 Neurological Conditions

Neurological conditions such as stroke and traumatic brain injury (TBI) often result in significant communication challenges, including aphasia, dysarthria, and cognitive-linguistic impairments. Speech-generating devices (SGDs) have become invaluable tools for individuals with severe speech deficits, particularly those with aphasia. These devices allow patients to communicate through text-to-speech conversion, enabling them to express their thoughts and needs effectively (Campbell et al., 2022). SGDs are a critical component of augmentative and alternative communication (AAC) systems. For example, a stroke survivor with non-fluent aphasia might use an SGD to select pre-programmed phrases or construct sentences using an intuitive interface. These tools not only restore a means of communication but also reduce frustration and enhance social participation. Beyond SGDs, neuroimaging techniques such as functional MRI and diffusion tensor imaging are being explored to understand the neural underpinnings of communication deficits and guide targeted rehabilitation strategies. By mapping brain activity and connectivity, clinicians can design personalized therapy programs that leverage the brain's neuroplasticity. Virtual reality (VR) is also gaining traction in neurological rehabilitation, offering immersive environments for practicing speech and language skills in simulated real-world scenarios (Campbell et al., 2022; Eshraghi et al., 2015).

10.4.2 Developmental Disorders

Developmental disorders, including autism spectrum disorder (ASD) and developmental language disorder, present unique challenges in communication. Children with ASD often struggle with expressive and receptive language, as well as pragmatic skills like turn-taking and understanding social cues. Technology has proven to be a powerful ally in addressing these challenges, with tablet-based apps emerging as a popular intervention. For instance, apps designed for children with ASD use visual supports, interactive games, and customizable features to improve vocabulary, sentence construction, and social communication skills. One widely used approach is the Picture Exchange Communication System (PECS), which has been shown through meta-analyses to significantly enhance communication abilities in individuals with ASD. These apps often incorporate gamification elements to maintain engagement and motivation, making therapy more enjoyable for young learners. Neuropsychological rehabilitation principles also guide interventions for developmental disorders. For example, structured and repetitive practice is crucial for reinforcing language skills in children with developmental language disorders. Apps and programs based on evidence-based techniques, such as Applied Behavior Analysis (ABA), provide systematic reinforcement to encourage desired behaviors and communication skills. Augmented reality (AR) and virtual reality (VR) technologies are also making strides in this area. AR tools can overlay digital prompts in real-world settings, while VR environments simulate social situations, helping children practice interactions in a controlled and supportive manner. These technologies not only improve language abilities but also build confidence and reduce anxiety associated with social communication (Campbell et al., 2022).

10.4.3 Degenerative Diseases

Degenerative diseases such as Parkinson's disease and amyotrophic lateral sclerosis (ALS) pose progressive challenges to communication. Voice banking technologies have become a beacon of hope for individuals with ALS, allowing them to preserve their personal speech before their condition advances. Voice banking involves recording a person's voice while they are still able to speak, creating a digital repository of phrases and sounds that can later be used in speech-generating devices. For example, an individual with ALS might record common expressions, personal anecdotes, and other unique vocal characteristics, enabling their SGD to produce speech that closely resembles their natural voice. This personalized approach not only maintains a sense of identity but also facilitates emotional connection with loved ones. In regions like Germany, assistive technology devices are routinely provided to individuals with ALS, ensuring access to these life-enhancing tools. Additionally, AI-powered speech recognition systems are being used to monitor speech changes in individuals with Parkinson's disease, enabling early intervention and tailored therapy. Wearable devices are also proving valuable in managing degenerative diseases. For instance, accelerometers and motion sensors can track the progression of motor symptoms, providing insights into how physical impairments impact communication abilities. These data-driven approaches empower clinicians to develop holistic care plans that address both motor and linguistic challenges (Campbell et al., 2022; Eshraghi et al., 2015).

10.4.4 Hearing Impairments

Hearing impairments, including those addressed by cochlear implants, require specialized rehabilitation to optimize communication outcomes. Cochlear implant training often integrates auditory-verbal therapy, which focuses on improving listening and spoken language skills. Apps designed for auditory training provide interactive exercises that help individuals with cochlear implants refine their ability to recognize sounds, distinguish speech, and understand spoken language. For example, a cochlear implant user might use an app to practice differentiating between similar-sounding words or identifying environmental noises. These exercises, combined with traditional therapy, enhance auditory processing and boost confidence in social interactions. However, the relationship between hearing impairments and conditions like autism spectrum disorder (ASD) remains an area of ongoing research and debate, highlighting the need for individualized approaches. Advanced technologies such as AR are being integrated into auditory training programs, creating engaging and immersive experiences for users. For instance, AR glasses can display visual cues synchronized with auditory stimuli, helping individuals associate sounds with corresponding visual information. These innovations not only improve auditory skills but also make rehabilitation more engaging and accessible (Campbell et al., 2022; Eshraghi et al., 2015).

10.5 Challenges and Ethical Considerations

The integration of advanced technologies into speech and language rehabilitation has ushered in a new era of possibilities, enabling more effective and personalized care for individuals with communication disorders. However, these advancements also bring significant challenges and ethical considerations that must be addressed to ensure equitable and responsible implementation. This section explores the practical and moral complexities surrounding these technologies, with a focus on neuropsychological rehabilitation and the latest innovations.

10.5.1 Challenges

- *High cost and accessibility of advanced technologies:* One of the most pressing challenges in adopting advanced speech and language rehabilitation technologies is their high cost. Devices like brain–computer interfaces (BCIs), augmented reality (AR) systems, and AI-driven speech recognition tools often come with prohibitive price tags, limiting their availability to individuals from lower-income backgrounds or underserved regions. This disparity is further exacerbated by geographic and socioeconomic barriers. In rural areas, for example, access to high-speed Internet—a prerequisite for many tele-rehabilitation platforms—remains limited. Similarly, healthcare providers in low-resource settings may lack the funding and infrastructure to implement cutting-edge technologies, leaving many patients without access to potentially life-changing interventions (Mennella et al., 2023). Addressing these gaps requires innovative funding models, such as public-private partnerships or subsidies, to make these tools more affordable and widely available.
- *Variability in user outcomes based on individual differences:* Another significant challenge is the variability in how individuals respond to different technologies. Factors such as age, cognitive ability, severity of the communication disorder, and familiarity with technology can all influence the effectiveness of an intervention (Mennella et al., 2023). For instance, older adults with limited technological literacy may struggle to use wearable devices or tele-rehabilitation platforms, while children with severe developmental disorders might require highly tailored solutions. This variability underscores the need for more research to identify which technologies work best for specific populations. Longitudinal studies and clinical trials are crucial to understanding the nuanced interactions between individual characteristics and therapeutic outcomes. By refining these insights, clinicians can better match patients with the tools and approaches that are most likely to benefit them (Beg & Verma, 2024).
- *Integration of technology with traditional therapy practices:* Incorporating advanced technologies into established therapeutic frameworks poses both logistical and cultural challenges. Many clinicians are accustomed to traditional

methods and may be hesitant to adopt new tools, particularly if they require significant changes to existing workflows or extensive training. Moreover, the integration process often reveals gaps in interoperability between different systems. For example, a speech therapist using AI-driven diagnostic tools might encounter difficulties in syncing data with wearable devices or electronic health records. Streamlining these processes requires not only technological innovation but also comprehensive training programs to equip clinicians with the skills and confidence needed to embrace new approaches (Mennella et al., 2024).

10.5.2 Ethical Considerations

- *Data privacy concerns with AI and wearable devices*: The use of AI and wearable devices in speech and language rehabilitation raises significant concerns about data privacy. These technologies often collect sensitive personal information, such as speech patterns, neural activity, and physiological metrics, which could be vulnerable to breaches or misuse if not adequately protected. Ensuring robust security measures is paramount. Encryption, secure storage, and regular audits should be standard practices for any organization deploying these technologies. Additionally, clear guidelines must be established to inform patients about how their data will be used, stored, and shared. Transparency in these processes is essential to building trust and ensuring compliance with ethical and legal standards (Al Kuwaiti et al., 2023; Mennella et al., 2024).
- *Ensuring equitable access to cutting-edge technologies:* Equitable access remains a central ethical concern in the adoption of advanced rehabilitation technologies. Disparities in income, geography, and healthcare infrastructure often determine who can benefit from these innovations. For instance, a family in a low-income community may not be able to afford an AI-powered speech therapy app, while patients in rural areas might lack access to the Internet required for tele-rehabilitation. Addressing these disparities requires systemic change. Policymakers and industry leaders must work together to develop strategies that ensure fair distribution of resources. This might include subsidizing costs for low-income families, investing in rural broadband infrastructure, or creating open-source tools that are accessible to all (Mennella et al., 2024).
- *Balancing technological reliance with human-centered therapy:* While technology has the potential to enhance rehabilitation outcomes, it is essential to strike a balance between technological reliance and human-centered care. Overemphasis on AI-driven tools or automated systems risks depersonalizing therapy, undermining the therapeutic relationship that is often crucial for patient engagement and success. Human therapists bring empathy, intuition, and adaptability to the therapeutic process—qualities that cannot be fully replicated by machines. Therefore, technology should be viewed as a complement to, rather than a replacement for, human expertise. Ensuring this balance requires thoughtful design and implementation of AI systems that prioritize collaboration with

clinicians, rather than attempting to replace them (Bekbolatova et al., 2024; Mennella et al., 2024).

10.6 Future Directions

The field of speech and language rehabilitation is advancing rapidly, driven by innovations that blend traditional methods with cutting-edge technology. These developments promise more personalized, effective, and accessible therapies for individuals with communication challenges. This section explores the emerging trends and future directions shaping the future of speech and language rehabilitation.

- *Neural speech decoding:* Neural speech decoding is at the forefront of technological advancements in speech rehabilitation. By leveraging brain-computer interface (BCI) technology, researchers are developing systems that translate brain activity directly into speech. Intracranial electrocorticographic (ECoG) recordings have been instrumental in uncovering the neural mechanisms underlying speech production, allowing dense electrode arrays to decode speech and reconstruct its acoustics. For individuals with severe physical disabilities or conditions like locked-in syndrome, neural speech decoding offers a transformative means of communication. These systems bypass traditional speech mechanisms, enabling users to express themselves through neural activity alone. As ECoG and BCI technologies continue to evolve, their integration into clinical settings holds immense promise for restoring communication abilities in profoundly impaired individuals (Luo et al., 2024).
- *Hybrid therapies:* A growing trend in speech and language rehabilitation is the integration of traditional therapy approaches with advanced digital tools. Hybrid therapies combine the personalized, human-centered aspects of traditional methods with the scalability and precision of technology. For instance, speech-generating devices (SGDs) are increasingly used alongside traditional speech therapy to enhance outcomes. These devices enable patients to practice communication skills in structured therapy sessions and real-world scenarios. Hybrid approaches also extend to combining cognitive and linguistic rehabilitation with neurofeedback techniques, creating a more holistic treatment paradigm. This synergy ensures that therapy addresses not only the mechanical aspects of speech but also the cognitive and emotional factors that influence communication (Luo et al., 2024).
- *Tele-rehabilitation platforms:* Tele-rehabilitation has emerged as a critical tool for expanding access to speech and language therapy. Remote platforms allow therapists to deliver high-quality care to patients in rural or underserved areas, overcoming geographic and logistical barriers. These platforms integrate video conferencing with interactive exercises, enabling real-time feedback and progress tracking. By incorporating AI and machine learning, tele-rehabilitation solutions can also personalize therapy sessions, adapting exercises to the individual's

needs and progress. This approach not only increases accessibility but also empowers patients to take an active role in their rehabilitation journey (Qiu et al., 2025).

- *Multimodal feedback systems:* The integration of visual, auditory, and tactile feedback in speech therapy is revolutionizing the way patients engage with rehabilitation. Multimodal feedback systems leverage multiple sensory modalities to enhance learning and retention. For example, a patient practicing articulation might receive auditory feedback from a speech recognition system, visual feedback from a spectrogram, and tactile feedback from a wearable device. These systems are particularly effective for individuals with complex communication needs, as they provide multiple pathways for reinforcing skills. By engaging different sensory modalities, multimodal feedback ensures that therapy is engaging, comprehensive, and tailored to the unique needs of each patient (Qiu et al., 2025).
- *Longitudinal studies:* While many speech and language rehabilitation technologies show promising short-term results, there is a pressing need for longitudinal research to evaluate their long-term efficacy. Understanding how these tools impact communication skills, quality of life, and overall independence over extended periods is crucial for optimizing their design and implementation. Longitudinal studies also provide insights into the sustainability of therapeutic gains and the factors that influence long-term outcomes. By identifying the conditions under which specific technologies are most effective, researchers can refine their approaches and maximize the benefits for diverse patient populations (Qiu et al., 2025; Zhuang et al., 2020).

10.7 Conclusion

The integration of advanced technologies into speech and language rehabilitation represents a profound shift in how communication disorders are addressed, offering new hope to individuals who face challenges in expressing themselves, understanding others, or navigating social interactions. These innovations, rooted in both traditional therapeutic principles and cutting-edge science, have reshaped the landscape of rehabilitation, making it more personalized, accessible, and effective. The chapter has explored the diverse applications of these technologies across neurological conditions, developmental disorders, degenerative diseases, and hearing impairments. From speech-generating devices for individuals with aphasia to tablet-based apps for children with autism spectrum disorder, the adaptability of these tools underscores their potential to meet a wide range of needs. Each application reflects the power of technology to complement and enhance human expertise, ensuring that individuals receive the support they need to overcome communication barriers.

Neuropsychological principles remain a cornerstone of effective rehabilitation, guiding the integration of technology into therapy. The brain's remarkable ability to adapt and reorganize itself, known as neuroplasticity, provides the foundation for many of these advancements. Whether through repetitive practice with multimodal feedback systems or the use of neural speech decoding to restore communication in individuals with severe physical impairments, the emphasis on leveraging the brain's potential has been central to the success of these interventions. However, the journey toward fully realizing the potential of speech and language rehabilitation technologies is not without its challenges. High costs and accessibility barriers remain significant hurdles, particularly in underserved or low-resource settings. While innovations such as tele-rehabilitation platforms and open-source tools aim to bridge these gaps, systemic changes in funding, infrastructure, and policy are needed to ensure that these benefits reach all who need them.

Ethical considerations also loom large as the field advances. Data privacy, equitable access, and the balance between technological reliance and human-centered care are critical areas of focus. AI-driven tools and wearable devices, while transformative, require robust safeguards to protect sensitive patient information and ensure fairness. Similarly, the adoption of these technologies must complement, rather than replace, the therapeutic relationship between clinicians and patients, which remains a cornerstone of effective care. Looking forward, the future of speech and language rehabilitation is bright with possibility. Neural speech decoding and brain–computer interfaces (BCIs) exemplify the profound impact of harnessing neural activity for communication, offering lifelines to individuals with severe impairments. The integration of traditional methods with digital tools, as seen in hybrid therapies, ensures that interventions remain holistic and adaptable to individual needs. Tele-rehabilitation platforms are expanding access to care, bringing high-quality therapy to remote and underserved communities, while multimodal feedback systems engage patients more comprehensively, making therapy both effective and enjoyable. The importance of longitudinal research cannot be overstated in this context. Understanding the long-term efficacy of these technologies is crucial for optimizing their design and implementation. By exploring how these tools impact communication skills, independence, and quality of life over extended periods, researchers can refine interventions to maximize their benefits. This emphasis on sustainability and adaptability will ensure that rehabilitation technologies continue to evolve in response to the diverse needs of patients.

Ultimately, the advancements in speech and language rehabilitation technologies are more than just tools; they represent a shift in how society views and addresses communication disorders. By combining scientific innovation with empathy and ethical responsibility, these technologies empower individuals to reclaim their voices and connect with the world around them. As collaboration among researchers, clinicians, policymakers, and technologists continues to grow, the field is poised to make even greater strides, transforming lives and redefining the boundaries of what is possible in rehabilitation. This chapter has highlighted the transformative potential of integrating neuropsychological principles with the latest technological advancements. It has also underscored the importance of addressing challenges and ethical considerations to ensure equitable and effective implementation. The journey ahead will require sustained effort, innovation, and a commitment to centering the needs and dignity of individuals with communication disorders. Through these collective efforts, the field of speech and language rehabilitation will continue to advance, offering hope and opportunity to those who need it most.

References

Al Kuwaiti, A., Nazer, K., Al-Reedy, A., Al-Shehri, S., Al-Muhanna, A., Subbarayalu, A. V., Al Muhanna, D., & Al-Muhanna, F. A. (2023). A review of the role of artificial intelligence in healthcare. *Journal of Personalized Medicine, 13*(6), Article 6. https://doi.org/10.3390/jpm13060951

Alduais, A., Alfadda, H., Allegretta, S., & Trivkovic, T. (2023). Pragmatic language impairment: A Scientometric review. *Applied Sciences, 13*(16), Article 16. https://doi.org/10.3390/app13169308

Arntz, A., Weber, F., Handgraaf, M., Lällä, K., Korniloff, K., Murtonen, K.-P., Chichaeva, J., Kidritsch, A., Heller, M., Sakellari, E., Athanasopoulou, C., Lagiou, A., Tzonichaki, I., Salinas-Bueno, I., Martínez-Bueso, P., Velasco-Roldán, O., Schulz, R.-J., & Grüneberg, C. (2023). Technologies in home-based digital rehabilitation: Scoping review. *JMIR Rehabilitation and Assistive Technologies, 10*, e43615. https://doi.org/10.2196/43615

Beg, M. J., & Verma, M. K. (2024). Exploring the potential and challenges of digital and AI-driven psychotherapy for ADHD, OCD, schizophrenia, and substance use disorders: A comprehensive narrative review. *Indian Journal of Psychological Medicine*, 02537176241300569. https://doi.org/10.1177/02537176241300569

Bekbolatova, M., Mayer, J., Ong, C. W., & Toma, M. (2024). Transformative potential of AI in healthcare: Definitions, applications, and navigating the ethical landscape and public perspectives. *Healthcare, 12*(2), 125. https://doi.org/10.3390/healthcare12020125

Campbell, P., Rooney, S., Nicoll, A., Brady, M. C., Smith, C. H., Deane, K. H., Herd, C. P., Tomlinson, C. L., Clarke, C. E., & Sackley, C. M. (2022). Speech and language therapy interventions for speech problems in Parkinson's disease. *The. Cochrane Database of Systematic Reviews, 2022*(6), CD015009. https://doi.org/10.1002/14651858.CD015009

Des Roches, C. A., & Kiran, S. (2017). Technology-based rehabilitation to improve communication after acquired brain injury. *Frontiers in Neuroscience, 11*, 382. https://doi.org/10.3389/fnins.2017.00382

Eshraghi, A. A., Nazarian, R., Telischi, F. F., Martinez, D., Hodges, A., Velandia, S., Cejas-Cruz, I., Balkany, T. J., Lo, K., & Lang, D. (2015). Cochlear implantation in children with autism Spectrum disorder. *Otology & Neurotology : Official Publication of the American Otological Society, American Neurotology Society [and] European Academy of Otology and Neurotology, 36*(8), e121–e128. https://doi.org/10.1097/MAO.0000000000000757

Lazcano-Herrera, A. G., Fuentes-Aguilar, R. Q., Chairez, I., Alonso-Valerdi, L. M., Gonzalez-Mendoza, M., & Alfaro-Ponce, M. (2022). Review on BCI virtual rehabilitation and remote technology based on EEG for assistive devices. *Applied Sciences, 12*(3), 12253. https://doi.org/10.3390/app122312253

Leech, K. A., Roemmich, R. T., Gordon, J., Reisman, D. S., & Cherry-Allen, K. M. (2022). Updates in motor learning: Implications for physical therapist practice and education. *Physical Therapy, 102*(1), pzab250. https://doi.org/10.1093/ptj/pzab250

Liégeois, F. J., Mahony, K., Connelly, A., Pigdon, L., Tournier, J.-D., & Morgan, A. T. (2013). Pediatric traumatic brain injury: Language outcomes and their relationship to the arcuate fasciculus. *Brain and Language, 127*(3), 388–398. https://doi.org/10.1016/j.bandl.2013.05.003

Ludlow, C. L., Hoit, J., Kent, R., Ramig, L. O., Shrivastav, R., Strand, E., Yorkston, K., & Sapienza, C. (2008). Translating principles of neural plasticity into research on speech motor control recovery and rehabilitation. *Journal of Speech, Language, and Hearing Research: JSLHR, 51*(1), S240–S258. https://doi.org/10.1044/1092-4388(2008/019)

Lundgren, K., Helm-Estabrooks, N., & Klein, R. (2010). Stuttering following acquired brain damage: A review of the literature. *Journal of Neurolinguistics, 23*(5), 447–454. https://doi.org/10.1016/j.jneuroling.2009.08.008

Luo, R., Huang, H., Yu, S., Zhang, X., & Xia, F. (2024). FairGT: A fairness-aware graph transformer. *Proceedings of the Thirty-Third International Joint Conference on Artificial Intelligence*, 449–457. https://doi.org/10.24963/ijcai.2024/50

Maier, M., Ballester, B. R., & Verschure, P. F. M. J. (2019). Principles of Neurorehabilitation after stroke based on motor learning and brain plasticity mechanisms. *Frontiers in Systems Neuroscience, 13*, 74. https://doi.org/10.3389/fnsys.2019.00074

Megan. (2019, September 12). *Speech therapy terms defined.* Tactus Therapy. https://tactustherapy.com/speech-therapy-terms-defined/

Mennella, C., Maniscalco, U., De Pietro, G., & Esposito, M. (2023). The role of artificial intelligence in future rehabilitation services: A systematic literature review. *IEEE Access, 11*, 11024–11043. IEEE Access. https://doi.org/10.1109/ACCESS.2023.3236084.

Mennella, C., Maniscalco, U., De Pietro, G., & Esposito, M. (2024). Ethical and regulatory challenges of AI technologies in healthcare: A narrative review. *Heliyon, 10*(4), e26297. https://doi.org/10.1016/j.heliyon.2024.e26297

Neuroplasticity in Clinical Practice. (n.d.). Physiopedia. Retrieved December 29, 2024, from https://www.physio-pedia.com/Neuroplasticity_in_Clinical_Practice

Pearce, L. M. N., Pryor, J., Redhead, J., Sherrington, C., & Hassett, L. (2024). Advanced Technology in a Real-World Rehabilitation Setting: Longitudinal observational study on clinician adoption and implementation. *Journal of Medical Internet Research, 26*(1), e60374. https://doi.org/10.2196/60374

Qiu, Y., Liu, H., & Zhao, M. (2025). A review of brain–computer Interface-based language decoding: From signal interpretation to intelligent communication. *Applied Sciences, 15*(1), Article 1. https://doi.org/10.3390/app15010392

Rasa, A. R. (2024). Artificial intelligence and its revolutionary role in physical and mental rehabilitation: A review of recent advancements. *BioMed Research International, 2024*(1), 9554590. https://doi.org/10.1155/bmri/9554590

Social Communication Disorder. (n.d.). American Speech-Language-Hearing Association; American Speech-Language-Hearing Association. Retrieved February 3, 2025, from https://www.asha.org/practice-portal/clinical-topics/social-communication-disorder/?srsltid=AfmBOoqfH3nSOiEaeEiMFIn5ehUm6X4HX2AVFG1ElFXm_hRNeMohBe53

Spaccavento, S., Caliendo, S., Galetta, R., Picciola, E., Losavio, E., & Glueckauf, R. (2024). Pragmatic communication deficit and functional outcome in patients with right- and left-brain damage: A pilot study. *Brain Sciences, 14*(4), 387. https://doi.org/10.3390/brainsci14040387

Speech Disorder—An overview | ScienceDirect Topics. (n.d.). Retrieved February 3, 2025, from https://www.sciencedirect.com/topics/neuroscience/speech-disorder

Telepractice. (n.d.). American Speech-Language-Hearing Association; American Speech-Language-Hearing Association. Retrieved February 3, 2025, from https://www.asha.org/practice-portal/professional-issues/telepractice/?srsltid=AfmBOoqMqDVSGqJc3blsyOyV5eC6MTpzFAHxkgUEh2__rxj5WUbCKY_l

What is the Role of Speech Therapy in Brain Injury Recovery?. (2023, June 23). *GSPP rehabilitation*. https://pennrehab.org/what-is-the-role-of-speech-therapy-in-brain-injury-recovery/

Zhuang, M., Wu, Q., Wan, F., & Hu, Y. (2020). State-of-the-art non-invasive brain–computer interface for neural rehabilitation: A review. *Journal of Neurorestoratology, 8*(1), 12–25. https://doi.org/10.26599/JNR.2020.9040001

Chapter 11
Motor Rehabilitation and Biofeedback

11.1 Introduction

Motor rehabilitation is a vital part of recovery for individuals affected by neurological conditions, injuries, and musculoskeletal disorders. It aims to restore strength, coordination, and independence, using targeted interventions that often draw on advanced technologies. Biofeedback, which provides real-time insights into physiological states, has become a valuable tool in this field. By helping patients monitor and adjust their movements, biofeedback enhances motor control and speeds up recovery. This section explores the key theories, motor training processes, neuropsychological impacts, and the role of technology in motor rehabilitation. Motor rehabilitation is grounded in several well-established theories. Motor learning theory emphasizes that motor skills are acquired through practice and repetition, leading to cognitive and neural changes that enable mastery of movements (Lagriffoul, 2022; Levin & Demers, 2021). Similarly, schema theory explains how movements are stored as generalized patterns in the brain, which can be adapted for new tasks through training. Underpinning these theories is the concept of neuroplasticity, the brain's ability to reorganize and form new neural connections after injury or disease. This remarkable adaptability is the foundation of many rehabilitation strategies, allowing individuals to recover lost functions through targeted training (Levin & Demers, 2021; Winstein et al., 2014).

Motor training focuses on three main elements: strength, coordination, and skill acquisition. Strengthening exercises, such as resistance training, rebuild muscle power and endurance, while coordination is enhanced through balance training, agility drills, and functional exercises. Skill acquisition is achieved through repetitive, goal-oriented tasks designed to address specific motor deficits. For example, constraint-induced movement therapy (CIMT), which encourages the use of an impaired limb by restricting the stronger one, has been shown to improve upper limb function in stroke survivors. These approaches rely on high-intensity,

K. J. Reddy, *Innovations in Neurocognitive Rehabilitation*,
https://doi.org/10.1007/978-3-031-88117-6_11

task-specific protocols, which have demonstrated significant success in motor recovery. Biofeedback adds another layer of precision to motor rehabilitation by enabling patients to actively participate in their recovery. Through devices that monitor physiological signals, such as muscle activity (EMG), brainwaves (EEG), or movement patterns, patients receive immediate feedback to guide their performance. For instance, EMG biofeedback helps patients activate specific muscle groups, promoting better strength and coordination. Similarly, EEG biofeedback, also known as neurofeedback, targets brain activity to foster neuroplasticity and improve motor function. Kinematic feedback provides real-time data on joint angles and motion, enabling more accurate retraining of complex movements. These technologies empower patients by increasing their awareness of their bodies and their ability to control them (Doğan-Aslan et al., 2012; Huang et al., 2006; Kwakkel et al., 2015).

Beyond physical recovery, motor rehabilitation plays a critical role in neuropsychological rehabilitation. Intensive motor practice drives neuroplastic changes, allowing the brain to reorganize itself and develop compensatory strategies. This process is particularly important for stroke survivors and individuals with traumatic brain injuries, who often rely on compensatory movements to navigate daily activities. Addressing these compensations through targeted training can lead to improved function and greater independence. Rehabilitation programs also increasingly integrate cognitive exercises with motor training, recognizing the interconnectedness of physical and mental recovery. For example, combining memory tasks with motor exercises has been shown to enhance overall brain function in patients recovering from neurological conditions (Levin & Demers, 2021; Winstein et al., 2014).

The integration of technology into motor rehabilitation has revolutionized the field, offering new opportunities for engagement and precision. Virtual reality (VR) and augmented reality (AR) create immersive environments where patients can practice simulated tasks, making therapy more interactive and motivating. For example, VR-based gait training has shown promising results in helping stroke survivors regain mobility. Robotics and exoskeletons provide precise, repetitive motion training, a key factor in promoting neuroplasticity. These devices are particularly beneficial for individuals with severe impairments, enabling them to engage in high-intensity practice that manual therapy alone cannot achieve. Wearable sensors further enhance rehabilitation by providing real-time feedback on performance, allowing for personalized adjustments to therapy plans (Banyai & Brişan, 2024; Kwakkel et al., 2015; Lagriffoul, 2022). Tele-rehabilitation is another transformative innovation, bringing specialized care to patients in remote or underserved areas. Through telemedicine platforms, patients can receive guidance from professionals without the need for in-person visits, significantly reducing barriers to access. This approach is not only more convenient but also more affordable, making rehabilitation accessible to a broader population. Additionally, brain-computer interfaces (BCIs) are opening new frontiers in rehabilitation by allowing patients to control assistive devices with their thoughts. These technologies hold immense potential for individuals with severe motor impairments, offering new avenues for communication and mobility (Jaswal et al., 2023; Young et al., 2021).

Despite these advancements, challenges remain. The high cost of advanced biofeedback devices and robotics limits their availability, particularly in low-resource settings. Variability in patient outcomes underscores the need for personalized rehabilitation plans that consider individual differences in recovery potential. Ethical concerns, such as ensuring data privacy and maintaining a balance between technology and human care, are also critical to address as these tools become more widespread (Banyai & Brişan, 2024; Winstein et al., 2014). The future of motor rehabilitation lies in integrating theoretical insights with innovative technologies to create holistic, patient-centered approaches. Artificial intelligence (AI) has the potential to revolutionize biofeedback systems by providing adaptive, predictive feedback tailored to each individual's needs. Home-based rehabilitation systems, powered by wearable devices and telemedicine, are making therapy more accessible and convenient, enabling patients to continue their recovery in the comfort of their homes. Multimodal approaches that combine biofeedback with cognitive and pharmacological interventions offer comprehensive solutions to address the complex challenges of motor recovery (Huang et al., 2006; Jaswal et al., 2023).

11.2 Principles of Motor Rehabilitation

11.2.1 Neuroplasticity and Recovery

Motor rehabilitation is a pivotal process in helping individuals regain movement, independence, and functionality after neurological injuries or conditions. It incorporates theoretical principles, intensive training protocols, and technological innovations to enhance outcomes for patients. This section synthesizes key insights on the theories underpinning motor rehabilitation, the processes involved in motor training, its neuropsychological implications, and the role of emerging technologies.

At the heart of motor rehabilitation lie several foundational theories. Motor learning theory emphasizes that motor skills are acquired through repetitive practice and meaningful experiences, which drive neural and cognitive adaptations to improve performance. Neuroplasticity, the brain's remarkable capacity to reorganize itself, allows individuals to relearn motor functions even after significant neurological damage. This principle highlights the importance of targeted, high-intensity training to stimulate the brain's adaptive mechanisms. Cortical remapping, another critical concept, refers to the brain's ability to reorganize functional areas so that intact regions can take over the roles of damaged ones. This process is essential for recovery after injuries such as stroke or traumatic brain injury. Synaptic plasticity, characterized by changes in the strength of neural connections, further supports the learning and memory mechanisms underlying motor recovery. By promoting long-term potentiation (strengthening of neural connections) and long-term depression (weakening of inefficient connections), rehabilitation fosters functional improvements (Dimyan & Cohen, 2011; "Neuroplasticity in Stroke and Brain Injury," 2024).

Motor training, the practical application of these theories, focuses on three key pillars: strength development, coordination improvement, and skill acquisition. Strength development is often achieved through resistance exercises and progressive overload, which rebuild muscular endurance and function. For instance, post-stroke patients may engage in strength training to counteract muscle atrophy and improve limb control. Coordination, an equally critical aspect, is enhanced through balance training, agility drills, and functional exercises that improve proprioception—the body's ability to sense and control its position in space (R. Yang et al., 2024). Rehabilitation exercises often involve activities like step training or agility tasks, which help patients refine their movements. Skill acquisition, the third pillar, emphasizes task-specific and goal-oriented practice. This approach enables patients to focus on functional activities that align with their recovery goals, such as grasping objects or walking on uneven terrain. Techniques like constraint-induced movement therapy (CIMT) exemplify this principle by forcing patients to rely on their weaker limb while restricting the stronger one. This intensive practice encourages neural reorganization and improves motor performance, particularly in individuals recovering from stroke ("Neuroplasticity in Stroke and Brain Injury," 2024).

The role of motor rehabilitation extends beyond physical recovery to encompass neuropsychological benefits. Intensive motor training drives neuroplastic changes in both damaged and undamaged brain regions, enabling functional reorganization. For example, compensatory strategies—where patients develop alternative movements to overcome deficits—are often addressed through rehabilitation, allowing patients to refine and optimize their motor skills. This process also promotes functional independence, enabling patients to perform daily activities and participate more fully in their lives. Moreover, the integration of cognitive and motor exercises in rehabilitation programs recognizes the interconnected nature of these functions. Cognitive-motor integration, such as pairing memory tasks with physical exercises, has been shown to enhance overall recovery and brain function (Dimyan & Cohen, 2011; "Neuroplasticity in Stroke and Brain Injury," 2024).

Technological advancements have revolutionized motor rehabilitation, introducing tools that make therapy more precise, engaging, and effective. Virtual reality (VR) and augmented reality (AR) are at the forefront of this innovation, offering immersive environments where patients can practice real-world tasks in simulated settings. For example, VR-based gait training allows stroke survivors to improve their walking ability while receiving real-time feedback. These technologies not only enhance motor learning but also increase patient motivation and adherence to therapy. Robotics and exoskeletons have also transformed rehabilitation by enabling repetitive, precise motion training. These devices are particularly valuable for individuals with severe motor impairments, as they provide high-intensity practice that might be difficult to achieve manually ("Neuroplasticity in Stroke and Brain Injury," 2024; Valkenborghs et al., 2017).

Wearable sensors further enhance rehabilitation by providing real-time data on a patient's performance. By tracking metrics like joint angles, muscle activity, and movement velocity, these devices allow therapists to personalize interventions and adjust them as needed. Brain-computer interfaces (BCIs) represent an exciting frontier in rehabilitation. These systems enable patients to control assistive devices using their thoughts, offering unprecedented opportunities for individuals with severe mobility limitations. BCIs have been particularly impactful in enabling communication and motor recovery in patients with conditions like amyotrophic lateral sclerosis (ALS) or spinal cord injuries. Other innovative techniques, such as noninvasive brain stimulation, have shown promise in enhancing neuroplasticity. Methods like transcranial magnetic stimulation (TMS) and transcranial direct current stimulation (tDCS) modulate brain activity in targeted areas, promoting recovery by facilitating neural reorganization. These approaches are often used in combination with traditional motor training to amplify their effects ("Neuroplasticity in Stroke and Brain Injury," 2024; Valkenborghs et al., 2017; Y.-W. Yang et al., 2020).

Tele-rehabilitation has emerged as a game-changer in expanding access to motor rehabilitation services. By leveraging telemedicine platforms, patients in remote or underserved areas can receive guidance from specialists without the need for in-person visits. This approach significantly reduces logistical barriers while maintaining high-quality care. Home-based rehabilitation systems, powered by wearable devices and remote monitoring, further enable patients to continue their recovery in familiar environments. These systems have been shown to increase patient engagement and compliance, making them a vital tool in modern rehabilitation practices. Despite these advancements, challenges remain. The high cost of advanced technologies, such as robotics and BCIs, limits their availability in many healthcare settings, particularly in low-resource regions. Ethical considerations, such as data privacy and the potential over-reliance on technology, also require careful attention. Additionally, variability in patient outcomes underscores the need for personalized approaches that account for individual differences in recovery potential. Addressing these challenges will be critical to ensuring the equitable and effective implementation of these innovations ("Neuroplasticity in Stroke and Brain Injury," 2024; Takeuchi & Izumi, 2013; Valkenborghs et al., 2017).

11.2.2 Motor Learning Theories

Motor rehabilitation plays a central role in helping individuals recover movement, independence, and functional ability following neurological injuries or disorders. Combining insights from motor learning theories, structured motor training processes, and innovative technologies, rehabilitation seeks to maximize the brain's capacity for recovery through neuroplasticity and targeted interventions. This section integrates the latest evidence to explore the theoretical foundations of motor rehabilitation, its practical applications, and the transformative potential of technological advancements. Motor learning is a cornerstone of rehabilitation, driven by several key theories and mechanisms. Practice and repetition have long been recognized as essential to strengthening neural connections, fostering use-dependent plasticity in the brain. Repeated performance of specific movements reinforces motor circuits, improving accuracy and efficiency over time. Feedback mechanisms, both internal and external, are equally critical. Providing augmented feedback, such as visual or auditory cues, enhances the refinement of motor skills, a principle validated by numerous studies. Goal-oriented training, where tasks align with a patient's functional objectives, has also been shown to boost engagement and outcomes, particularly in individuals with developmental coordination disorders or stroke (Dimyan & Cohen, 2011; "Neuroplasticity in Stroke and Brain Injury," 2024; Takeuchi & Izumi, 2013).

Beyond these core theories, rehabilitation leverages additional motor learning mechanisms to facilitate recovery. Use-dependent learning, primarily involving the cerebellum, drives improvements through repetitive practice, while reinforcement learning relies on rewards and corrections to optimize movement strategies. Sensorimotor adaptation enables the automatic adjustments required to maintain posture and coordination, a function critical for individuals regaining balance or mobility after injury. Together, these mechanisms provide a comprehensive framework for designing effective interventions tailored to individual needs (Takeuchi & Izumi, 2013).

Motor training, the practical application of motor learning theories, focuses on three key components: strength development, coordination improvement, and skill acquisition. Strengthening exercises, such as resistance training and progressive overload, are widely used to rebuild muscular endurance and control. These activities are particularly beneficial for patients recovering from stroke or traumatic brain injury, where muscle atrophy or spasticity often limits function. Coordination improvement focuses on enhancing proprioception and neuromuscular control through balance training, agility drills, and functional exercises. For example, activities like step drills or reaching tasks help refine spatial awareness and movement precision, laying the groundwork for more complex motor skills (Y.-W. Yang et al., 2020).

Skill acquisition, a critical component of motor training, emphasizes task-specific and goal-oriented practice. Techniques like constraint-induced movement therapy (CIMT) exemplify this approach. CIMT involves restricting the use of an unaffected limb to encourage intensive practice with the impaired limb, driving neural reorganization and improved motor performance. Research has consistently demonstrated its efficacy in improving upper limb function in stroke survivors. High-intensity, repetitive, and meaningful practice has emerged as the gold standard for fostering recovery through neuroplastic changes (Dimyan & Cohen, 2011; Takeuchi & Izumi, 2013).

Motor rehabilitation extends beyond physical recovery, playing a critical role in neuropsychological rehabilitation. Intensive motor practice promotes structural and functional neuroplasticity in both damaged and intact brain regions, enabling compensatory mechanisms to emerge. For example, stroke patients often develop alternative movement strategies to overcome deficits, but targeted rehabilitation can refine these strategies, leading to more efficient and effective movement patterns. Restoring motor skills also fosters functional independence, allowing individuals to perform daily activities with greater ease and confidence. Moreover, modern rehabilitation increasingly integrates cognitive training alongside motor exercises, recognizing the interplay between cognitive and motor functions. Cognitive-motor integration, such as pairing memory tasks with physical exercises, has been shown to enhance overall brain function and recovery outcomes (Dimyan & Cohen, 2011; Nudo et al., 1996; Taub et al., 2002).

The field of motor rehabilitation has been transformed by technological advancements, which have introduced innovative tools and systems to enhance therapy. Virtual reality (VR) and augmented reality (AR) create immersive environments where patients can engage in simulated tasks that mimic real-life scenarios. These technologies make therapy more engaging and motivating while offering precise feedback on performance. For instance, VR-based gait training has shown significant promise in improving mobility in stroke survivors. Robotics and exoskeletons provide another avenue for enhancing motor recovery. By delivering repetitive, accurate motion training, these devices support neuroplasticity and enable high-intensity practice for individuals with severe impairments (Taub et al., 2002). Wearable sensors have also emerged as valuable tools in rehabilitation, providing real-time feedback on joint angles, muscle activity, and movement velocity. These systems allow therapists to monitor progress and adjust interventions dynamically, offering a level of personalization that was previously unattainable (Nudo et al., 1996). Brain-computer interfaces (BCIs) represent the cutting edge of rehabilitation technology. BCIs enable patients to control assistive devices using their thoughts, offering new possibilities for communication and mobility in individuals with severe motor impairments or conditions like amyotrophic lateral sclerosis (ALS). These innovations not only improve outcomes but also empower patients by increasing their sense of agency and involvement in their recovery. In addition to these tools, noninvasive brain stimulation techniques such as transcranial magnetic stimulation (TMS) and transcranial direct current stimulation (tDCS) are being explored as adjuncts to traditional rehabilitation. By modulating brain activity in targeted regions, these methods enhance neuroplasticity and facilitate recovery. Combined with high-intensity motor training, they hold significant potential for improving functional outcomes in patients with neurological injuries ("Neuroplasticity in Stroke and Brain Injury," 2024; Y.-W. Yang et al., 2020).

Tele-rehabilitation has further expanded access to care, particularly for patients in remote or underserved areas. Through telemedicine platforms, individuals can receive guidance and support from specialists without the need for in-person visits, reducing logistical barriers and increasing convenience. Home-based rehabilitation systems, powered by wearable devices and remote monitoring, enable patients to continue therapy in familiar environments. These approaches not only improve accessibility but also foster greater adherence and engagement, making them a vital component of modern rehabilitation practices (Takeuchi & Izumi, 2013).

Despite these advancements, challenges remain in implementing these technologies widely. High costs and limited accessibility pose significant barriers, particularly in low-resource settings. Ethical considerations, such as ensuring data privacy and avoiding over-reliance on technology, are also critical as these tools become more integrated into clinical practice. Variability in patient outcomes highlights the need for personalized approaches that account for individual differences in recovery potential. Addressing these challenges will be essential to ensuring that the benefits of technological innovation reach all patients. The future of motor rehabilitation lies in integrating theoretical insights with advanced technologies to create holistic, patient-centered approaches. Artificial intelligence (AI) has the potential to revolutionize biofeedback systems by providing adaptive, predictive feedback tailored to individual needs. Multimodal approaches that combine motor training with cognitive and pharmacological interventions offer comprehensive solutions to address the multifaceted nature of recovery (Taub et al., 2002; Y.-W. Yang et al., 2020).

11.3 Biofeedback in Motor Rehabilitation

11.3.1 What Is Biofeedback?

Biofeedback is an innovative therapeutic approach that empowers individuals to take greater control over their physiological processes, many of which are typically involuntary. By monitoring and providing real-time feedback about these processes, biofeedback helps patients develop awareness and voluntary control over bodily functions, making it a powerful tool for managing various physical and mental health conditions. This section explores the principles of biofeedback, its mechanisms, practical applications, and its integration into broader healthcare practices. At its core, biofeedback relies on the principle that awareness is the first step toward control. During a biofeedback session, sensors or devices placed on the body measure physiological signals such as muscle activity, brain waves, or movement patterns. These signals are then translated into visual, auditory, or tactile feedback that patients can interpret and act upon. The process uses operant conditioning principles, where positive reinforcement is applied when patients achieve the desired physiological changes. For instance, if a patient successfully relaxes a specific muscle group, they might see a visual indicator change on a monitor, reinforcing the behavior (*Biofeedback*, n.d.-a; Giggins et al., 2013).

One of the most widely used forms of biofeedback is electromyography (EMG), which measures muscle activity. EMG biofeedback is particularly effective for conditions such as chronic pain, muscle tension, or recovery from injury. By providing real-time information about muscle contractions, EMG biofeedback helps patients identify patterns of tension and learn to consciously relax targeted muscles. Another common modality is electroencephalography (EEG), which measures electrical activity in the brain. EEG biofeedback, often referred to as neurofeedback, is used to treat conditions such as anxiety, attention-deficit/hyperactivity disorder (ADHD), and insomnia by training individuals to regulate their brainwave patterns. Motion tracking is another valuable form of biofeedback that monitors movement patterns. This modality is particularly useful in rehabilitation settings, where patients recovering from injury or surgery can refine their movements based on real-time feedback about joint angles, velocity, and posture. Across all modalities, the goal is to help patients become aware of their physiological responses, learn to control them during biofeedback sessions, and eventually apply these skills in daily life without the need for equipment (*Biofeedback*, n.d.-a; Giggins et al., 2013).

The applications of biofeedback are extensive, spanning both physical and mental health conditions. For example, it has been effectively used to manage anxiety and depression by helping patients control their physiological stress responses. By learning to reduce heart rate and muscle tension, patients often report feeling calmer and more in control. In conditions such as ADHD, biofeedback helps children and adults improve focus and attention by training them to modulate their brainwave activity through neurofeedback. Chronic pain management is another area where biofeedback has shown promise. By reducing muscle tension and improving posture, patients can often experience significant relief from conditions like migraines, fibromyalgia, and lower back pain. Biofeedback is also widely used to treat incontinence, where it helps patients strengthen pelvic floor muscles and regain bladder control. Additionally, it is employed to manage insomnia by training patients to relax their bodies and lower arousal levels before sleep. This non-invasive approach offers an appealing alternative to medication for individuals seeking to improve their sleep quality. For individuals with movement disorders or undergoing physical rehabilitation, motion biofeedback provides precise insights into how to optimize their movements, leading to improved outcomes and a faster recovery process (*Biofeedback*, n.d.-a; *Biofeedback*, n.d.-b; *Biofeedback as a Treatment of Chronic Pain*, 2024; Giggins et al., 2013).

The process of biofeedback treatment is both structured and interactive. During a typical session, a healthcare provider attaches sensors to the patient's body to measure the targeted physiological signals. These signals are then displayed in a way that is easy to understand, such as a graph or a series of lights on a monitor. The feedback allows patients to observe their physiological responses in real time and practice strategies to change them. For instance, a patient might use deep breathing or progressive muscle relaxation to lower their heart rate or reduce tension in a specific muscle group. Over time, with practice and guidance from their provider, patients often become proficient in making these changes independently, even outside of the biofeedback setting. While biofeedback offers significant benefits, its success largely depends on the active participation and motivation of the individual. Unlike medications that work passively, biofeedback requires patients to engage in the learning process and apply the techniques regularly. For many, this sense of empowerment—knowing they can directly influence their health—is a key part of biofeedback's appeal. Research consistently supports its effectiveness across various conditions, though it is often most effective when used alongside other treatments as part of a comprehensive care plan (*Biofeedback*, n.d.-a; *Biofeedback*, n.d.-b; *Biofeedback as a Treatment of Chronic Pain*, 2024).

One of the unique advantages of biofeedback is its adaptability across diverse patient populations and conditions. For example, children with ADHD may benefit from neurofeedback sessions designed to improve their focus and reduce impulsivity. Athletes recovering from injury might use motion biofeedback to refine their form and prevent future injuries. In chronic pain patients, EMG biofeedback helps to identify and reduce muscular tension, offering relief without the need for pain medication. This versatility makes biofeedback a valuable tool in integrative healthcare approaches, where it complements other therapies to address complex, multifaceted conditions. Despite its many strengths, biofeedback has some limitations. It is not typically used as a stand-alone therapy but is often integrated into broader treatment plans that include behavioral therapy, medication, or physical rehabilitation. Additionally, the number of sessions required for noticeable improvement can vary widely depending on the individual and the condition being treated. Some patients may see benefits after just a few sessions, while others may require months of regular practice (*Biofeedback*, n.d.-a).

Moreover, while biofeedback is generally considered safe, it is important for individuals to work with a qualified provider who has experience in the specific modality being used. Technological advancements have further expanded the potential of biofeedback, making it more accessible and effective. Wearable biofeedback devices, for instance, allow patients to monitor their physiological responses in real time, even outside of clinical settings. These devices, such as heart rate monitors or wearable EMG sensors, enable patients to practice biofeedback techniques in their daily lives, fostering greater consistency and long-term improvement. As technology continues to evolve, biofeedback systems are becoming increasingly user-friendly, with mobile apps and gamified interfaces that enhance engagement and adherence to treatment plans (*Biofeedback*, n.d.-a; *Biofeedback*, n.d.-b; *Biofeedback as a Treatment of Chronic Pain*, 2024).

11.3.2 Benefits of Biofeedback

Biofeedback is an innovative therapeutic approach that empowers individuals to understand and control physiological processes, particularly those related to muscle activity, posture, and motor coordination. By providing real-time feedback on bodily functions, biofeedback has become a valuable tool in rehabilitation and treatment for various conditions. Its benefits span multiple domains, including enhanced movement awareness, improved muscle activation and relaxation, and superior motor control. One of the key advantages of biofeedback is its ability to enhance awareness of movement and posture. By providing real-time feedback on physiological signals such as muscle activity or joint alignment, biofeedback fosters a deeper understanding of bodily movements. This increased awareness can lead to improved muscle control, as patients learn to voluntarily control muscles that were previously difficult to activate. For example, studies have shown that biofeedback aids in restoring muscle function after stroke or injury by helping individuals regain precise control over impaired muscle groups (*Biofeedback*, n.d.-b).

Furthermore, biofeedback encourages better posture by providing cues about muscle tension and alignment, enabling individuals to maintain correct postural positions during daily activities. Enhanced proprioception, the body's sense of its position in space, is another significant benefit. This improvement is particularly valuable for patients working on balance and coordination, such as those recovering from neurological injuries or older adults at risk of falls. Biofeedback also facilitates targeted muscle activation and relaxation, enabling individuals to gain control over specific muscle groups. EMG biofeedback, which measures and provides feedback on muscle activity, is particularly effective in this area. For example, patients recovering from musculoskeletal injuries or dealing with chronic pain conditions can use biofeedback to precisely activate weakened muscles while relaxing overactive or tense ones. This targeted approach not only helps in rebalancing muscle usage but also reduces the risk of further injury or chronic tension (Giggins et al., 2013; Seo, 2024).

Additionally, biofeedback's role in managing chronic pain cannot be overstated. By teaching individuals to consciously relax tense muscles, biofeedback has been shown to alleviate symptoms of conditions like fibromyalgia, lower back pain, and tension headaches. Another significant benefit of biofeedback lies in its ability to improve motor control and coordination. Traditional rehabilitation methods often focus on static exercises, but biofeedback introduces a dynamic, task-oriented approach. This modern methodology provides real-time feedback during functional movements, such as walking, gripping, or reaching. For example, dynamic biofeedback systems offer feedback on variables like hand trajectory, finger–thumb coordination, or grip strength, allowing patients to refine these movements and achieve better functional outcomes. Such systems are particularly effective in improving gait function, joint range of motion, and motor recovery in individuals with conditions like stroke, cerebral palsy, or spinal cord injuries. Research also supports the use of biofeedback-assisted therapy in sports, where athletes use it to fine-tune movement patterns, enhance performance, and prevent injuries (*Biofeedback*, n.d.-a; Huang et al., 2006; Sears, 2024).

The benefits of biofeedback extend to diverse patient populations and treatment settings. For individuals with neurological disorders, biofeedback is instrumental in retraining the brain and body to work together more effectively. Stroke survivors, for instance, often struggle with impaired motor skills. By providing precise feedback during movement retraining, biofeedback helps these patients regain control over their affected limbs, improve coordination, and restore functional independence. Similarly, for patients with chronic pain conditions, biofeedback not only reduces pain levels but also addresses the root causes by encouraging proper muscle use and relaxation. In addition to its rehabilitative applications, biofeedback is increasingly used as a preventive tool. Athletes, for example, use biofeedback to monitor and optimize their movement patterns, reducing the likelihood of overuse injuries or muscular imbalances. By offering detailed insights into muscle activation and joint alignment, biofeedback provides a pathway to healthier, more efficient movement, both in high-performance sports and everyday activities (*Biofeedback*, n.d.-a; Huang et al., 2006; Sears, 2024; Seo, 2024).

Despite its many advantages, biofeedback is typically not used as a stand-alone therapy. Instead, it is most effective when integrated into a comprehensive treatment plan that includes physical therapy, behavioral interventions, and, when necessary, medical management. For instance, patients recovering from a stroke might combine biofeedback with constraint-induced movement therapy (CIMT) to maximize the recovery of upper limb function. Similarly, individuals with chronic pain might pair biofeedback with mindfulness-based stress reduction techniques to address both the physical and psychological aspects of their condition. The number of biofeedback sessions required varies widely depending on the individual's needs, the condition being treated, and the progress made during therapy (*Biofeedback*, n.d.-a).

Technological advancements have further enhanced the effectiveness and accessibility of biofeedback. Modern biofeedback systems now offer more intuitive and engaging interfaces, such as wearable devices and mobile apps, that patients can use outside clinical settings. For example, wearable EMG devices allow patients to monitor muscle activity during daily tasks, providing continuous feedback that encourages adherence to therapeutic exercises (Bates, 2022; *Biofeedback*, n.d.-b). These portable systems are particularly useful for individuals managing long-term conditions, as they enable consistent practice and monitoring without the need for frequent clinic visits.

Looking ahead, the future of biofeedback lies in integrating advanced technologies like artificial intelligence (AI) and machine learning. These innovations can provide more personalized and adaptive feedback, tailoring interventions to the unique needs of each patient. For instance, AI-driven biofeedback systems could analyze patterns in muscle activation and suggest targeted adjustments in real time, optimizing the rehabilitation process. Moreover, as research continues to explore the potential of biofeedback, its applications are likely to expand into new areas, including cognitive rehabilitation, stress management, and even preventive healthcare for at-risk populations (Peng & Stirling, 2023; Sears, 2024).

11.3.3 Types of Biofeedback Used in Rehabilitation

Biofeedback is a versatile and effective tool in rehabilitation, offering tailored solutions to improve motor function, cognitive recovery, and physiological control. By providing real-time feedback on physiological processes, biofeedback helps patients consciously adjust and optimize their movements and functions. Among the various types of biofeedback modalities used in rehabilitation, electromyography (EMG), electroencephalography (EEG), and kinematic feedback are the most prominent, with additional options like heart rate variability (HRV) and real-time ultrasound imaging (RTUS) further expanding the scope of this approach.

EMG biofeedback is one of the most widely utilized biofeedback methods in rehabilitation. This modality measures muscle activity and tension, offering real-time feedback that helps patients improve muscle strength, coordination, and relaxation. EMG biofeedback is particularly valuable in the rehabilitation of musculoskeletal and neurological conditions. For instance, it is frequently used in knee, hip, and shoulder recovery programs, as well as for managing chronic conditions like lower back pain. By helping patients monitor and control specific muscle groups, EMG biofeedback enhances voluntary muscle contraction and activation, enabling more precise movement patterns. A common application includes retraining weakened or poorly activated muscles after surgery or injury. For example, individuals recovering from anterior cruciate ligament (ACL) reconstruction benefit from EMG biofeedback to restore quadriceps strength and function (Giggins et al., 2013; Spencer et al., 2021).

EEG biofeedback, often referred to as neurofeedback, focuses on brain activity, making it an essential tool for motor and cognitive rehabilitation. By measuring electrical activity in the brain, EEG biofeedback enables patients to regulate their brainwave patterns, promoting neuroplasticity and functional recovery. This technique has been particularly effective in addressing motor impairments in neurological conditions. For example, studies have shown that EEG biofeedback can improve upper extremity function in children with cerebral palsy and walking ability in individuals with multiple sclerosis. Additionally, neurofeedback has applications beyond motor recovery, such as improving cognitive function in patients with attention-deficit/hyperactivity disorder (ADHD) or mitigating symptoms of anxiety and depression (Giggins et al., 2013; Kopańska et al., 2022).

Kinematic feedback is another critical modality that uses motion tracking technology to analyze and optimize movement patterns. This type of biofeedback provides detailed information about joint angles, velocity, and spatial movement, offering patients and therapists precise insights into motor control and coordination. Kinematic feedback is especially beneficial in rehabilitation settings, where correcting abnormal movement patterns is essential for recovery. For example, patients recovering from stroke often experience impaired gait patterns. By using kinematic feedback during walking exercises, they can adjust their posture, stride length, and joint movements to achieve a more natural gait. The ability to track and refine movements in real time empowers patients to develop more efficient motor strategies, facilitating faster and more effective recovery (Giggins et al., 2013; Spencer et al., 2021).

In addition to these primary modalities, other biofeedback techniques play valuable roles in rehabilitation. **Heart Rate Variability (HRV) Biofeedback** is widely used to assess and improve autonomic nervous system function. By measuring fluctuations in heart rate, HRV biofeedback helps patients manage stress and enhance relaxation, which can indirectly support motor recovery by reducing muscle tension and improving focus during rehabilitation exercises. **Real-Time Ultrasound Imaging (RTUS)**, another innovative biofeedback method, provides visual feedback on soft tissue structures, allowing patients to observe and adjust their muscle activation and movement in real time. RTUS is particularly useful for retraining deep core muscles, such as the transverse abdominis, in individuals with lower back pain or postural instability (Casellato et al., 2019; Huang et al., 2006; Pizzoli et al., 2021).

Biomechanical biofeedback, which focuses on external measurements of movement and forces, further expands the capabilities of rehabilitation programs. By incorporating tools like force plates or wearable sensors, biomechanical biofeedback systems provide detailed feedback on parameters such as ground reaction forces, joint torques, and movement efficiency. This approach is particularly effective for athletes recovering from injury or individuals undergoing gait retraining after a stroke or amputation. The ability to monitor and adjust biomechanical variables enhances performance and reduces the risk of re-injury. The versatility of biofeedback is evident in its applications across diverse clinical populations. For example, in patients recovering from spinal cord injuries, EMG and kinematic biofeedback help restore motor control and improve functional independence. Similarly, EEG biofeedback is being increasingly used in cognitive rehabilitation programs to enhance attention, memory, and executive function in individuals recovering from brain injuries. These technologies are also valuable in chronic pain management, where biofeedback helps patients identify and relax overactive muscle groups, reducing pain and improving quality of life (Bowman et al., 2021; Giggins et al., 2013).

The integration of biofeedback into rehabilitation programs is not only effective but also empowering for patients. By providing real-time insights into physiological processes, biofeedback fosters a sense of agency and control over recovery. For instance, patients with Parkinson's disease can use biofeedback to monitor and correct tremors or posture during daily activities, enhancing their confidence and independence. Similarly, individuals recovering from orthopedic surgeries can use biomechanical biofeedback to ensure proper weight distribution and alignment during movement, preventing complications and promoting faster recovery. Despite its many benefits, biofeedback is most effective when integrated into a comprehensive treatment plan. It is rarely used as a stand-alone therapy but rather as a complement to traditional rehabilitation approaches, such as physical therapy, occupational therapy, or cognitive-behavioral interventions. The number of biofeedback sessions required varies widely depending on the condition, severity, and individual response to treatment. Some patients may experience significant improvements after just a few sessions, while others may require ongoing biofeedback training to achieve their goals (Casellato et al., 2019; Giggins et al., 2013; Ovadia-Blechman et al., 2022; Pizzoli et al., 2021).

Technological advancements are continuing to enhance the capabilities of biofeedback systems. Wearable devices and mobile applications have made biofeedback more accessible, allowing patients to practice and monitor their progress outside clinical settings. For example, portable EMG sensors can be used during daily activities to track muscle activity, providing continuous feedback that reinforces proper movement patterns. Similarly, AI-driven biofeedback systems are being developed to provide more personalized and adaptive feedback, tailoring interventions to the unique needs of each patient. These innovations are not only improving outcomes but also making biofeedback a more convenient and user-friendly option for patients and providers alike (Bowman et al., 2021; Casellato et al., 2019).

11.4 Technology-Enhanced Motor Rehabilitation

11.4.1 Wearable Devices

Wearable devices and smart sensors have revolutionized the field of rehabilitation by providing continuous monitoring and real-time feedback to both patients and clinicians. These innovations have significantly enhanced rehabilitation practices, particularly for individuals with motor impairments such as those caused by spinal cord injuries (SCI). Among these technologies, exoskeletons represent a groundbreaking example of wearable devices designed to restore mobility and improve functional outcomes. One of the primary advantages of wearable devices is their ability to deliver continuous monitoring of motor functions. Smart sensors embedded within these devices can track various physiological signals, such as muscle activity, joint angles, and movement velocity, allowing clinicians to assess progress and adjust therapy in real time (Siviy et al., 2022). For patients, this feedback fosters a better understanding of their movement patterns, encouraging active participation in their rehabilitation process.

Exoskeleton technology has become one of the most notable applications of wearable devices in rehabilitation. These robotic systems are designed to support individuals with severe motor impairments, such as SCI, by providing powered motion to the hips and knees. By enabling basic locomotion movements—including standing, walking, turning, and even climbing stairs—exoskeletons play a vital role in helping individuals regain mobility and independence. They are particularly effective in reducing reliance on wheelchairs and promoting upright posture, which has been associated with numerous physical and psychological benefits (Lingampally et al., 2024).

Clinical studies have demonstrated the functional benefits of using exoskeletons in rehabilitation. Users of these devices have shown objective improvements in key performance metrics, such as increased walking distance, velocity, and duration. These improvements not only enhance mobility but also contribute to better overall physical health. For example, standing and walking upright using an exoskeleton can help mitigate secondary health complications associated with prolonged immobility, including muscle atrophy, spasticity, impaired cardiovascular function, and poor bone density. Additionally, improved muscle tone and joint alignment have been observed in patients who regularly engage with exoskeleton-assisted therapy. The use of wearable devices like exoskeletons also aligns with the principles of personalized rehabilitation, as these technologies can be tailored to the specific needs of each individual. For optimal outcomes, robotic devices are often adjusted based on the patient's physical condition, goals, and progress, ensuring a user-centered approach to rehabilitation (Bove, 2019; Siviy et al., 2022). This customization enhances the device's effectiveness and increases the likelihood of patient adherence to therapy.

Despite their benefits, the adoption of exoskeletons remains limited, with relatively few patients choosing to use them regularly. This hesitancy can be attributed to several factors, including psychological and physiological barriers. Some users may experience discomfort or find the devices cumbersome, while others may struggle to integrate the exoskeleton into their body image. For instance, there is uncertainty about how these devices can be seamlessly incorporated into a user's mental representation of their body, which may impact their willingness to engage with the technology. Additionally, the interaction between the exoskeleton and the patient's autonomic and interoceptive signals—such as heart rate, breathing patterns, and muscle fatigue—plays a crucial role in determining the user's overall experience and satisfaction (Lingampally et al., 2024; Siviy et al., 2022).

To address these challenges, researchers are exploring embodied approaches to improve the interaction between wearable devices and their users. By focusing on how exoskeletons can be better integrated into a patient's bodily experience, these approaches aim to enhance the user's sense of control and comfort while using the device. For example, more intuitive and adaptive interfaces are being developed to align the device's movements with the user's natural motor patterns, fostering a smoother and more cohesive experience. Additionally, advancements in sensor technology and machine learning are enabling wearable devices to provide more personalized and adaptive feedback, further improving user engagement and outcomes (Emish & Young, 2024).

Beyond exoskeletons, other wearable devices are also making significant contributions to rehabilitation. Smart sensors embedded in wearable garments or accessories, such as wristbands or shoe insoles, provide valuable data on movement patterns, posture, and balance. These devices are particularly useful for patients undergoing gait retraining or recovering from lower limb injuries, as they offer continuous feedback on walking mechanics and alignment. Furthermore, wearable devices equipped with haptic feedback systems can deliver tactile cues to guide patients during rehabilitation exercises, enhancing their motor learning and coordination. In the context of SCI rehabilitation, wearable devices are increasingly being integrated into technology-based gait training programs, which combine robotic assistance with real-time feedback to optimize recovery outcomes. For example, patients using wearable sensors during treadmill training can receive instant feedback on their walking speed, stride length, and weight distribution, allowing them to make adjustments and improve their performance over time (Lingampally et al., 2024; Martinez-Hernandez et al., 2021; Siviy et al., 2022).

These programs not only enhance motor recovery but also promote neuroplasticity by stimulating the nervous system through repetitive, task-specific movements. While wearable devices hold immense promise, their widespread adoption requires addressing several practical and psychological barriers. For instance, the cost of advanced robotic systems like exoskeletons can be prohibitive for many patients, limiting access to these life-changing technologies. Additionally, healthcare providers must ensure that patients receive adequate training and support to use these devices effectively, as improper usage may lead to frustration or diminished outcomes. Ongoing research and innovation are essential to developing more affordable, accessible, and user-friendly wearable devices that cater to the diverse needs of rehabilitation patients (Lingampally et al., 2024; Seneviratne et al., 2017).

11.4.2 Virtual Reality (VR) and Augmented Reality (AR)

Virtual Reality (VR) and Augmented Reality (AR) technologies are transforming rehabilitation by creating immersive and interactive environments that enhance recovery outcomes, particularly in gait training for stroke survivors. By simulating real-world tasks and providing real-time feedback, these technologies offer a range of benefits that traditional rehabilitation methods cannot match. This section highlights the advantages of VR and AR in stroke rehabilitation and their potential to improve motor recovery and patient engagement. One of the standout benefits of VR and AR is their ability to provide engaging and motivating environments for rehabilitation. Stroke survivors often face challenges with adherence to repetitive exercises, but the playful and interactive nature of VR tasks increases intrinsic motivation and participation. For example, patients might find themselves walking through a virtual park or crossing busy streets, making therapy sessions more enjoyable and stimulating. This heightened engagement can lead to better adherence to rehabilitation protocols, which is crucial for recovery (Kim & Kaneko, 2023; Morizio et al., 2022).

VR-based rehabilitation also supports task-oriented practice, a cornerstone of effective motor recovery. By replicating real-world scenarios, VR allows patients to practice functional movements in controlled settings, enhancing the transferability of skills learned during therapy. For example, a stroke survivor can practice stepping over virtual obstacles or navigating uneven terrain, which prepares them for similar challenges in their daily life. The customizable nature of virtual environments ensures that these tasks can be tailored to each patient's specific needs and progress, further optimizing outcomes. A critical advantage of VR is its ability to provide real-time feedback. Many VR systems offer immediate visual or auditory feedback on movement patterns, enabling patients to adjust their gait and posture as they practice. This feedback accelerates motor learning by reinforcing correct movements and helping patients identify and correct errors. The incorporation of multiple sensory inputs—visual, auditory, and sometimes haptic feedback—further enhances the learning process by engaging various sensory modalities (de Rooij et al., 2021; Kim & Kaneko, 2023; Morizio et al., 2022).

Data collection and assessment are additional strengths of VR-based rehabilitation. VR systems can collect detailed metrics on gait parameters, movement patterns, and task performance, providing valuable insights for clinicians to track progress and adjust therapy plans. For instance, a clinician can analyze data on stride length, walking speed, or weight distribution to identify areas needing improvement and monitor changes over time. This level of precision and documentation surpasses what is typically achievable in traditional rehabilitation settings. Combining VR with robot-assisted gait training (RAGT) further enhances its effectiveness. RAGT provides physical support and assistance during walking, while VR adds a layer of engagement and feedback. This combination has been shown to improve walking distance, speed, and balance in stroke survivors, making it a powerful tool for gait rehabilitation. For example, patients can walk on a treadmill with robotic assistance while navigating a virtual cityscape, integrating physical and cognitive training into a single session (Kim & Kaneko, 2023; Winter et al., 2020).

One of the most exciting aspects of VR is its potential for home-based rehabilitation. Advances in technology have made VR systems more accessible and portable, allowing patients to continue their therapy outside clinical settings. Home-based VR programs can increase the duration and consistency of rehabilitation, which are key factors in achieving long-term recovery. Moreover, these programs can be tailored to individual needs and progress, ensuring that patients remain challenged and engaged over time. While the benefits of VR are substantial, there are challenges that need to be addressed. One notable issue is cybersickness, a type of motion sickness caused by discrepancies between visual motion in the virtual environment and the physical sensations of the body. Symptoms such as nausea, dizziness, and disorientation can limit the usability of VR for some patients. However, researchers are working on minimizing these effects by designing virtual environments that align more closely with natural movements and using head-mounted displays that accurately track users' head motions (Aderinto et al., 2023; de Rooij et al., 2021; Morizio et al., 2022).

Another challenge is ensuring the widespread accessibility and cost-effectiveness of VR-based rehabilitation programs. While the technology holds immense potential, its adoption is often limited by high costs and the need for specialized equipment and training. Future research should focus on developing affordable VR systems that can be easily integrated into standard rehabilitation practices, particularly in underserved or resource-limited settings. The usability and sense of presence offered by VR also play a crucial role in its effectiveness. Studies have shown that patients report high levels of presence in virtual environments, which enhances their engagement and motivation during therapy. The intuitive design of modern VR devices has made them more user-friendly, even for older adults or those with cognitive impairments. These improvements in usability make VR a viable option for a wide range of patients (de Rooij et al., 2021; *Virtual Reality for Individuals Affected by Stroke*, n.d.; Winter et al., 2020).

11.4.3 Robotics in Rehabilitation

Robotic devices have become an integral part of rehabilitation, offering promising solutions for enhancing recovery in patients with motor impairments, particularly those recovering from strokes. For post-stroke patients with upper limb impairments, robotic arms have shown potential in facilitating repetitive motion training, improving arm function, and supporting activities of daily living (ADLs). However, recent research has also highlighted the complexities and limitations of robotic-assisted rehabilitation, prompting discussions about its role in clinical practice.

One of the most significant advantages of robotic devices in rehabilitation is their ability to facilitate **repetitive motion training**. Stroke recovery often requires repetitive, task-specific movements to promote neuroplasticity and motor learning—key principles supported by evidence-based theories [1, 5]. Robotic arms excel in delivering high-frequency, consistent, and precise movements tailored to a patient's motor abilities. This consistency is particularly beneficial for individuals with moderate to severe upper limb impairments, enabling them to engage in therapeutic exercises that may otherwise be challenging to perform manually. A landmark 2018 Cochrane review analyzed 45 trials involving 1619 participants and found that **robot-assisted arm training significantly improved arm function (standardized mean difference [SMD] 0.32) and ADL scores (SMD 0.31)** compared to usual care. These findings highlight the efficacy of robotics in addressing motor impairments. However, when robot-assisted training was compared to conventional therapy of the same intensity and duration, no significant differences in outcomes were observed (Rodgers et al., 2019). This suggests that while robotic devices offer certain advantages, their effectiveness may not necessarily surpass that of traditional approaches when therapy is delivered at comparable intensities.

The **RATULS trial (Randomized Assessment of Utility and Liver Safety)**, a large multicenter randomized controlled trial, further examined the efficacy of robot-assisted training. It compared robot-assisted therapy to an enhanced upper limb therapy program (EULT) and usual care in stroke survivors with moderate to severe upper limb functional limitations. At three months, there were no significant differences in upper limb function between the three groups, including robot-assisted training, EULT, and usual care. These findings imply that while robotic rehabilitation can be effective, it may not be universally superior to alternative interventions, particularly enhanced therapy programs that emphasize task-oriented and intensive practice. The implications of these findings are critical for shaping the future of robotic rehabilitation. The RATULS trial emphasized the potential benefits of **enhanced upper limb therapy programs (EULT)**, which demonstrated significant improvements in upper limb impairment, mobility, and ADL scores compared to usual care. These results suggest that traditional or manual therapy approaches, when enhanced and delivered intensively, can yield comparable or even superior outcomes to robotic interventions (Rodgers et al., 2019).

While the current evidence highlights some limitations of robotic-assisted training, it also underscores the need for further research to optimize its application in clinical settings. A key area of focus is translating the **improvements in upper limb impairment observed with robotic devices into functional gains** that directly impact daily life and independence. For example, while patients may demonstrate improved motor control during robotic-assisted exercises, these improvements do not always translate to enhanced performance in ADLs such as dressing, eating, or writing (Chen et al., 2023; Rodgers et al., 2019).

Future research should also explore how robotic devices can be better integrated into **personalized rehabilitation programs**. Robotics have the potential to complement existing therapy by addressing specific needs, such as providing support for patients with severe impairments who may be unable to participate in conventional therapy initially. By customizing robotic interventions to target individual deficits and goals, clinicians can maximize the benefits of these technologies while addressing their limitations. The role of robotics in promoting **neuroplasticity** and motor learning also remains an area of interest. Robotic devices are uniquely positioned to deliver task-specific, high-intensity practice—factors known to drive neuroplastic changes in the brain. For instance, repetitive movement patterns facilitated by robotic arms may help rewire neural pathways and restore lost motor functions. However, integrating feedback mechanisms, such as visual or tactile feedback, could further enhance the effectiveness of robotic rehabilitation by engaging patients more actively in their recovery process (Grosmaire et al., 2022; Rodgers et al., 2019).

The high cost of robotic systems and the need for specialized training to operate them can limit their availability, particularly in resource-constrained settings. Developing more affordable and user-friendly robotic devices could help bridge this gap and make these technologies more accessible to a broader population of patients. Emerging trends in robotic rehabilitation also highlight the potential for combining robotics with other technologies, such as virtual reality (VR) and wearable sensors. For example, VR environments can be paired with robotic arms to create immersive and interactive therapy experiences, enhancing patient motivation and engagement. Similarly, wearable sensors can provide real-time feedback on movement quality, enabling clinicians to monitor progress and adjust robotic interventions dynamically (Grosmaire et al., 2022; Hasan et al., 2024; Zhang et al., 2018).

11.5 Applications of Motor Rehabilitation and Biofeedback

11.5.1 Neurological Rehabilitation

Neurological rehabilitation is an essential component of care for individuals living with complex neurological conditions such as stroke, Parkinson's disease (PD), and multiple sclerosis (MS). These conditions often present with multifaceted motor, cognitive, and sensory impairments that necessitate personalized, multidisciplinary approaches. The field of neurorehabilitation continues to evolve, incorporating innovative technologies and evidence-based practices to optimize recovery and improve quality of life for patients.

11.6 Stroke Rehabilitation

Stroke rehabilitation focuses on restoring motor, cognitive, and functional independence in patients who have experienced cerebrovascular events. Given the high prevalence of chronic stroke, significant efforts have been made to explore new approaches for improving motor outcomes. Traditional rehabilitation often emphasizes task-specific training and motor recovery, but recent research has expanded to include advanced techniques and innovative interventions. For instance, studies have investigated the relationship between prefrontal and premotor cortical connections and their influence on residual motor function in chronic stroke patients. Functional cortico-muscular coupling, assessed through tools such as functional MRI and high-density EEG, has also been explored as a marker of motor function and recovery potential. These insights are crucial for designing interventions that target specific neural pathways. Novel therapies, such as **perturbation-based balance training programs**, are being developed to improve reactive step quality and balance in chronic stroke patients. These programs simulate real-life scenarios, helping patients regain dynamic balance control and reduce fall risk. Additionally, the application of **oropharyngeal muscle exercises** has been examined for stroke patients with obstructive sleep apnea syndrome, addressing both respiratory and swallowing functions that are often compromised post-stroke (Tamburin et al., 2019).

11.7 Parkinson's Disease Rehabilitation

Parkinson's disease is characterized by progressive motor symptoms, such as bradykinesia, rigidity, and tremor, as well as non-motor symptoms like cognitive decline and mood disturbances. Neurorehabilitation for PD addresses this wide spectrum of impairments, with a strong emphasis on improving gait, balance, and functional mobility. Recent research has explored the use of **body weight support combined with treadmill training** to enhance gait in PD patients. This approach offers a safe environment for practicing walking, enabling patients to focus on stride length, cadence, and posture. However, it is essential to consider contraindications such as severe postural instability, balance disorders, or orthostatic hypotension, which may limit the feasibility of certain interventions. Beyond motor rehabilitation, neurorehabilitation for PD often incorporates cognitive exercises and dual-task training, addressing the interaction between motor and cognitive domains. For example, therapies that combine physical tasks with memory or problem-solving exercises have shown promise in improving both gait and executive function in PD patients (Tamburin et al., 2019).

11.8 Multiple Sclerosis Rehabilitation

Multiple sclerosis presents unique challenges due to its diverse and unpredictable symptoms, including motor and sensory impairments, cognitive deficits, and pharmaco-resistant pain. Neurorehabilitation for MS aims to mitigate these symptoms while promoting functional independence and quality of life. Innovative interventions, such as **tablet-based home training programs**, are being developed to improve dexterity and fine motor skills in MS patients. These programs use gamified tasks to engage patients in repetitive, task-specific exercises that target hand function. Additionally, technologies like **functional near-infrared spectroscopy (fNIRS)** and advanced imaging techniques are being employed to assess the neural correlates of training-induced plasticity, providing deeper insights into the effectiveness of rehabilitation strategies. Pain management is another critical aspect of MS rehabilitation. Non-invasive brain stimulation techniques, such as transcranial magnetic stimulation (TMS), are being explored as adjunct therapies to alleviate neuropathic pain and enhance motor recovery (Giggins et al., 2013; Tamburin et al., 2019).

11.9 Common Themes Across Conditions

Despite the unique challenges posed by each condition, several overarching themes emerge in neurological rehabilitation (*Neurorehabilitation*, n.d.; *Optimum Physical Therapy and Sports Rehab Neurological Rehabilitation*, n.d.; Tamburin et al., 2019):

1. **Advanced technologies:** The integration of functional MRI, high-density EEG, and near-infrared spectroscopy allows researchers and clinicians to explore the effects of neurorehabilitation on neural plasticity and motor recovery. These tools provide valuable biomarkers for assessing treatment efficacy and tailoring interventions to individual needs.
2. **Timing of interventions:** The timing of rehabilitation interventions—whether during the acute or chronic phase of a neurological condition—plays a critical role in determining outcomes. Personalized strategies are being developed to address the varying needs of patients at different stages of recovery.
3. **Personalized medicine:** Advances in imaging and neurophysiological assessment are driving the development of personalized rehabilitation programs. By correlating CNS lesions with clinical features and outcomes, clinicians can design targeted therapies that maximize recovery potential.
4. **Evidence-based practice:** The field is increasingly emphasizing systematic reviews, meta-analyses, and consensus-building initiatives to refine evidence-based approaches in neurorehabilitation. These efforts aim to standardize care and identify the most effective interventions across diverse patient populations.
5. **Patient-centered approaches:** Rehabilitation programs are being tailored to address not only physical impairments but also emotional and social well-being. Incorporating patient preferences and goals into therapy plans enhances engagement and satisfaction.

11.9.1 Orthopedic Rehabilitation

Orthopedic rehabilitation is essential for restoring function, mobility, and strength following musculoskeletal injuries and surgeries such as anterior cruciate ligament (ACL) reconstruction. A structured, evidence-based approach ensures effective recovery and long-term joint stability. ACL rehabilitation encompasses multiple phases, each designed to address specific aspects of recovery, from regaining mobility to re-establishing proprioception and strength.

- **Immediate post-surgery phase:** Rehabilitation typically begins immediately after ACL reconstruction, with the primary goals of achieving full knee extension (straightening), reducing swelling, and establishing normal gait patterns. Early mobility exercises are crucial to prevent stiffness and promote healing. Patients are encouraged to focus on restoring full extension of the knee, as deficits in this range of motion can lead to long-term functional impairments. During this phase, managing pain and swelling is a priority. Common strategies include using ice, elevation, and rest to minimize inflammation and discomfort. Quadriceps sets, which involve contracting the quadriceps muscle without moving the leg, are foundational exercises introduced early in the recovery process. These exercises activate the quadriceps and prevent muscle atrophy, forming the basis for more advanced rehabilitation activities. Weight-bearing is gradually reintroduced using crutches to ensure the knee is protected while allowing safe mobility. Patients are advised to bear weight as tolerated on the surgical leg, progressing from partial to full weight-bearing as their pain and confidence improve (*ACL Reconstruction Surgery*, n.d.; "ACL Rehabilitation Phase 1 | MOON Knee ACL Research," n.d., p. 1; blackrockortho, 2024).
- **Mobility and strength development:** The transition to later phases of ACL rehabilitation focuses on restoring joint mobility and building strength. Achieving and maintaining full range of motion (ROM), particularly knee extension, remains a key objective throughout the recovery process. Flexion exercises are also introduced to improve knee bending while protecting the surgical graft. Strengthening exercises target not only the quadriceps but also the hamstrings, gluteal muscles, and core, which are essential for stabilizing the knee during dynamic movements. These exercises are performed in a progressive manner, moving from non-weight-bearing to weight-bearing activities. For example, straight-leg raises and mini squats help improve muscle strength and control without placing excessive strain on the knee joint (*Anterior Cruciate Ligament (ACL) Rehabilitation*, n.d.; Saka, 2014).
- **Proprioception and neuromuscular training:** Regaining proprioception—the body's ability to sense joint position and movement—is a critical aspect of ACL rehabilitation. Proprioception is vital for balance, coordination, and preventing re-injury. Training typically begins when patients experience minimal pain, regain confidence, and demonstrate a willingness to engage in more dynamic exercises. Proprioceptive exercises often involve balance boards, single-leg stance drills, and dynamic stability exercises. These activities challenge the neuromuscular system, helping patients develop the reflexive control needed to stabilize the knee during sports or high-impact activities (Chaput et al., 2022; Saka, 2014).
- **Evidence-based approach and phased progression:** Rehabilitation for ACL reconstruction is divided into multiple phases, each with specific goals and milestones. For example, the early phases emphasize reducing swelling, restoring

ROM, and reactivating the quadriceps, while later phases focus on dynamic strength, agility, and return-to-sport activities. The MOON Knee Group and other evidence-based protocols provide structured guidelines for progressing through these phases, ensuring that patients achieve optimal recovery outcomes. Success in rehabilitation depends on completing each phase before advancing to the next. Skipping steps or rushing the process can compromise the surgical graft and increase the risk of re-injury. As such, individualized care is crucial, with patients working closely with their rehabilitation team to address specific needs and limitations (*Anterior Cruciate Ligament (ACL) Rehabilitation - Physiopedia*, n.d.; Chaput et al., 2022; Piedade et al., 2023).

- **Pain management and patient support:** Throughout the rehabilitation process, effective pain management remains a cornerstone of recovery. Patients often experience discomfort, swelling, and bruising after surgery, which can be managed through a combination of physical therapy techniques and home care strategies. Education on the importance of adhering to rehabilitation protocols is essential for maintaining motivation and achieving long-term success ("ACL Rehabilitation Phase 1 | MOON Knee ACL Research," n.d., p. 1; blackrockortho, 2024).
- **Advanced rehabilitation and return to function:** As patients progress to advanced stages of rehabilitation, the focus shifts to activities that mimic real-life movements and sport-specific tasks. Plyometric exercises, agility drills, and cutting maneuvers are introduced to prepare patients for the demands of their desired activities. Return-to-sport testing, which assesses strength, balance, and dynamic control, helps determine readiness for resuming athletic activities. Research underscores the importance of a multi-systems approach to human movement after ACL reconstruction. This includes addressing the nervous system, musculoskeletal system, and psychological readiness to ensure a holistic recovery. For example, integrating dual-task training—where patients perform cognitive tasks alongside physical exercises—can improve both physical and mental readiness for high-demand activities ("ACL Rehabilitation Phase 1 | MOON Knee ACL Research," n.d.; *Anterior Cruciate Ligament (ACL) Rehabilitation – Physiopedia*, n.d.; Piedade et al., 2023).

11.9.2 Sports Performance and Injury Prevention

Proprioceptive training is a critical component of sports performance enhancement and injury prevention, offering athletes a refined sense of body awareness, balance, and motor control. By targeting the body's ability to perceive its position and movement in space, proprioceptive training not only improves athletic skills but also reduces the likelihood of injuries, particularly in high-impact and dynamic sports. This section explores the multifaceted benefits, methodologies, and future directions of proprioceptive training based on recent research and practice.

11.9.2.1 The Importance of Proprioception

Proprioception, often referred to as the "sixth sense," enables individuals to sense and interpret the position, movement, and orientation of their bodies without visual cues. It integrates sensory input from joints, muscles, and the nervous system, forming the foundation for coordinated movement and balance. For athletes, enhanced proprioception translates to greater control over their movements, improved reaction times, and the ability to navigate complex physical environments effectively (Aman et al., 2015; Yilmaz & Dedeli, 2016).

11.10 Proprioceptive Training and Athletic Performance

Proprioceptive training has been shown to significantly enhance athletic performance by improving motor skills, somatosensory integration, and overall sensorimotor function. Key benefits include (Cooper, 2019; Yilmaz & Dedeli, 2016):

1. **Enhanced balance and stability**: Proprioceptive exercises improve trunk control, gait speed, and functional mobility, which are essential for athletes engaged in dynamic sports such as basketball, soccer, and gymnastics.
2. **Refined motor control**: Athletes undergoing proprioceptive training exhibit improved accuracy in movement execution, leading to more efficient and effective athletic performance.
3. **Sport-specific gains**: Studies have highlighted unique benefits in various sports. For instance, core proprioceptive training has been shown to enhance executive functions in young female basketball players, while improving shoulder stability and throwing performance in water polo athletes.
4. **Long-term neuromuscular adaptations**: Prolonged proprioceptive training fosters lasting improvements in joint position sense, particularly in the ankle and knee, reducing the risk of re-injury and supporting long-term athletic performance.

11.11 Proprioceptive Training in Injury Prevention

Injury prevention is one of the most significant applications of proprioceptive training. By enhancing joint stability and dynamic neuromuscular control, athletes can reduce the likelihood of both contact and non-contact injuries, particularly in the lower limbs (Aman et al., 2015; Cooper, 2019; Daryabor et al., 2023; *How Digital Balance Boards Enhance Stability with Biofeedback and Obje*, 2024; Yilmaz & Dedeli, 2016). For example:

1. **Joint stabilization**: Proprioceptive exercises strengthen the neuromuscular pathways that stabilize joints, minimizing the risk of ligament injuries such as ACL tears.
2. **Dynamic balance and control**: Training programs that incorporate balance boards, single-leg stance drills, and agility exercises help athletes develop the reflexive responses needed to maintain stability during rapid directional changes.
3. **Targeted prevention in high-risk sports**: In fencers, for example, proprioceptive training has been shown to improve dynamic control, reducing the risk of injury during complex movements. Similarly, knee proprioception is crucial for athletes in sports like skiing and tennis, where joint stability under dynamic loads is critical.

11.12 Training Methodologies and Tools

Proprioceptive training programs typically combine passive and active movements with and without visual feedback to enhance sensory integration. Common tools and methodologies include:

1. **Balance training devices**: Digital balance boards and wobble boards provide biofeedback, allowing athletes to refine their stability and monitor their progress in real time.
2. **Dynamic neuromuscular exercises**: Activities such as single-leg squats, plyometric drills, and reaction-based tasks challenge the body's ability to maintain control in dynamic scenarios.
3. **Sport-specific simulations**: Customized training programs replicate the demands of specific sports, ensuring that athletes develop proprioceptive skills relevant to their performance environments.
4. **Biofeedback integration**: Electromyography (EMG) biofeedback tools are increasingly used to monitor muscle activity and guide proprioceptive exercises, providing athletes with detailed insights into their movement patterns.

11.13 Challenges and Ethical Considerations

11.13.1 Challenges

The implementation of advanced biofeedback technologies holds great promise for enhancing rehabilitation and performance outcomes. However, several significant challenges must be addressed to facilitate their widespread adoption and equitable distribution. These challenges span economic, technical, and operational domains, reflecting the complexity of integrating advanced technologies into healthcare and wellness practices.

- High cost of advanced devices: One of the most significant barriers to implementing advanced biofeedback technologies is their high cost. Many of these devices require sophisticated hardware, specialized sensors, and complex software, making them prohibitively expensive for individuals and organizations in resource-constrained settings (Cui et al., 2024; Peake et al., 2018). This financial hurdle limits the availability of biofeedback technologies to a narrow segment of the population, typically those in high-income regions or institutions with substantial resources. As a result, patients in low-income or underserved areas often miss out on the benefits these technologies can offer.
- Limited accessibility in low-resource settings: The cost and complexity of biofeedback devices disproportionately affect low-resource settings, where healthcare infrastructure and funding are already limited. Clinics in such regions often lack the necessary equipment, technical expertise, and training programs to implement biofeedback technologies effectively (Kerr et al., 2011; Peake et al., 2018). This disparity exacerbates existing inequalities in healthcare access, leaving many patients without opportunities to benefit from innovative therapeutic tools.
- Variability in patient outcomes: Patient responses to biofeedback interventions can vary widely due to factors such as individual health conditions, motivation, and prior experience with technology. For example, patients who are less familiar with digital tools may struggle to engage with biofeedback devices, reducing the efficacy of interventions. Additionally, certain conditions may not respond as well to biofeedback, creating inconsistencies in outcomes across different patient populations (Peake et al., 2018). Addressing this variability requires a deeper understanding of patient-specific factors and tailored approaches to intervention design.
- Need for standardization: A lack of standardized protocols and guidelines for using biofeedback technologies further complicates their implementation. Currently, practices and methodologies vary widely across different conditions, patient populations, and healthcare settings, making it difficult to compare results and replicate successful interventions (Peake et al., 2018). Standardization is crucial to ensure consistency in application, enhance the reliability of outcomes, and facilitate the integration of biofeedback into broader healthcare systems.
- Technological complexity and training requirements: Many advanced biofeedback devices are technically complex, requiring specialized training for effective use. Healthcare professionals may need extensive instruction to operate these devices, interpret data, and integrate them into clinical workflows. This training demand can pose a barrier to adoption, particularly in settings where staffing shortages or time constraints are prevalent. For patients, the complexity of the devices can lead to frustration, reducing engagement and adherence to therapy (Cheng & Lin, 2023; Peake et al., 2018).
- Data integration and analysis: Advanced biofeedback devices generate vast amounts of data, including physiological signals, motion patterns, and performance metrics. While this data can provide valuable insights, it also presents challenges in terms of management, integration, and analysis (Boikanyo et al.,

2023). Many healthcare providers lack robust systems to process and interpret this data effectively, limiting its utility in guiding treatment decisions. Developing streamlined data management solutions is essential to maximize the value of biofeedback technologies.

- Validation and evidence base: Although many biofeedback technologies show promise, rigorous scientific validation is often lacking. The effectiveness of these devices must be demonstrated through high-quality studies and clinical trials to build confidence among clinicians and patients (Boikanyo et al., 2023). Establishing a strong evidence base is critical for overcoming skepticism and facilitating the integration of biofeedback technologies into standard clinical practice.

11.13.2 Ethical Considerations

The use of wearable and connected devices in research has expanded rapidly, offering unprecedented opportunities to collect real-time data and enhance health interventions. However, these technologies raise critical ethical considerations that researchers and developers must address to ensure participant rights and data security. This section explores the major ethical challenges associated with wearable device research and provides guidance for creating ethical frameworks to navigate these issues.

One of the most significant ethical concerns in wearable device research is ensuring the privacy and security of participant data. Wearable devices often collect sensitive information, including physiological metrics, location data, and behavioral patterns. Breaches of this data can have serious consequences, as demonstrated by recent security incidents exposing millions of fitness tracker records online. The involvement of third-party companies in data storage and analysis adds an additional layer of complexity, as researchers may have limited control over how data is stored, transferred, and secured. To address these risks, researchers must prioritize robust data protection measures. Encryption protocols, secure data storage, and clear policies regarding data sharing are essential for minimizing vulnerabilities. Moreover, researchers should carefully vet third-party companies to ensure they comply with stringent security standards and avoid the use of opaque "black box" algorithms that hinder accountability and transparency (IABAC®, 2023; Karapetyan, 2024; Silva, 2024).

Participants in wearable device studies often have limited knowledge or control over how their data will be used, particularly when third-party companies are involved. This lack of transparency raises questions about data ownership and informed consent. Researchers may find themselves constrained by commercial agreements, which can limit their ability to guarantee the ethical use of participant data. To mitigate these concerns, researchers must adopt practices that ensure participants fully understand how their data will be collected, stored, and used. Informed consent processes should include detailed explanations of data usage, potential risks, and the entities involved in managing the data. Clear communication of these details helps build trust between researchers and participants while upholding ethical research practices (Silva, 2024; Sui et al., 2023).

The integration of wearable devices into research often involves partnerships with commercial entities, which may prioritize profit over research integrity. These competing interests can blur the boundaries between research objectives and corporate agendas, potentially skewing study outcomes or compromising participant welfare. To maintain the integrity of research, it is crucial to establish clear boundaries between commercial and research activities. Researchers should strive to retain control over study design, data analysis, and publication of results. Additionally, transparency about any financial relationships between researchers and device manufacturers is essential to avoid conflicts of interest (Sui et al., 2023).

The transfer and storage of data collected by wearable devices present significant ethical challenges. Data breaches during transmission or storage can compromise participant confidentiality, while reliance on proprietary data storage systems can limit researchers' ability to ensure data security. Furthermore, wearable devices often generate vast quantities of data, creating logistical challenges in managing and analyzing this information responsibly. Developing robust systems for data integration and analysis is critical for addressing these challenges. Ethical frameworks should include guidelines for securely transferring and storing data, as well as protocols for managing large datasets. Employing data anonymization techniques can further enhance privacy while maintaining the utility of the data for research purposes (IABAC®, 2023; Karapetyan, 2024; Sui et al., 2023).

Ensuring informed consent is a cornerstone of ethical research, particularly when using wearable devices. Participants must be made fully aware of the scope of data collection, the risks involved, and the entities that will have access to their data. However, the technical complexity of wearable devices can make it challenging to communicate these details effectively. To address this issue, researchers should simplify the language used in consent forms and provide visual aids or interactive tools to explain data collection processes. Ongoing communication with participants is also vital, ensuring they remain informed about how their data is being used throughout the study (Karapetyan, 2024; Sui et al., 2023).

While wearable devices offer innovative solutions for health research, there is a risk of over-reliance on these technologies at the expense of traditional methods. Wearable devices may not capture the full scope of human experiences, and their use could exclude populations without access to technology or digital literacy. To maintain inclusivity and rigor in research, researchers should strive for a balanced approach that integrates wearable devices with traditional methods. Combining quantitative data from wearables with qualitative insights from interviews or surveys can provide a more comprehensive understanding of participant experiences. The lack of standardized protocols for using wearable devices in research poses challenges for consistency and comparability across studies. Variability in device types, data collection methods, and analysis techniques can hinder the development of generalizable conclusions (Karapetyan, 2024; Silva, 2024; Sui et al., 2023).

Creating standardized guidelines for wearable device research is essential for advancing the field. These guidelines should address device selection, data collection protocols, and ethical considerations, ensuring consistency and reliability across diverse research contexts. Developing robust ethical frameworks is crucial for addressing the challenges associated with wearable device research. These frameworks should encompass data privacy, security, informed consent, and equitable access to technology. Collaboration between researchers, device manufacturers, and policymakers is essential to establish comprehensive guidelines that protect participant rights while leveraging the benefits of wearable technologies (Lewczak et al., 2024; Sui et al., 2023).

11.14 Future Directions

Future directions in rehabilitation technology highlight the potential of advanced tools and systems to transform care delivery, improve accessibility, and personalize treatment. One promising area is **AI-powered biofeedback**, which integrates artificial intelligence into biofeedback systems to provide adaptive and predictive feedback for patients. By analyzing large datasets in real time, AI enables the delivery of personalized and anticipatory insights, enhancing patient outcomes and optimizing therapy. Similarly, the development of **home-based rehabilitation** tools, such as wearable sensors and mobile applications, is making rehabilitation more accessible. These user-friendly and affordable devices allow patients to continue their therapy in the comfort of their homes, particularly benefiting those in remote or underserved areas (Barzegar Khanghah et al., 2023).

A **multimodal approach** to rehabilitation is gaining traction, combining biofeedback with pharmacological therapies and cognitive interventions. This comprehensive method addresses both physical and cognitive challenges, offering a holistic framework for recovery. Moreover, **tele-rehabilitation** is expanding the reach of care by leveraging telemedicine platforms for real-time monitoring and remote interaction between patients and clinicians. Advanced tele-rehabilitation systems not only improve accessibility but also maintain the quality of care, making rehabilitation more inclusive and convenient. Innovations in **sensor technology** continue to enhance the scope of rehabilitation. Motion-capturing sensors, vital parameter monitors, and activity trackers are increasingly integrated into wearable devices and smartphones, enabling the collection and analysis of real-time patient data. These advancements allow for better assessment of patient progress and more tailored interventions. Furthermore, the application of **machine learning** in rehabilitation offers the potential to analyze vast amounts of data and predict patient outcomes. Intelligent systems powered by AI can adapt dynamically to a patient's progress, ensuring that treatment plans are personalized and evolve over time (Arntz et al., 2023; Barzegar Khanghah et al., 2023).

While these technological advancements promise significant benefits, they also bring ethical challenges. Concerns surrounding **data privacy, security, and**

ownership must be addressed, especially as big data and AI become central to rehabilitation practices. It is critical to strike a balance between leveraging technology and preserving human expertise to ensure ethical and effective care. Efforts toward **standardization** are equally important, with a focus on developing protocols and guidelines for implementing advanced technologies across various conditions and populations. Standardization would enable consistent application and facilitate the comparison of outcomes across studies. Collaborative research is another key future direction. Interdisciplinary partnerships between AI developers, clinicians, and policymakers are essential to ensure that AI integration in rehabilitation is both reliable and actionable. These collaborations will help address real-world clinical challenges while fostering innovation grounded in patient needs. Moreover, **ethical and responsible AI integration** must remain a priority, with transparency, accountability, and fairness guiding the development and deployment of these technologies. This ensures that AI applications in rehabilitation are not only effective but also respectful of patient rights and societal values (Arntz et al., 2023; Barzegar Khanghah et al., 2023; Rasa, 2024).

11.15 Conclusion

Rehabilitation is entering a transformative era driven by technological advancements, cutting-edge research, and an evolving understanding of the human body's capacity for recovery. Across neurological, orthopedic, motor, and sports rehabilitation, innovations such as biofeedback, wearable devices, virtual and augmented reality, robotics, and artificial intelligence are redefining therapeutic possibilities. These technologies not only enhance recovery but also empower patients by offering more personalized, accessible, and engaging interventions. One of the most promising advancements is biofeedback, which provides real-time physiological insights to patients, enabling them to actively participate in their recovery. Its diverse applications—from improving motor control in stroke survivors to enhancing athletic performance—underscore its adaptability and effectiveness. For instance, EMG biofeedback aids muscle activation, while EEG biofeedback helps regulate brain activity for cognitive and motor improvements. Combined with robotics, wearable sensors, and virtual reality, biofeedback is becoming a cornerstone of modern rehabilitation strategies.

Wearable devices and smart sensors have made rehabilitation more accessible, especially in remote and underserved areas. These technologies enable continuous monitoring, provide real-time feedback, and support home-based therapy, which is crucial for maintaining consistency in care. For example, exoskeletons assist individuals with spinal cord injuries in regaining mobility, while wearable sensors offer valuable insights into movement patterns during recovery. Virtual and augmented reality have also proven transformative by creating immersive environments that enhance motivation and engagement in therapy, particularly for gait training in stroke survivors. Robotic-assisted rehabilitation, especially for post-stroke upper limb recovery, demonstrates how precise, repetitive motion training can facilitate neuroplasticity and motor recovery. Although challenges remain in translating gains in impairment reduction into improved daily function, robotic devices show promise as a valuable adjunct to traditional therapy. Similarly, proprioceptive training has proven essential in sports rehabilitation, helping athletes refine movement patterns, improve balance, and reduce injury risks.

Despite these advancements, significant challenges remain. High costs, limited accessibility in resource-constrained settings, and the complexity of implementing advanced technologies present barriers to equitable healthcare. Ethical considerations, particularly around data privacy, ownership, and security, are also increasingly relevant as rehabilitation becomes more reliant on connected devices and AI-driven systems. Addressing these issues requires collaboration among researchers, clinicians, policymakers, and technology developers to create inclusive, transparent, and effective solutions. Future directions in rehabilitation emphasize integrating artificial intelligence, machine learning, and multimodal approaches to personalize care and predict outcomes. AI-powered biofeedback and wearable devices are already enabling tailored interventions that adapt dynamically to patient progress. Tele-rehabilitation, supported by advancements in telemedicine and digital platforms, offers the potential to extend the reach of care and ensure continuity even in remote areas. Meanwhile, efforts toward standardizing protocols and guidelines will improve the consistency and reliability of these technologies across diverse conditions and populations. At the heart of these developments is the need to balance innovation with human-centered care. Technology must complement, not replace, the expertise and compassion of healthcare providers. By leveraging the strengths of both technology and traditional methods, rehabilitation can become more inclusive, adaptable, and effective.

References

ACL Reconstruction Surgery: Procedure and Recovery | HSS. (n.d.). *Hospital for special surgery*. Retrieved February 4, 2025, from https://www.hss.edu/condition-list_acl-surgery.asp

ACL Rehabilitation Phase 1 | MOON Knee ACL Research. (n.d.). [Video recording]. In *Https://acltear.info/*. Retrieved February 4, 2025, from https://acltear.info/anterior-cruciate-ligament-rehabilitation/acl-rehabilitation-phase-1/

Aderinto, N., Olatunji, G., Abdulbasit, M. O., Edun, M., Aboderin, G., & Egbunu, E. (2023). Exploring the efficacy of virtual reality-based rehabilitation in stroke: A narrative review of current evidence. *Annals of Medicine, 55*(2), 2285907. https://doi.org/10.1080/07853890.2023.2285907

Aman, J. E., Elangovan, N., Yeh, I.-L., & Konczak, J. (2015). The effectiveness of proprioceptive training for improving motor function: A systematic review. *Frontiers in Human Neuroscience, 8*, 1075. https://doi.org/10.3389/fnhum.2014.01075

Anterior Cruciate Ligament (ACL) Rehabilitation—Physiopedia. (n.d.). Retrieved February 4, 2025, from https://www.physio-pedia.com/Anterior_Cruciate_Ligament_(ACL)_Rehabilitation

Arntz, A., Weber, F., Handgraaf, M., Lällä, K., Korniloff, K., Murtonen, K.-P., Chichaeva, J., Kidritsch, A., Heller, M., Sakellari, E., Athanasopoulou, C., Lagiou, A., Tzonichaki, I., Salinas-

Bueno, I., Martínez-Bueso, P., Velasco-Roldán, O., Schulz, R.-J., & Grüneberg, C. (2023). Technologies in home-based digital rehabilitation: Scoping review. *JMIR Rehabilitation and Assistive Technologies, 10*, e43615. https://doi.org/10.2196/43615

Banyai, A. D., & Brişan, C. (2024). Robotics in physical rehabilitation: Systematic review. *Healthcare, 12*(17), Article 17. https://doi.org/10.3390/healthcare12171720

Barzegar Khanghah, A., Fernie, G., & Roshan Fekr, A. (2023). A novel approach to tele-rehabilitation: Implementing a biofeedback system using machine learning algorithms. *Machine Learning with Applications, 14*, 100499. https://doi.org/10.1016/j.mlwa.2023.100499

Bates, K. (2022, February 2). Biofeedback for low back pain & its impacts on motor learning. *mTrigger*. https://www.mtrigger.com/biofeedback-lbp-motor-learning/

Biofeedback. (n.d.-a). *Cleveland clinic*. Retrieved February 4, 2025, from https://my.clevelandclinic.org/health/treatments/13354-biofeedback

Biofeedback. (n.d.-b). *Physiopedia*. Retrieved February 4, 2025, from https://www.physio-pedia.com/Biofeedback

Biofeedback as a Treatment of Chronic Pain. (2024). https://www.southcarolinablues.com/web/public/brands/medicalpolicy/external-policies/biofeedback-as-a-treatment-of-chronic-pain/

blackrockortho. (2024, August 20). *The road to recovery: What to expect after ACL reconstruction surgery*. Blackrock Orthopedics. https://blackrockortho.com/the-road-to-recovery-what-to-expect-after-acl-reconstruction-surgery/.

Boikanyo, K., Zungeru, A. M., Sigweni, B., Yahya, A., & Lebekwe, C. (2023). Remote patient monitoring systems: Applications, architecture, and challenges. *Scientific African, 20*, e01638. https://doi.org/10.1016/j.sciaf.2023.e01638

Bove, L. A. (2019). Increasing patient engagement through the use of wearable technology. *The Journal for Nurse Practitioners, 15*(8), 535–539. https://doi.org/10.1016/j.nurpra.2019.03.018

Bowman, T., Gervasoni, E., Arienti, C., Lazzarini, S. G., Negrini, S., Crea, S., Cattaneo, D., & Carrozza, M. C. (2021). Wearable devices for biofeedback rehabilitation: A systematic review and meta-analysis to design application rules and estimate the effectiveness on balance and gait outcomes in neurological diseases. *Sensors, 21*(10), Article 10. https://doi.org/10.3390/s21103444

Casellato, C., Ambrosini, E., Galbiati, A., Biffi, E., Cesareo, A., Beretta, E., Lunardini, F., Zorzi, G., Sanger, T. D., & Pedrocchi, A. (2019). EMG-based vibro-tactile biofeedback training: Effective learning accelerator for children and adolescents with dystonia? A pilot crossover trial. *Journal of Neuroengineering and Rehabilitation, 16*(1), 150. https://doi.org/10.1186/s12984-019-0620-y

Chaput, M., Ness, B. M., Lucas, K., & Zimney, K. J. (2022). A multi-systems approach to human movement after ACL reconstruction: The nervous system. *International Journal of Sports Physical Therapy, 17*(1), 47–59. https://doi.org/10.26603/001c.30020

Chen, Z.-J., He, C., Xu, J., Zheng, C.-J., Wu, J., Xia, N., Hua, Q., Xia, W.-G., Xiong, C.-H., & Huang, X.-L. (2023). Exoskeleton-assisted anthropomorphic movement training for the upper limb after stroke: The EAMT randomized trial. *Stroke, 54*(6), 1464–1473. https://doi.org/10.1161/STROKEAHA.122.041480

Cheng, C.-F., & Lin, C. J. (2023). Building a low-cost wireless biofeedback solution: Applying design science research methodology. *Sensors, 23*(6), Article 6. https://doi.org/10.3390/s23062920

Cooper, M. (2019, August 1). Reconstituting proprioception training for sports performance. *SimpliFaster*. https://simplifaster.com/articles/proprioception-training-sports-performance/

Cui, F., Wang, Y., Lei, S., & Shi, Y. (2024). CardboardHRV: Bridging virtual reality and biofeedback with a cost-effective heart rate variability system. *Extended Abstracts of the CHI Conference on Human Factors in Computing Systems*, 1–6. https://doi.org/10.1145/3613905.3650782

Daryabor, A., Naghizad, J., Jamehbozorgi, A., Kalantari, K., & Rezaei, M. (2023). The evaluation of electromyography biofeedback on proprioception and balance in healthy young athletes. *Journal of Modern Rehabilitation, 17*(3). https://doi.org/10.18502/jmr.v17i3.13074

de Rooij, I. J. M., van de Port, I. G. L., Punt, M., Abbink-van Moorsel, P. J. M., Kortsmit, M., van Eijk, R. P. A., Visser-Meily, J. M. A., & Meijer, J.-W. G. (2021). Effect of virtual reality gait training on participation in survivors of subacute stroke: A randomized controlled trial. *Physical Therapy, 101*(5), pzab051. https://doi.org/10.1093/ptj/pzab051

Dimyan, M. A., & Cohen, L. G. (2011). Neuroplasticity in the context of motor rehabilitation after stroke. *Nature Reviews. Neurology, 7*(2), 76–85. https://doi.org/10.1038/nrneurol.2010.200

Doğan-Aslan, M., Nakipoğlu-Yüzer, G. F., Doğan, A., Karabay, İ., & Özgirgin, N. (2012). The effect of electromyographic biofeedback treatment in improving upper extremity functioning of patients with hemiplegic stroke. *Journal of Stroke and Cerebrovascular Diseases, 21*(3), 187–192. https://doi.org/10.1016/j.jstrokecerebrovasdis.2010.06.006

Emish, M., & Young, S. D. (2024). Remote wearable neuroimaging devices for health monitoring and neurophenotyping: A scoping review. *Biomimetics, 9*(4), 237. https://doi.org/10.3390/biomimetics9040237

Giggins, O. M., Persson, U. M., & Caulfield, B. (2013). Biofeedback in rehabilitation. *Journal of Neuroengineering and Rehabilitation, 10*, 60. https://doi.org/10.1186/1743-0003-10-60

Grosmaire, A.-G., Pila, O., Breuckmann, P., & Duret, C. (2022). Robot-assisted therapy for upper limb paresis after stroke: Use of robotic algorithms in advanced practice. *Neuro Rehabilitation, 51*(4), 577–593. https://doi.org/10.3233/NRE-220025

Hasan, S. K., Bhujel, S. B., & Niemiec, G. S. (2024). Emerging trends in human upper extremity rehabilitation robot. *Cognitive Robotics, 4*, 174–190. https://doi.org/10.1016/j.cogr.2024.09.001

How Digital Balance Boards Enhance Stability with Biofeedback and Obje. (2024, October 3). *Techcare store*. https://techcareinnovation.com/blogs/blog/how-digital-balance-boards-enhance-stability-with-biofeedback-and-objective-measurement

Huang, H., Wolf, S. L., & He, J. (2006). Recent developments in biofeedback for neuromotor rehabilitation. *Journal of Neuroengineering and Rehabilitation, 3*(1), 11. https://doi.org/10.1186/1743-0003-3-11

IABAC®. (2023, October 17). *Data ethics in the Internet of Things (IoT): Balancing convenience and privacy*. IABAC®. https://iabac.org/blog/data-ethics-in-the-internet-of-things-iot-balancing-convenience-and-privacy

Jaswal, S., Lo, J., Sithamparanathan, G., & Nowrouzi-Kia, B. (2023). The era of technology in healthcare: An evaluation of telerehabilitation on patient outcomes—A systematic review and meta-analysis protocol. *Systematic Reviews, 12*, 76. https://doi.org/10.1186/s13643-023-02248-8

Karapetyan, M. (2024, June 25). *Ethics of data sharing and digital privacy*. Viterbi Conversations in Ethics. https://vce.usc.edu/volume-7-issue-2/ethics-of-data-sharing-and-digital-privacy/

Kerr, A. L., Cheng, S.-Y., & Jones, T. A. (2011). Experience-dependent neural plasticity in the adult damaged brain. *Journal of Communication Disorders, 44*(5), 538–548. https://doi.org/10.1016/j.jcomdis.2011.04.011

Kim, M., & Kaneko, F. (2023). Virtual reality-based gait rehabilitation intervention for stroke individuals: A scoping review. *Journal of Exercise Rehabilitation, 19*(2), 95. https://doi.org/10.12965/jer.2346114.057

Kopańska, M., Ochojska, D., Mytych, W., Lis, M. W., & Banaś-Ząbczyk, A. (2022). Development of a brain wave model based on the quantitative analysis of EEG and EEG biofeedback therapy in patients with panic attacks during the COVID-19 pandemic. *Scientific Reports, 12*(1), 14908. https://doi.org/10.1038/s41598-022-19068-w

Kwakkel, G., Veerbeek, J. M., van Wegen, E. E. H., & Wolf, S. L. (2015). Constraint-induced movement therapy after stroke. *The Lancet Neurology, 14*(2), 224–234. https://doi.org/10.1016/S1474-4422(14)70160-7

Lagriffoul, F. (2022). A schema-based robot controller complying with the constraints of biological systems. *Frontiers in Neurorobotics, 16*. https://doi.org/10.3389/fnbot.2022.836767

Levin, M. F., & Demers, M. (2021). Motor learning in neurological rehabilitation—PubMed. *Disability and Rehabilitation, 43*(24), 3445–3453. https://doi.org/10.1080/09638288.2020.1752317

Lewczak, Z., Mitchell, M., Lewczak, Z., & Mitchell, M. G. (2024). Wearable technology and chronic illness: Balancing justice and care ethics. *Cureus, 16*(11). https://doi.org/10.7759/cureus.73686

Lingampally, P. K., Ramanathan, K. C., Shanmugam, R., Cepova, L., & Salunkhe, S. (2024). Wearable assistive rehabilitation robotic devices—A comprehensive review. *Machines, 12*(6), Article 6. https://doi.org/10.3390/machines12060415

Martinez-Hernandez, U., Metcalfe, B., Assaf, T., Jabban, L., Male, J., & Zhang, D. (2021). Wearable assistive robotics: A perspective on current challenges and future trends. *Sensors (Basel, Switzerland), 21*(20), 6751. https://doi.org/10.3390/s21206751

Morizio, C., Compagnat, M., Boujut, A., Labbani-Igbida, O., Billot, M., & Perrochon, A. (2022). Immersive virtual reality during robot-assisted gait training: Validation of a new device in stroke rehabilitation. *Medicina, 58*(12), 1805. https://doi.org/10.3390/medicina58121805

Neuroplasticity in stroke and brain injury: Shaping modern rehabilitation…. (2024). *MedLink Neurology*. Retrieved December 29, 2024, from https://www.medlink.com/news/neuroplasticity-in-stroke-and-brain-injury-shaping-modern-rehabilitation-practices

Neurorehabilitation. (n.d.). Retrieved February 4, 2025, from https://www.henryford.com/services/rehabilitation/outpatient/neurorehabilitation

Nudo, R. J., Birute, W. M., Sifuentes, F., & Milliken, G. W. (1996). Neural substrates for the effects of rehabilitative training on motor recovery after ischemic infarct | Science. *Science, 272*(5629), 1791–1794. https://doi.org/10.1126/science.272.5269.1791

Optimum Physical Therapy and Sports Rehab Neurological Rehabilitation | Bringing You Back. (n.d.). Retrieved February 4, 2025, from https://optimumoutpatientrehabilitation.com/77459-physical-therapy-specialties-services/neurological-rehabilitation/

Ovadia-Blechman, Z., Tarrasch, R., Velicki, M., & Chalutz Ben-Gal, H. (2022). Reducing test anxiety by device-guided breathing: A pilot study. *Frontiers in Psychology, 13*, 678098. https://doi.org/10.3389/fpsyg.2022.678098

Peake, J. M., Kerr, G., & Sullivan, J. P. (2018). A critical review of consumer wearables, mobile applications, and equipment for providing biofeedback, monitoring stress, and sleep in physically active populations. *Frontiers in Physiology, 9*, 743. https://doi.org/10.3389/fphys.2018.00743

Peng, X., & Stirling, L. (2023). Effects of biofeedback on muscle effort reduction when holding positions with a powered upper limb exoskeleton. *Proceedings of the Human Factors and Ergonomics Society Annual Meeting, 67*(1), 790–794. https://doi.org/10.1177/21695067231192634

Piedade, S. R., Leite Arruda, B. P., de Vasconcelos, R. A., Parker, D. A., & Maffulli, N. (2023). Rehabilitation following surgical reconstruction for anterior cruciate ligament insufficiency: What has changed since the 1960s?—State of the art. *Journal of ISAKOS, 8*(3), 153–162. https://doi.org/10.1016/j.jisako.2022.10.001

Pizzoli, S. F. M., Marzorati, C., Gatti, D., Monzani, D., Mazzocco, K., & Pravettoni, G. (2021). A meta-analysis on heart rate variability biofeedback and depressive symptoms. *Scientific Reports, 11*(1), 6650. https://doi.org/10.1038/s41598-021-86149-7

Rasa, A. R. (2024). Artificial intelligence and its revolutionary role in physical and mental rehabilitation: A review of recent advancements. *BioMed Research International, 2024*(1), 9554590. https://doi.org/10.1155/bmri/9554590

Rodgers, M. M., Alon, G., Pai, V. M., & Conroy, R. S. (2019). Wearable technologies for active living and rehabilitation: Current research challenges and future opportunities. *Journal of Rehabilitation and Assistive Technologies Engineering, 6*, 2055668319839607. https://doi.org/10.1177/2055668319839607

Saka, T. (2014). Principles of postoperative anterior cruciate ligament rehabilitation—PMC. *World Journal of Orthopedics, 5*(4), 450–459. https://doi.org/10.5312/wjo.v5.i4.450

Sears, B. (2024). *Biofeedback therapy: Control involuntary actions in your body*. Verywell Health. https://www.verywellhealth.com/biofeedback-therapy-5211371

Seneviratne, S., Hu, Y., Nguyen, T., & Lan, G. (2017). A survey of wearable devices and challenges. *IEEE Communications Surveys & Tutorials, 99*(1), 1. https://doi.org/10.1109/COMST.2017.2731979

Seo. (2024, February 2). What is muscle biofeedback? *ScienceBeam*. https://sciencebeam.com/what-is-muscle-biofeedback/

da Silva, J. P. (2024). *Privacy data ethics of wearable digital health technology | Center for Digital Health | Engineering | Brown University*. https://cdh.brown.edu/news/2023-05-04/ethics-wearables

Siviy, C., Baker, L. M., Quinlivan, B. T., Porciuncula, F., Swaminathan, K., Awad, L. N., & Walsh, C. J. (2022). Opportunities and challenges in the development of exoskeletons for locomotor assistance. *Nature Biomedical Engineering, 7*(4), 456. https://doi.org/10.1038/s41551-022-00984-1

Spencer, J., Wolf, S. L., & Kesar, T. M. (2021). Biofeedback for post-stroke gait retraining: A review of current evidence and future research directions in the context of emerging technologies. *Frontiers in Neurology, 12*, 637199. https://doi.org/10.3389/fneur.2021.637199

Sui, A., Sui, W., Liu, S., & Rhodes, R. (2023). Ethical considerations for the use of consumer wearables in health research. *DIGITAL HEALTH, 9*, 20552076231153740. https://doi.org/10.1177/20552076231153740

Takeuchi, N., & Izumi, S.-I. (2013). Rehabilitation with poststroke motor recovery: A review with a focus on neural plasticity. *Stroke Research and Treatment, 2013*, 128641. https://doi.org/10.1155/2013/128641

Tamburin, S., Smania, N., Saltuari, L., Hoemberg, V., & Sandrini, G. (2019). Editorial: New advances in neurorehabilitation. *Frontiers in Neurology, 10*, 1090. https://doi.org/10.3389/fneur.2019.01090

Taub, E., Uswatte, G., & Elbert, T. (2002). New treatments in neurorehabiliation founded on basic research. *Nature Reviews Neuroscience, 3*(3), 228–236. https://doi.org/10.1038/nrn754

Valkenborghs, S. R., Visser, M. M., Dunn, A., Erickson, K. I., Nilsson, M., Callister, R., & van Vliet, P. (2017). AExaCTT – Aerobic Exercise and Consecutive Task-specific Training for the upper limb after stroke: Protocol for a randomised controlled pilot study. *Contemporary Clinical Trials Communications, 7*, 179–185. https://doi.org/10.1016/j.conctc.2017.07.009

Virtual Reality for Individuals Affected by Stroke. (n.d.). *Physiopedia*. Retrieved December 28, 2024, from https://www.physio-pedia.com/Virtual_Reality_for_Individuals_Affected_by_Stroke

Winstein, C., Lewthwaite, R., Blanton, S. R., Wolf, L. B., & Wishart, L. (2014). Infusing motor learning research into neurorehabilitation practice: A historical perspective with case exemplar from the accelerated skill acquisition program. *Journal of Neurologic Physical Therapy : JNPT, 38*(3), 190–200. https://doi.org/10.1097/NPT.0000000000000046

Winter, S. C., Obara, L. M., & McMahon, S. (2020). Intimate partner violence: A key correlate of women's physical and mental health in informal settlements in Nairobi, Kenya. *PLOS One, 15*(4), e0230894. https://doi.org/10.1371/journal.pone.0230894

Yang, R., Yuan, Q., Zhang, W., Cai, H., & Wu, Y. (2024). Application of artificial intelligence in rehabilitation science: A scientometric investigation utilizing citespace. *SLAS Technology, 29*(4). https://doi.org/10.1016/j.slast.2024.100162

Yang, Y.-W., Pan, W.-X., & Xie, Q. (2020). Combined effect of repetitive transcranial magnetic stimulation and physical exercise on cortical plasticity. *Neural Regeneration Research, 15*(11), 1986. https://doi.org/10.4103/1673-5374.282239

Yilmaz, A., & Dedeli, O. (2016). Assessment of anxiety, depression, loneliness and stigmatization in patients with tuberculosis. *Acta Paulista de Enfermagem, 29*(5), 549–557. https://doi.org/10.1590/1982-0194201600076

Young, M. J., Lin, D. J., & Hochberg, L. R. (2021). Brain-computer interfaces in neurorecovery and neurorehabilitation. *Seminars in Neurology, 41*(2), 206. https://doi.org/10.1055/s-0041-1725137

Zhang, K., Chen, X., Liu, F., Tang, H., Wang, J., & Wen, W. (2018). System framework of robotics in upper limb rehabilitation on poststroke motor recovery. *Behavioural Neurology*. https://doi.org/10.1155/2018/6737056

Chapter 12
Ethical and Privacy Issues in Technology

12.1 Introduction

Technology is transforming the way we approach health care and rehabilitation, particularly in the field of neuropsychological rehabilitation. Tools like wearable devices, virtual reality (VR), and artificial intelligence (AI) are making it possible to tailor treatments to individual needs, monitor progress in real time, and even predict recovery patterns. For patients recovering from conditions like strokes, traumatic brain injuries, or degenerative diseases, such as Alzheimer's, these innovations are more than just tools—they're lifelines offering hope for a better quality of life.

Take Jeni, for example. Jeni, a high-school teacher in her 50 s, suffered a traumatic brain injury after a car accident. As part of her recovery, she was introduced to an AI-powered cognitive rehabilitation program. It provided personalized exercises to help her regain focus, memory, and problem-solving skills. The technology played a pivotal role in her recovery, helping her feel more in control of her progress. But months later, Jeni learned that her personal data, including sensitive details about her rehabilitation, had been shared with third-party organizations without her explicit consent. What had initially felt like a breakthrough in her recovery left her feeling betrayed and vulnerable.

Jeni's story highlights the dual-edged nature of technology in neuropsychological rehabilitation. On the one hand, these advancements offer incredible possibilities to improve outcomes and empower patients. On the other hand, they raise important ethical questions: Who owns the data these tools collect? How can we ensure that patients like Jeni are informed about how their information will be used? And how do we protect their privacy and dignity in a world where data is a valuable commodity? The current chapter explores these challenges and opportunities. It looks at the ethical principles that should guide the use of technology, delves into

K. J. Reddy, *Innovations in Neurocognitive Rehabilitation*,
https://doi.org/10.1007/978-3-031-88117-6_12

the privacy risks patients face, and offers strategies to balance innovation with responsibility.

12.1.1 Technology and Transformation

Neuropsychological rehabilitation has come a long way, thanks to advancements in technology. Wearable devices, such as smartwatches or biosensors, allow clinicians to track a patient's physical and cognitive activity in real time. For someone recovering from a stroke, this might mean monitoring heart rate, movement, or even subtle patterns in brain activity that signal cognitive recovery. Similarly, VR-based rehabilitation programs immerse patients in controlled environments, helping them relearn essential skills like walking, navigating a busy street, or engaging in social interactions. These tools not only accelerate recovery but also build confidence, allowing patients to practice in a safe, supportive space. AI, meanwhile, is playing a transformative role by analyzing complex data to uncover trends that clinicians might miss. For example, AI algorithms can identify subtle improvements in memory or reaction time, providing insights that inform personalized treatment plans. These tools are not just about efficiency, they are about creating more meaningful, patient-centered care. But with all this potential comes a set of ethical challenges. As these tools collect more detailed and intimate data about patients, we must grapple with questions about privacy, consent, and the fairness of their use (Klingler et al., 2017).

12.1.2 The Ethical Challenges of Technology

One of the most fundamental principles in health care is **autonomy**—ensuring that patients have control over their care and the choices they make. For technology to support autonomy, patients need to fully understand how tools like AI or wearables work. They need to know what data is being collected, how it will be used, and who will have access to it. But studies show that many patients and even clinicians lack this understanding. For example, social media platforms used for health research often collect vast amounts of data without users being fully aware of what they've agreed to (Ford et al., 2021).

This lack of transparency raises concerns about **informed consent**. In Jeni's case, her consent for data collection was buried in lengthy, jargon-filled terms and conditions. Like many patients, she agreed because she trusted the tool and her healthcare team. But when that trust was broken, the emotional toll was significant—an overlooked consequence of poor ethical practices.

Another critical principle is **beneficence**, or the duty to do good. Technologies in neuropsychological rehabilitation should enhance patient well-being, but they can also unintentionally cause harm. For instance, an unvalidated cognitive app might provide ineffective or even misleading results, causing frustration or delaying recovery. Similarly, AI systems trained on biased data sets can lead to disparities in treatment recommendations, perpetuating inequities in health care (Ford et al., 2021; Klingler et al., 2017).

The principle of **justice**, which ensures fairness and equity, is also at risk in the current landscape. While technology has the potential to make care more accessible, it often remains out of reach for those in underserved communities. High costs, limited internet access, and language barriers mean that many patients cannot benefit from these advancements. As a result, the digital divide risks deepening existing health disparities rather than bridging them (Ford et al., 2021).

12.1.3 Data Privacy Concerns

In neuropsychological rehabilitation, privacy is more than just an abstract concept—it is a cornerstone of trust. The data collected by wearables, apps, and other tools often includes deeply personal information, from brain activity patterns to emotional responses during therapy. One of the biggest concerns is data ownership. Who owns the data generated by these tools—the patient, the clinician, or the company that created the device? In many cases, patients unknowingly give up ownership when they agree to use a device. Companies may then share or sell this data, as social media platforms often do, for purposes far removed from patient care. Data security is another pressing issue. High-profile data breaches in healthcare highlight the risks of storing sensitive information. For neuropsychological patients, a breach could expose details about their cognitive and emotional health, leading to stigma or even discrimination. Finally, there's the issue of transparency. Many tools, especially those powered by AI, operate as "black boxes," where even developers cannot fully explain how decisions are made. This lack of transparency undermines trust and can make it difficult for clinicians to justify using such tools in patient care (Ford et al., 2021).

12.1.4 Bridging the Gap with Innovation Responsibility

Addressing these challenges requires a collaborative, patient-centered approach. One promising strategy is the creation of dedicated "technology ethics boards" to oversee how data is collected, stored, and used. These boards can also advise on issues like informed consent and equitable access. A bottom-up approach to ethics, as suggested in research on social media and health technologies, can also help. By involving patients, clinicians, and developers in conversations about the ethical implications of new tools, we can ensure that technology is designed with real-world concerns in mind. Regulatory frameworks like GDPR and HIPAA provide valuable guidelines for safeguarding privacy. But as technology evolves, these regulations must adapt to address emerging challenges, such as AI bias and cross-border data sharing. Technology has the potential to revolutionize neuropsychological rehabilitation, offering tools that empower patients and enhance recovery. But as Jeni's experience reminds us, innovation must go hand in hand with responsibility. By prioritizing ethical practices, protecting privacy, and ensuring equitable access, we can create a future where technology transforms rehabilitation not just in function but also in dignity and trust (Ford et al., 2021).

12.1.5 Ethical Principles in Healthcare Technology

The ethical use of technology in health care is grounded in several core principles:

1. **Autonomy: Supporting Patient Decision-Making**: The ethical principle of autonomy is foundational in health care, emphasizing that patients must retain the ability to make informed decisions about their care. In the context of neuropsychological rehabilitation, this principle becomes particularly critical, as many patients may experience cognitive or emotional impairments that challenge their ability to exercise autonomy. Technologies designed for rehabilitation, such as AI-powered diagnostic tools and wearable devices, must prioritize patient empowerment by providing clear, accessible information about their functionality and implications. A patient recovering from a stroke might be introduced to a virtual reality (VR) therapy program. While the program could offer significant benefits for relearning motor skills or practicing real-world tasks, it is essential that the patient understands how the therapy works, what data is being collected during sessions, and how that data will be used. Studies show that many patients and even healthcare providers struggle to understand the complexities of digital tools, particularly when it comes to advanced technologies like artificial intelligence. To ensure autonomy, healthcare providers must simplify technical explanations, offer visual aids, and involve caregivers when necessary. This approach fosters trust, supports informed decision-making, and respects patients' rights to choose treatments that align with their personal values and goals (*Ethics in Health Care*, 2023; Varkey, 2021).
2. **Beneficence: Promoting Patient Well-Being**: Beneficence underscores the responsibility of healthcare providers and technology developers to act in the best interests of patients. In neuropsychological rehabilitation, this principle is exemplified by the use of technologies like brain–computer interfaces (BCIs) and personalized cognitive rehabilitation apps, which are designed to enhance recovery and improve quality of life. A patient with traumatic brain injury might benefit from an AI-driven tool that monitors cognitive recovery and offers tailored recommendations for exercises. These tools can significantly improve outcomes by providing insights that clinicians may not otherwise detect. However, beneficence also requires that these technologies undergo rigorous validation to ensure their efficacy and safety. While the primary goal is to promote well-being,

this principle also demands a careful evaluation of the potential unintended consequences of new innovations. Poorly designed tools may produce inaccurate results, leading to frustration, delayed recovery, or harm. By focusing on evidence-based practices and ongoing monitoring of technological effectiveness, healthcare providers can uphold the principle of beneficence (*Ethics in Health Care*, 2023; *Understanding the Four Main Ethical Principles in Healthcare*, 2024; Varkey, 2021).

3. **Non-Maleficence: Avoiding Harm**: Closely linked to beneficence, the principle of non-maleficence emphasizes the importance of minimizing risks and avoiding harm to patients. This principle is particularly relevant in the use of healthcare technologies, where the potential for harm can manifest in various ways, including adverse effects, data breaches, and biased algorithms. One significant area of concern is the security of patient data. Wearable devices, mobile apps, and AI systems often collect vast amounts of sensitive information, from cognitive performance metrics to emotional health data. If this information is not adequately protected, it could be exposed in data breaches or misused by third parties. For example, a cyberattack on a rehabilitation device could compromise sensitive patient information, leading to emotional distress, reputational damage, or even discrimination (*Ethics in Health Care*, 2023; Varkey, 2021). Another challenge lies in the potential biases of AI algorithms. If an AI system is trained on data sets that do not represent diverse populations, it may produce biased outcomes that disproportionately harm certain groups of patients. For example, an AI-powered cognitive assessment tool might misdiagnose or underdiagnosed conditions in patients from underrepresented backgrounds due to biases in its training data (*Understanding the Four Main Ethical Principles in Healthcare*, 2024). To uphold non-maleficence, developers and clinicians must prioritize rigorous testing, robust security measures, and continuous monitoring to identify and address potential harms before they reach patients (*Ethics in Health Care*, 2023).
4. **Justice: Ensuring Equitable Access**: The principle of justice emphasizes fairness and equality in the distribution of healthcare resources. In neuropsychological rehabilitation, this means ensuring that technologies like virtual reality systems, wearable devices, and AI tools are accessible to all individuals, regardless of their socioeconomic status, geographic location, or ethnicity. While technology has the potential to democratize health care, it can also exacerbate existing disparities. For example, expensive rehabilitation tools may be inaccessible to patients in low-income communities, while patients in rural areas may lack the internet connectivity needed to use telehealth platforms or AI-driven apps. Additionally, language barriers and cultural biases in technology design can further exclude marginalized groups (*Ethics in Health Care*, 2023; *Understanding the Four Main Ethical Principles in Healthcare*, 2024). Addressing these disparities requires a commitment to creating inclusive, affordable, and accessible technologies. Subsidized programs, open-source software, and tools designed with diverse populations in mind are essential steps toward achieving justice in neuropsychological rehabilitation. Policymakers and healthcare organizations must also advocate for equitable distribution of resources and

provide training to ensure that all patients can fully benefit from these innovations (*Understanding the Four Main Ethical Principles in Healthcare*, 2024).

5. **Confidentiality: Protecting Patient Information**: Confidentiality is a cornerstone of ethical healthcare practice, requiring the protection of patients' sensitive data from unauthorized access or misuse. In neuropsychological rehabilitation, where the data collected can include intimate details about a patient's cognitive and emotional health, safeguarding confidentiality is particularly important. One of the most pressing challenges in this area is the question of data ownership. Many patients are unaware that the data collected by their wearable devices or apps is often owned by the technology companies that develop them, rather than the patients themselves. This lack of ownership can lead to misuse, such as data being sold to third parties without the patient's knowledge or consent (*Understanding the Four Main Ethical Principles in Healthcare*, 2024; Varkey, 2021). Moreover, the increasing sophistication of data collection tools has made it more difficult to ensure true anonymity. Even when data is de-identified, advanced re-identification techniques can compromise patient privacy. For example, patterns in brain activity collected by a rehabilitation device could potentially be linked back to an individual, posing risks to their privacy and dignity (Varkey, 2021). To uphold confidentiality, healthcare providers and developers must adopt robust privacy measures. This includes encrypting data, anonymizing information to the greatest extent possible, and clearly communicating to patients how their data will be used. Regulatory frameworks like the General Data Protection Regulation (GDPR) and the Health Insurance Portability and Accountability Act (HIPAA) provide valuable guidelines, but adherence to these regulations is only the beginning. A culture of transparency and respect for patient privacy must guide every aspect of technology use in health care (*Ethics in Health Care*, 2023; *Understanding the Four Main Ethical Principles in Healthcare*, 2024).

12.2 Privacy Issues in Healthcare Technology

As healthcare systems adopt more digital solutions, patient data becomes increasingly vulnerable to breaches, misuse, or exploitation. Privacy concerns are particularly pertinent when considering the following issues:

1. **Data Security**: Tools such as wearable devices, brain–computer interfaces, and AI-powered cognitive assessment platforms often collect information about a patient's mental health, behavior, and recovery progress. While these technologies offer immense benefits, they also make patients vulnerable to data breaches, unauthorized access, and cyberattacks. For instance, imagine Sarah, a young woman recovering from a traumatic brain injury. She uses a wearable device to monitor her cognitive functions and progress. While the device provides valuable insights for her care team, a cyberattack on the platform compromises her

data, exposing intimate details about her cognitive health. Such incidents can erode trust in healthcare systems and leave patients feeling vulnerable and violated. To mitigate these risks, healthcare providers and technology developers must implement robust safeguards. This includes encrypting patient data, conducting regular security audits, and ensuring compliance with data protection standards such as the Health Insurance Portability and Accountability Act (HIPAA) (Chiruvella & Guddati, 2021). Additionally, healthcare organizations must stay ahead of evolving cyber threats by updating their security measures and educating their staff about best practices. By prioritizing data security, we can protect patients like Sarah from unnecessary harm and maintain their trust in technological advancements.

2. **Informed Consent**: Informed consent is a cornerstone of ethical health care and an essential component of using technology in neuropsychological rehabilitation. Patients must fully understand how their data will be collected, used, and shared to make informed decisions about their care. This principle is especially important in the context of advanced technologies, where the complexity of data usage can create confusion and uncertainty. Consider John, a stroke survivor using an AI-driven cognitive rehabilitation platform. John agreed to use the platform but was unaware that his data might be shared with third parties for research purposes. When he later learned about this, he felt betrayed and concerned about how his private information might be used. This highlights the importance of transparent communication and obtaining explicit consent from patients. Dynamic consent models offer a promising solution. Unlike one-time consent forms, dynamic consent allows patients to adjust their preferences as new data uses or stakeholders emerge (Robert, 2024). This approach ensures that patients remain in control of their data and can make decisions that align with their evolving needs and values. By adopting such models, healthcare providers can build trust and empower patients to take an active role in their rehabilitation journey.
3. **Data Ownership and Control**: Who owns the data collected during neuropsychological rehabilitation? This question has sparked significant debate in recent years, as technology companies, healthcare providers, and patients all have stakes in the information generated by digital tools. At its core, the issue of data ownership is about control—who has the right to access, edit, share, or delete the data? For patients like Jeni, who is recovering from a brain injury and uses a wearable device to track her recovery, having control over her data is critical. Without clear ownership rights, Jeni might find herself unable to access her data or prevent it from being used in ways she does not approve of. Patients should have greater control over their health data, including the ability to review, edit, or delete it when necessary (Robert, 2024). Technology developers and healthcare organizations must prioritize transparency by providing patients with clear information about who owns their data and how it will be used. By empowering patients with control, we can address ethical concerns and ensure that technology serves their best interests.
4. **Data Sharing**: Data sharing has the potential to revolutionize neuropsychological rehabilitation by enabling clinicians, researchers, and policymakers to col-

laborate and develop more effective treatments. For example, sharing anonymized patient data across institutions can lead to breakthroughs in understanding brain recovery and cognitive rehabilitation. However, data sharing also raises significant privacy concerns, particularly when patients are not adequately informed about how their data will be used. Consider a case where a patient's anonymized data is shared with researchers studying new rehabilitation techniques. While the intention is noble, if the patient was not informed about this secondary use of their data, it could violate their trust and autonomy. This example underscores the need for clear guidelines around data sharing and respect for patient preferences (Chiruvella & Guddati, 2021). Healthcare organizations must establish transparent policies for data sharing and ensure that patients are aware of how their information might be used. By respecting patient autonomy and adhering to ethical principles, we can strike a balance between advancing research and protecting individual privacy (Robert, 2024).

5. **Anonymization and De-identification**: One way to address privacy concerns is through anonymization and de-identification, processes that strip personal identifiers from patient data to protect their identity. In neuropsychological rehabilitation, these techniques are essential for enabling data sharing while safeguarding patient privacy. However, ensuring that data cannot be re-identified remains a significant challenge, particularly as technology advances. For example, even when data is anonymized, patterns in brain activity or behavioral responses could potentially be traced back to an individual with the right tools. This risk is particularly concerning in the era of big data, where algorithms can analyze massive data sets to uncover subtle connections (Chiruvella & Guddati, 2021; Robert, 2024). To address these challenges, developers must adopt robust anonymization techniques and continuously evaluate their effectiveness. Additionally, patients should be informed about the limitations of anonymization and given the choice to opt-out of data sharing if they are uncomfortable with the risks. By prioritizing transparency and security, we can uphold patient trust while enabling valuable research.

12.3 Ethical Frameworks for Addressing Challenges

12.3.1 The Four Principles Framework

The Four Principles Framework, developed by Beauchamp and Childress, provides a foundational approach for ethical decision-making in health care, including neuropsychological rehabilitation. It is built around four key principles—autonomy, beneficence, non-maleficence, and justice—which guide clinicians and technology developers in making ethical choices.

1. **Autonomy**: Respecting patient autonomy is a cornerstone of ethical health care. In neuropsychological rehabilitation, where patients often face cognitive or

emotional challenges, it is essential to empower them to make informed decisions about their treatment. For instance, a patient using a brain–computer interface (BCI) for cognitive recovery should understand how the device works, what data it collects, and how that data might be used. Without this knowledge, their ability to give informed consent is compromised (Chiruvella & Guddati, 2021; Williamson & Prybutok, 2024). To respect autonomy, healthcare providers must engage patients in open, accessible communication, simplifying complex technological explanations and ensuring that patients are active participants in their care. For patients recovering from conditions like stroke or traumatic brain injury, this respect for autonomy reinforces their dignity and control during what is often a vulnerable time in their lives.

2. **Beneficence**: The principle of beneficence emphasizes acting in the patient's best interest. In neuropsychological rehabilitation, this might involve integrating advanced technologies like virtual reality (VR) or AI-powered cognitive assessment tools to enhance recovery outcomes. For example, VR environments can help patients practice real-world skills, such as navigating public spaces, in a safe and controlled setting (Chiruvella & Guddati, 2021; Williamson & Prybutok, 2024). However, beneficence also requires careful consideration of potential risks and benefits. Not all technologies are equally effective, and unvalidated tools may hinder progress rather than promote it. Clinicians must critically evaluate the technologies they use to ensure that they genuinely enhance patient well-being.
3. **Non-Maleficence**: Non-maleficence, or the principle of "do no harm," requires healthcare providers to minimize risks and avoid harm to patients. This principle is particularly relevant in the context of neuropsychological rehabilitation technologies, where potential risks include data breaches, algorithmic bias, or physical discomfort caused by wearable devices (Chiruvella & Guddati, 2021; Williamson & Prybutok, 2024). Consider the case of an AI-powered cognitive rehabilitation app. If the app's algorithms are based on biased data, it might deliver suboptimal or even harmful recommendations for patients from underrepresented populations. To uphold non-maleficence, developers and clinicians must test technologies thoroughly, address potential biases, and continuously monitor their performance.
4. **Justice**: Justice focuses on ensuring equitable access to healthcare resources, including technologies. In neuropsychological rehabilitation, this means addressing disparities in access to advanced tools like wearable devices or AI-driven platforms. Patients in low-income or rural areas often lack access to these technologies, exacerbating existing health inequities (Chiruvella & Guddati, 2021; Williamson & Prybutok, 2024). Healthcare organizations can promote justice by subsidizing costs for underserved populations, developing low-cost alternatives, and advocating for policies that expand access to digital tools. Ensuring that all patients benefit from technological advancements is essential for achieving fairness in neuropsychological rehabilitation.

12.3.2 The Bioethical Approach

The bioethical approach expands on the Four Principles Framework by emphasizing the balance between respecting patient autonomy and embracing technological innovation. This approach encourages healthcare providers to explore the moral implications of using advanced technologies in treatment and care. In neuropsychological rehabilitation, this might involve considering how a new tool impacts a patient's sense of independence. For instance, while a wearable device that tracks cognitive activity might provide valuable data for clinicians, it could also make a patient feel overly monitored or controlled. The bioethical approach requires healthcare providers to weigh these factors carefully, ensuring that technological innovation enhances, rather than diminishes, the patient's quality of life (Chiruvella & Guddati, 2021).

12.3.3 Fair Information Practice Principles (FIPPs)

The Fair Information Practice Principles (FIPPs) offer a framework for collecting, using, and sharing personal data in ways that respect privacy rights. These principles are especially relevant in neuropsychological rehabilitation, where technologies often collect sensitive health data.

- **Authority**: Under FIPPs, personal data should only be collected and used if authorized. For example, a clinician using an AI tool to track a patient's recovery must ensure that data collection aligns with legal and ethical standards (Paul et al., 2023).
- **Minimization**: Minimization emphasizes collecting only the data necessary to achieve specific, authorized purposes. In neuropsychological rehabilitation, this might mean limiting data collection to cognitive and behavioral metrics directly related to a patient's recovery, rather than gathering unrelated personal information (Paul et al., 2023).
- **Individual Participation**: FIPPs encourage involving individuals in decisions about their personal data. Patients should be informed about how their data will be used and given opportunities to provide input. For instance, a stroke survivor using a cognitive rehabilitation platform should have the ability to review and modify their consent preferences as new data-sharing practices emerge (Paul et al., 2023).
- **Purpose Specification and Use Limitation**: Patients must be informed about why their data is being collected and how it will be used. For example, if data from a wearable device is shared with researchers, patients should know the specific purpose of the research and how it aligns with their treatment goals. By limiting data use to compatible purposes, FIPPs help build trust between patients and providers (Paul et al., 2023).

12.3.4 Ethical AI Frameworks

As artificial intelligence becomes increasingly integrated into neuropsychological rehabilitation, ethical AI frameworks are essential for ensuring responsible use. These frameworks focus on transparency, fairness, accountability, and explainability (Chiruvella & Guddati, 2021). For example, an AI-powered diagnostic tool might analyze patterns in brain activity to identify cognitive impairments. While the tool could offer valuable insights, it is crucial for patients and clinicians to understand how its algorithms work and whether its recommendations are unbiased. Transparency ensures that patients can trust the technology, while accountability ensures that developers and providers are responsible for its outcomes. Ethical AI frameworks also emphasize the importance of fairness. For instance, an algorithm trained on a diverse data set is more likely to deliver equitable results for patients from different backgrounds. By adhering to these principles, healthcare providers can use AI tools ethically and effectively (Chiruvella & Guddati, 2021).

12.4 Case Studies

12.4.1 Case Studies in Ethical Issues in Neuropsychological Rehabilitation Technology

Examining real-world scenarios through case studies provides a powerful way to understand and address the ethical dilemmas posed by healthcare technologies. Neuropsychological rehabilitation, where technology plays an increasingly prominent role, is no exception. By analyzing cases involving automation, AI, and data-driven tools, we can illuminate key ethical challenges and explore strategies for fostering ethical practices. Automated Healthcare App: Balancing Transparency and Paternalism: Imagine a mobile app designed to guide patients recovering from traumatic brain injuries. The app uses artificial intelligence to recommend daily cognitive exercises, offer reminders for medication, and track a patient's progress. On the surface, this seems like an ideal solution—automating rehabilitation tasks while reducing the workload for both patients and healthcare providers. However, ethical challenges quickly emerge. The app's algorithms are opaque, offering little transparency into how it makes recommendations or whether its suggestions align with evidence-based practices. Patients like Sarah, a 29-year-old recovering from a car accident, might feel that their autonomy is undermined when the app overrides their preferences, such as advising them to avoid certain activities they feel capable of performing. This case underscores the tension between paternalism—making decisions "for" the patient's benefit—and respecting patient autonomy. Developers of healthcare apps must prioritize transparency by explaining how algorithms work, allowing patients to make informed decisions about their care. Moreover, tools like dynamic consent models could give users greater control over the app's settings and recommendations, fostering a sense of collaboration rather than control (*Case Study*, 2019; Mathews et al., 2023).

12.4.2 Dynamic Sound Identification: Rights and Downstream Responsibility

In neuropsychological rehabilitation, sound identification systems can be used to aid patients with auditory processing difficulties. These AI-driven tools analyze environmental sounds to help patients recognize and respond to auditory cues, such as the sound of a doorbell or a fire alarm. While these tools have significant therapeutic potential, they also raise ethical concerns about representational harm and downstream responsibility. For example, if the sound identification system is trained on a limited data set, it might misidentify sounds in diverse cultural or environmental contexts. Consider Michael, a patient recovering from a stroke who relies on such a system to navigate his daily life. If the system fails to recognize critical auditory signals due to biases in its training data, it could place Michael in harm's way. This case highlights the importance of diversity in training data sets and the need for accountability in AI development. Developers must ensure their tools are inclusive and adaptable to varied real-world scenarios, minimizing risks for vulnerable populations in rehabilitation (*Case Study*, 2019).

12.4.3 AI in Hiring and Fairness in Access

While not directly tied to neuropsychological rehabilitation, the ethical issues raised by AI-powered hiring systems provide valuable parallels. These systems have been criticized for perpetuating biases in decision-making, such as favoring certain demographics over others. In the context of neuropsychological rehabilitation, similar biases could emerge in AI tools used for cognitive assessments. For instance, an AI system that evaluates a patient's progress might produce skewed results if its algorithms are trained on homogenous data sets that do not account for cultural or linguistic diversity. This could lead to disparities in treatment plans or rehabilitation outcomes. To address this, healthcare providers and developers must ensure that AI systems are designed with fairness and equity in mind. By incorporating diverse data sets and regularly auditing algorithms for biases, we can create tools that support equitable access to high-quality care for all patients (*Case Study*, 2019; Parsi & Elster, 2015).

12.4.4 *Public Sector Data Analytics: Transparency and Accountability*

Data analytics plays an increasingly central role in public health and neuropsychological rehabilitation. Consider a scenario where patient data collected from wearable devices is aggregated to identify trends in rehabilitation outcomes. While this approach could lead to valuable insights for improving care, it also raises concerns about privacy, secrecy, and accountability. A stroke survivor uses a wearable device to monitor her daily activity and progress. She later learns that her anonymized data was included in a research study without her explicit consent. Although the data was used for a beneficial purpose, Jeni feels that her trust was violated. This case illustrates the need for transparency in data collection and usage. Patients must be informed about how their data will be used, and mechanisms such as dynamic consent models can allow them to opt in or out of data-sharing initiatives. Furthermore, healthcare organizations must establish clear accountability for how data is handled and shared, ensuring that patient preferences are respected at every step (*Case Study PDFs*, 2019; Cirone, 2005; Mathews et al., 2023).

12.4.5 *Telehealth and Automation: Navigating Consequences*

Telehealth platforms have become a cornerstone of neuropsychological rehabilitation, offering patients convenient access to therapy sessions and cognitive assessments. However, the automation inherent in telehealth systems can sometimes lead to ethical dilemmas. Consider John, a 45-year-old recovering from a brain injury, who relies on telehealth for weekly therapy sessions. The platform's AI system flags John's responses as "uncooperative" based on its analysis of his facial expressions and tone of voice, leading to an automated recommendation to adjust his treatment plan. John's therapist, relying on the AI's input, implements the changes without verifying their accuracy. This case raises concerns about the fallibility of AI systems and their impact on patient care. Automated decisions can lack the nuance and context that human clinicians provide, potentially leading to harmful consequences. To mitigate these risks, telehealth platforms should incorporate safeguards such as human oversight and the ability for patients to contest automated decisions. Ensuring that clinicians remain central to decision-making processes is critical for maintaining the quality of care (Afzal & Arshad, 2021; Parsi & Elster, 2015).

12.4.6 Vaccine Refusal and Broader Ethical Dilemmas

Although not specific to neuropsychological rehabilitation, the ethical issues surrounding vaccine refusal during pandemics offer important lessons. These cases often involve conflicts between individual autonomy and public health. For example, mandatory vaccination policies may protect the broader community but infringe on personal freedoms.

In the context of rehabilitation, similar dilemmas arise when balancing individual rights with broader societal benefits. For instance, if a rehabilitation tool collects data for public health research, how do we balance the patient's right to privacy with the potential societal benefits of the research? These cases highlight the need for ethical frameworks that carefully weigh competing priorities and foster trust among all stakeholders (Mathews et al., 2023; Parsi & Elster, 2015).

12.5 Emerging Trends and Future Directions

12.5.1 Blockchain for Healthcare Data

Blockchain technology has emerged as a powerful tool for addressing privacy and security concerns in health care. By offering decentralized, immutable records, blockchain provides patients and providers with greater control over data ownership and access. A patient recovering from a traumatic brain injury might use multiple digital tools, such as wearable devices and telehealth platforms, all of which generate sensitive data. Blockchain can securely link these disparate data sources, ensuring that the patient retains ownership and control while minimizing risks of unauthorized access. Additionally, blockchain's transparency and traceability enhance trust between patients and providers. Looking ahead, integrating blockchain into neuropsychological rehabilitation could improve data-sharing practices, streamline care coordination, and empower patients to take an active role in their recovery (Aastha, 2023; Bathula et al., 2024; Doughman, 2023).

12.5.2 AI and Data Privacy

As AI becomes increasingly integral to neuropsychological rehabilitation, concerns about data privacy are coming to the forefront. AI-powered tools, such as cognitive assessment platforms and virtual assistants, often rely on large datasets to deliver personalized care. However, the use of these tools raises questions about how patient data is collected, stored, and analyzed. Transparent algorithms that respect patient privacy are essential. Emerging regulations like the General Data Protection Regulation (GDPR) are paving the way for more robust privacy protections, setting standards for how AI tools manage sensitive health information. In the future, neuropsychological rehabilitation programs may adopt privacy-preserving AI techniques, such as federated learning, which trains machine-learning models on decentralized data without transferring it to a central server. This approach significantly reduces privacy risks while still enabling advanced AI functionalities (Bathula et al., 2024; Nathella, 2024).

12.5.3 Universal Healthcare Access

One of the most pressing challenges in healthcare technology is the digital divide, which disproportionately affects underserved populations. For example, rural areas or low-income communities often lack access to telehealth platforms, wearable devices, or advanced rehabilitation tools, limiting their ability to benefit from technological advancements. In neuropsychological rehabilitation, addressing this divide is critical. Future developments should focus on creating affordable, accessible technologies that reach all patients, regardless of their socioeconomic or geographic status. Subsidized programs, community-based technology hubs, and low-cost devices are some strategies that could bridge this gap. Ensuring universal access is not just an ethical imperative but also a practical one—broadening access to rehabilitation technologies can improve overall public health outcomes and reduce long-term healthcare costs (Bathula et al., 2024; Nathella, 2024).

12.5.4 Integration of Blockchain and AI

The integration of blockchain and artificial intelligence marks a paradigm shift in healthcare technology. Together, these tools can address critical challenges in securing electronic health records, automating compliance processes, and improving care delivery. AI-driven algorithms could analyze patient data to predict recovery trajectories, while blockchain ensures that the data used in these analyses remains secure and tamper-proof. This fusion enhances transparency, privacy, and efficiency, ultimately leading to better patient outcomes. By combining blockchain's robust security features with AI's analytical power, healthcare providers can create systems that are not only innovative but also ethically sound (Bathula et al., 2024; Nathella, 2024).

12.5.5 Data Protection Measures

As the volume of data generated by healthcare technologies continues to grow, robust data protection measures are becoming more important than ever. In neuropsychological rehabilitation, tools like wearable devices, IoT platforms, and telehealth systems collect vast amounts of sensitive information. Protecting this data requires adherence to regulations such as HIPAA and the implementation of advanced IT security protocols. Healthcare organizations are increasingly investing in encryption, multi-factor authentication, and real-time monitoring systems to safeguard patient information. These measures are essential not only for compliance but also for maintaining trust between patients and providers. By prioritizing data protection, neuropsychological rehabilitation programs can ensure that technological innovations do not come at the expense of patient privacy (Aastha, 2023; Ghadi et al., 2024).

12.5.6 Innovative Technologies: Precision Medicine and IoT

Precision medicine, which tailors treatments to individual patient profiles, is gaining traction in neuropsychological rehabilitation. This approach relies on multimodal healthcare data analysis, incorporating inputs from wearable devices, genetic testing, and clinical assessments to create highly personalized care plans. At the same time, the Internet of Things (IoT) is transforming rehabilitation by enabling continuous monitoring through connected devices. For example, a wearable EEG monitor could track a patient's brain activity during cognitive exercises, providing real-time feedback to both the patient and their care team. However, the widespread use of IoT devices also necessitates strong authentication, encryption, and secure communication protocols to protect sensitive data. As these technologies advance, their integration into neuropsychological rehabilitation has the potential to significantly enhance recovery outcomes while addressing long-standing challenges in care delivery (Aastha, 2023; Nathella, 2024).

12.5.7 AI-Driven Solutions and Federated Learning

AI-driven tools are automating various aspects of neuropsychological rehabilitation, from compliance monitoring to personalized therapy recommendations. For example, generative AI (GenAI) solutions can help bridge gaps in fragmented regulatory frameworks, ensuring that rehabilitation programs adhere to ethical and legal standards. Federated learning, a novel approach to AI training, further enhances privacy by enabling models to learn from decentralized data. This means that sensitive patient data never leaves the device where it was generated, reducing the risk of breaches while still allowing for advanced machine-learning capabilities. These innovations represent a promising future for neuropsychological rehabilitation, where AI-driven tools and privacy-preserving techniques work hand in hand to improve care delivery (Bathula et al., 2024; Ghadi et al., 2024).

12.6 Conclusion

Technology has transformed neuropsychological rehabilitation, offering new hope to patients recovering from strokes, brain injuries, and neurodegenerative conditions. Tools like artificial intelligence, virtual reality, wearable devices, and telehealth have empowered patients and clinicians alike, making rehabilitation more personalized, accessible, and effective. But with this progress comes responsibility. The ethical and privacy challenges posed by these innovations cannot be ignored. How we navigate these challenges will define the future of care. At the heart of ethical technology use are the patients themselves—their autonomy, well-being, and trust. Throughout this chapter, we explored how respecting these values through principles like beneficence, non-maleficence, justice, and confidentiality can ensure that technology enhances, rather than undermines, the human experience of recovery. Empowering patients with informed choices, safeguarding their privacy, and addressing inequities in access are essential to keeping the patient's dignity at the center of innovation.

Real-world stories illustrate the delicate balance we must achieve. A wearable device might transform a patient's recovery journey but also raise questions about who owns their data. An AI-powered assessment tool might offer valuable insights but also carry risks of bias if not designed inclusively. These cases remind us that technology is not inherently good or bad—it is how we design, use, and regulate it that determines its impact.

Looking to the future, emerging technologies like blockchain, federated learning, and precision medicine hold great promise. They offer tools to improve data security, expand access to care, and deliver even more personalized treatment. But progress must be patient-centered. Bridging the digital divide, ensuring transparency in how technologies work, and fostering collaboration between technologists, clinicians, ethicists, and patients will be crucial for success. Ultimately, the goal of neuropsychological rehabilitation is more than just restoring cognitive and physical abilities—it is about helping people regain their independence, confidence, and humanity. Technology can be a powerful ally in this journey, but only if we remain vigilant in aligning its use with the core values of compassion, fairness, and trust. By prioritizing the person behind the data, we can ensure that technological advancements not only improve outcomes but also honor the lived experiences of every patient. In this way, technology becomes not just a tool for recovery, but a means to restore hope and dignity.

References

Aastha. (2023, August 9). *Blockchain in healthcare data management*. Silicon Valley InnovationCenter. https://siliconvalley.center/blog/blockchain-in-healthcare-data-management

Afzal, S., & Arshad, A. (2021). Ethical issues among healthcare workers using electronic medical records: A systematic review. *Computer Methods and Programs in Biomedicine Update, 1*, 100030. https://doi.org/10.1016/j.cmpbup.2021.100030

Bathula, A., Gupta, S. K., Merugu, S., Saba, L., Khanna, N. N., Laird, J. R., Sanagala, S. S., Singh, R., Garg, D., Fouda, M. M., & Suri, J. S. (2024). Blockchain, artificial intelligence, and healthcare: The tripod of future—a narrative review. *Artificial Intelligence Review, 57*(9), 238. https://doi.org/10.1007/s10462-024-10873-5

Case Study PDFs. (2019, January 24). *Princeton dialogues on AI and ethics*. https://aiethics.princeton.edu/case-studies/case-study-pdfs/

Chiruvella, V., & Guddati, A. K. (2021). Ethical issues in patient data ownership. *Interactive Journal of Medical Research, 10*(2), e22269. https://doi.org/10.2196/22269

Cirone, C. (2005). *Cases in medical ethics*. Markkula Center for Applied Ethics. https://www.scu.edu/ethics/focus-areas/bioethics/resources/cases-in-medical-ethics-student-led-discussions/

Doughman, S. (2023, December 5). *How blockchain can enhance the security of healthcare data*. World Economic Forum. https://www.weforum.org/stories/2023/12/healthcare-data-breaches-blockchain-cybersecurity/

Ethics in Health Care: Improving Patient Outcomes. (2023, January 19). https://publichealth.tulane.edu/blog/ethics-in-healthcare/

Ford, E., Shepherd, S., Jones, K., & Hassan, L. (2021). Toward an ethical framework for the text mining of social media for health research: A systematic review. *Frontiers in Digital Health, 2*. https://doi.org/10.3389/fdgth.2020.592237

Ghadi, Y. Y., Mazhar, T., Shahzad, T., Amir Khan, M., Abd-Alrazaq, A., Ahmed, A., & Hamam, H. (2024). The role of blockchain to secure internet of medical things. *Scientific Reports, 14*, 18422. https://doi.org/10.1038/s41598-024-68529-x

Klingler, C., Silva, D. S., Schuermann, C., Reis, A. A., Saxena, A., & Strech, D. (2017). Ethical issues in public health surveillance: A systematic qualitative review. *BMC Public Health, 17*(1), 295. https://doi.org/10.1186/s12889-017-4200-4

Mathews, D., Abernethy, A., Butte, A. J., Ginsburg, P., Kocher, B., Novelli, C., Sandy, L., Smee, J., Fabi, R., Ii, A. C. O., Sherkow, J. S., Sullenger, R. D., Freiling, E., & Balatbat, C. (2023). Telehealth and mobile health: Case study for understanding and anticipating emerging science and technology. *NAM Perspectives*. https://doi.org/10.31478/202311e.

Nathella, G. (2024, July 17). *Data privacy in healthcare: balancing innovation with patient security | Healthcare IT Today*. HealthcareITToday. https://www.healthcareittoday.com/2024/07/17/data-privacy-in-healthcare-balancing-innovation-with-patient-security/

Parsi, K., & Elster, N. (2015). Why can't we be friends? A case-based analysis of ethical issues with social media in health care. *AMA Journal of Ethics, 17*(11), 1009–1018. https://doi.org/10.1001/journalofethics.2015.17.11.peer1-1511

Paul, M., Maglaras, L., Ferrag, M. A., & Almomani, I. (2023). Digitization of healthcare sector: A study on privacy and security concerns. *ICT Express, 9*(4), 571–588. https://doi.org/10.1016/j.icte.2023.02.007

Robert, S. (2024). Ethical considerations in health informatics: Privacy, consent and data ownership. *Journal of Health & Medical Informatics, 15*(1). https://doi.org/10.37421/2157-7420.2024.15.509

Understanding the Four Main Ethical Principles in Healthcare: Autonomy, Beneficence, Justice, and Non-maleficence | Simbo AI – Blogs. (2024, November 1). https://www.simbo.ai/blog/understanding-the-four-main-ethical-principles-in-healthcare-autonomy-beneficence-justice-and-non-maleficence-3072231/

Varkey, B. (2021). Principles of clinical ethics and their application to practice. *Medical Principles and Practice, 30*(1), 17–28. https://doi.org/10.1159/000509119

Williamson, S. M., & Prybutok, V. (2024). Balancing privacy and progress: A review of privacy challenges, systemic oversight, and patient perceptions in AI-driven healthcare. *Applied Sciences, 14*(2), Article 2. https://doi.org/10.3390/app14020675

Chapter 13
Accessibility and Affordability of Technologies

13.1 Introduction

The rapid evolution of healthcare and rehabilitation technologies has transformed the way medical care is delivered and experienced. Breakthroughs in artificial intelligence (AI), telehealth, wearable devices, and advanced diagnostic tools have not only enabled earlier and more accurate diagnoses but also improved treatment outcomes and optimized healthcare delivery processes. For individuals facing chronic conditions or rehabilitation needs, these innovations promise a future where quality care is both personalized and effective. However, this technological progress has also raised profound challenges around who can access and benefit from these advancements. Accessibility, affordability, and equity remain at the heart of the global discourse on healthcare technology (Iecovich & Carmel, 2009; Varadarajan, 2024). Technological advancements often carry significant costs. Cutting-edge medical devices, such as robotic surgical systems or AI-driven diagnostic tools, are frequently developed in high-income countries where healthcare budgets can accommodate such investments. However, for individuals and health systems in low- and middle-income countries (LMICs), these costs present insurmountable barriers. In many cases, the price of acquiring, operating, and maintaining these technologies far exceeds what these regions can afford. Even within high-income nations, low-income and marginalized populations often face similar challenges, as high out-of-pocket expenses or inadequate insurance coverage put these innovations beyond their reach (Khasnabis et al., 2020).

Beyond financial constraints, structural barriers further hinder access to healthcare technologies. In many underserved communities, particularly in LMICs, inadequate infrastructure—such as unreliable electricity, poor transportation networks, and limited internet connectivity—poses significant obstacles. Digital literacy is another pressing issue. For patients and providers alike, unfamiliarity with the operation and integration of new technologies can create gaps in adoption and

K. J. Reddy, *Innovations in Neurocognitive Rehabilitation*,
https://doi.org/10.1007/978-3-031-88117-6_13

utilization. Together, these factors illustrate that affordability alone does not guarantee accessibility; addressing structural and systemic inequities is equally crucial (*Availability, Accessibility, Acceptability, and Quality Framework*, n.d.; Jamal et al., 2023). The inequitable distribution of healthcare technologies also reflects a broader issue of concentrated technological development and knowledge. The phenomenon of "gatekeeping knowledge and technology" perpetuates disparities by centralizing advancements in urban centers or developed nations. Rural regions and LMICs often lack the resources to develop or acquire such technologies independently, leaving them reliant on external sources. This dependence exacerbates global health inequities, as those with fewer resources and capabilities face delays in accessing life-saving innovations (Jamal et al., 2023; Varadarajan, 2024).

These challenges are particularly evident in the field of neurorehabilitation, a domain that has witnessed significant technological advancements in recent years. From virtual reality-based therapy tools to AI-driven cognitive rehabilitation programs, these innovations hold immense promise for improving the quality of life for individuals with neurological conditions. Yet, the high costs associated with these technologies and the specialized training required for their implementation often limit their accessibility to a privileged few. As a result, individuals in underserved regions or those from disadvantaged socioeconomic backgrounds remain excluded from these benefits, perpetuating cycles of health disparity (Holloway & Dawes, 2016). Addressing these barriers requires a holistic and multi-stakeholder approach that prioritizes equity. Governments, private enterprises, non-profit organizations, and healthcare providers must collaborate to create policies and solutions that bridge the gaps in access and affordability. Public–private partnerships can play a transformative role by pooling resources and expertise to develop cost-effective technologies tailored to the needs of underserved populations. For example, initiatives like Gavi, the Vaccine Alliance, have demonstrated the power of collaboration in making essential medical interventions affordable and widely accessible in LMICs. Applying similar models to healthcare technologies could yield comparable success (Reibling & Wendt, 2012).

Open-source innovation also offers a promising pathway to democratize access to healthcare technologies. By reducing the costs associated with proprietary software and intellectual property, open-source platforms can enable the development and dissemination of affordable solutions. In fields like diagnostics and telehealth, open-source AI models have already begun to demonstrate their potential to bridge gaps in access. However, realizing the full benefits of open-source technology will require sustained investment, robust infrastructure, and community-driven efforts to ensure that these tools are practical and scalable (*Healthcare Programs and Services:* National Academies of Sciences et al., 2018; Reibling & Wendt, 2012). Subsidies and financial incentives are another critical lever for enhancing affordability. Governments and international organizations can provide financial support to manufacturers and distributors, encouraging them to lower prices for underserved markets. Tax breaks, grants, and other forms of economic support can create an environment where innovative solutions are not only developed but also made available at scale to those who need them most. Additionally, decentralized manufacturing models, such as 3D printing, can further reduce costs by enabling localized production and minimizing reliance on complex supply chains (Jamal et al., 2023).

Telehealth and mobile health (mHealth) platforms have emerged as transformative tools for improving healthcare access in remote and underserved areas. By leveraging existing mobile networks and devices, these platforms bypass many of the infrastructural barriers associated with traditional healthcare delivery. For instance, telehealth consultations allow patients in rural areas to connect with specialists in urban centers, while mHealth applications empower individuals to monitor and manage their health conditions in real time. These technologies have already demonstrated their potential to reduce costs and improve health outcomes, but scaling their adoption will require targeted efforts to enhance digital literacy and expand connectivity (Holloway & Dawes, 2016). While these strategies hold significant promise, the path toward equitable healthcare technology is not without challenges. Policymakers and stakeholders must navigate complex ethical, cultural, and logistical considerations to ensure that solutions are inclusive and sustainable. For example, cultural perceptions of technology and health care can influence adoption rates, necessitating community engagement and culturally sensitive approaches. Similarly, ensuring that healthcare technologies are designed with user accessibility in mind—particularly for individuals with disabilities or limited digital literacy—is essential for fostering widespread adoption (Holloway & Dawes, 2016; Reibling & Wendt, 2012).

The need for equitable healthcare technology is not only a technical or economic challenge but also a moral imperative. Bridging the gaps in accessibility and affordability is essential for reducing health disparities and improving outcomes for all populations. By prioritizing equity and inclusivity in the design, development, and distribution of healthcare technologies, we can create a future where innovation benefits everyone, regardless of geography, socioeconomic status, or technological proficiency (Iecovich & Carmel, 2009). This chapter will explore these themes in depth, drawing on case studies, policy analysis, and examples of grassroots innovation to illustrate the challenges and opportunities in making healthcare technologies accessible and affordable. Through this exploration, we aim to contribute to the global dialogue on how to harness technological progress to build a more equitable and inclusive healthcare landscape, where no one is left behind.

13.2 Global Disparities in Healthcare Technology Access

Access to healthcare technologies, particularly advanced imaging modalities like magnetic resonance imaging (MRI), is starkly unequal across the globe. High-income countries (HICs) dominate in the development and deployment of these technologies, while low- and middle-income countries (LMICs) face significant challenges in accessing and utilizing them. These disparities, deeply rooted in economic, infrastructural, and educational inequalities, have profound implications for global health outcomes (Hilabi et al., 2023; Morris & Becker, 2004; Valerio et al., 2024).

13.3 Geographic Inequalities

Healthcare technologies are disproportionately concentrated in HICs, leaving LMICs struggling to provide basic diagnostic and treatment tools. For instance, advanced imaging devices like MRI scanners are often unavailable in resource-limited areas due to high acquisition costs, insufficient infrastructure, and lack of trained personnel (Hilabi et al., 2023; Valerio et al., 2024). In 2017, the African region had just 0.7 MRI scanners per million people, the lowest globally, while the South-East Asian region had only 1.1. In contrast, the Asia-Pacific region had 5.4 MRI scanners per million people, underscoring the regional disparities. In West Africa, a region with over 370 million people, only 84 MRI facilities were operational in 2016. Notably, more than two-thirds of these facilities were located in Nigeria, leaving vast populations in neighboring countries without access to essential imaging services (Gore et al., 2023; Morin et al., 2023). These gaps highlight the urgent need for equitable distribution of healthcare technologies.

- **Barriers to accessibility in LMICs:** Several factors contribute to the limited availability of healthcare technologies in LMICs. High acquisition costs for equipment like MRI scanners are a significant barrier, compounded by the additional expenses of maintenance, operation, and continuous power supply. Many LMICs lack the infrastructure required to support such advanced technologies, including stable electricity and reliable internet connectivity. Shortages in trained healthcare professionals further exacerbate the problem. In Sub-Saharan Africa and the Indian subcontinent, the scarcity of radiologists and technologists limits the operational capacity of imaging centers. This shortage, coupled with low scanner density, results in delayed diagnoses and inadequate care for millions of patients (Chowdhury & Pick, 2019; Hilabi et al., 2023; Valerio et al., 2024).
- **Impacts on healthcare delivery:** The lack of advanced imaging services in LMICs hampers the diagnosis and treatment of both infectious and non-communicable diseases. Diseases like tuberculosis, COVID-19, and cancer often go undetected or are diagnosed too late due to the absence of accessible imaging technologies. Imaging is essential not only for primary prevention and early detection but also for monitoring and managing treatment responses. For example, in Sub-Saharan Africa, where non-communicable diseases are on the rise, limited access to imaging services significantly undermines healthcare outcomes. Without timely diagnostic capabilities, conditions that could have been managed effectively often progress to advanced stages, resulting in higher mortality rates and increased healthcare costs (Hilabi et al., 2023; Murali et al., 2024; Paik et al., 2023; Ukert et al., 2024).

13.4 Socioeconomic Barriers

The persistent socioeconomic barriers to healthcare technology access represent a critical factor perpetuating global health inequities. High costs associated with cutting-edge innovations, such as robotic surgical systems and AI-driven diagnostic tools, remain prohibitive for low-income populations. Insurance models in many nations fail to cover these advancements, compelling individuals and healthcare systems in low- and middle-income countries (LMICs) to bear the financial burden. This dynamic exacerbates disparities, ensuring that technological advancements predominantly benefit wealthier countries and communities (Ali et al., 2021; Duroseau et al., 2023).

The inequities surrounding COVID-19 vaccine distribution starkly illustrate these challenges. High-income countries (HICs) swiftly secured large quantities of vaccines, leveraging their financial and logistical capabilities. In contrast, LMICs encountered insurmountable obstacles, including limited affordability and inadequate infrastructure, which delayed mass vaccination efforts. As of October 2022, only 25% of the population in low-income countries had received at least one vaccine dose compared to 72% in high-income nations (Duroseau et al., 2023; Kana et al., 2023). Economic disparities were a principal driver of these inequities. LMICs often faced higher prices for vaccines than their HIC counterparts. For instance, while European nations negotiated vaccine prices of approximately $3.50 per dose, some low-income countries were charged between $5.25 and $7 per dose. Such inflated costs strained already limited healthcare budgets, compelling LMICs to make difficult trade-offs in resource allocation (Duroseau et al., 2023).

Compounding these financial barriers were logistical challenges inherent to LMIC healthcare systems. Rural regions frequently lacked the necessary infrastructure for vaccine storage and distribution. Maintaining the cold chain for temperature-sensitive vaccines proved nearly impossible in areas without reliable electricity. Similarly, weak transportation networks often left remote communities disconnected from centralized healthcare facilities, delaying vaccine delivery and administration. Healthcare infrastructure deficiencies also shaped the accessibility of vaccines in LMICs. Many rural healthcare facilities operated with inadequate funding, insufficient staffing, and minimal resources to manage mass immunization campaigns. Trained personnel to administer vaccines and handle potential adverse effects were often unavailable, further delaying rollouts and compromising the efficacy of public health interventions (Duroseau et al., 2023; Tagoe et al., 2021).

Misinformation and low health literacy in LMICs amplified vaccine hesitancy, particularly in rural areas. Distrust of vaccines and healthcare providers—often fueled by misinformation campaigns—eroded public confidence in vaccination initiatives. Without targeted education campaigns, the disparity in vaccination rates between urban and rural populations widened, underscoring the interplay between socioeconomic, cultural, and informational barriers to healthcare access (Ali et al., 2022).

Global initiatives addressing these inequities highlighted both progress and limitations in international cooperation. For example, the COVAX Initiative aimed to promote equitable vaccine access by pooling resources and distributing doses to participating countries. While COVAX delivered over 1.7 billion doses globally by 2023, its impact was constrained by supply chain disruptions, delayed deliveries, and funding shortages. These challenges limited its ability to close the gap between HICs and LMICs. The COVID-19 Vaccine Delivery Partnership (CoVDP), launched in January 2022, supplemented these efforts by providing in-country support for vaccine distribution in 92 low-income nations. The initiative prioritized logistical solutions, such as cold chain management and improved transportation. However, broader structural inequalities, including insufficient local healthcare infrastructure and inadequate funding for implementation, limited the reach and success of these programs (Duroseau et al., 2023; Sinuraya et al., 2024).

Efforts to waive intellectual property rights for vaccines further illustrated the complexities of addressing global health disparities. Many LMICs advocated for temporary patent suspensions to enable local vaccine production and reduce dependency on external suppliers. However, resistance from several HICs delayed these measures, highlighting the inherent tension between profit-driven pharmaceutical models and the ethical imperative to prioritize global health equity (Duroseau et al., 2023; Pattanshetty et al., 2023). The implications of these disparities extend beyond the immediate context of the pandemic. They underscore the broader role of socioeconomic factors in shaping access to healthcare technologies. The unequal distribution of COVID-19 vaccines exemplifies how financial, infrastructural, and informational barriers converge to exclude vulnerable populations from the benefits of medical advancements. These challenges reflect systemic inequities that must be addressed to achieve global health equity (Ali et al., 2022; Boro & Stoll, 2022). The pandemic also emphasized the importance of international solidarity in tackling global health challenges. Collaborative initiatives, such as resource pooling and technology sharing, have demonstrated the potential to mitigate disparities. However, meaningful progress requires sustained investment in healthcare systems, education, and infrastructure, particularly in LMICs. Policymakers, global health organizations, and private stakeholders must prioritize dismantling systemic barriers to ensure equitable access to healthcare technologies (Roy et al., 2024).

Lessons from the pandemic underscore the necessity of comprehensive, long-term strategies to build resilient healthcare systems. These strategies should include investments in decentralized manufacturing and local production capabilities, reducing dependence on external suppliers. Additionally, bolstering healthcare infrastructure in rural areas and enhancing the capacity of frontline healthcare workers can create systems that are better equipped to deliver essential services, particularly in crisis scenarios. The COVID-19 pandemic has exposed the fragility and inequities of global healthcare systems, offering a critical opportunity to reimagine healthcare technology distribution. By embedding equity and inclusivity into healthcare planning and policy, stakeholders can work toward a future where all populations—irrespective of socioeconomic or geographic context—benefit from medical advancements (Roy et al., 2024; Santangelo et al., 2024; Sinuraya et al., 2024).

13.5 Challenges of Affordability in Healthcare Technology

1. **High development costs:** The development and commercialization of medical technologies necessitate significant investments in research, regulatory approval, and manufacturing. These substantial upfront costs frequently translate into elevated end-user prices, rendering cutting-edge technologies inaccessible to a large proportion of low-income populations (Dutta & Dhar, 2024; E. Kim, 2023). Sophisticated medical tools such as robotic surgical systems and AI-driven diagnostic platforms demand extensive research, compliance with rigorous regulatory frameworks, and complex production processes. Consequently, their affordability is limited to well-funded healthcare institutions in high-income countries, leaving LMICs unable to integrate these innovations into their healthcare systems (Dutta & Dhar, 2024; Leevy, 2006). Regulatory approval costs exacerbate these challenges. Medical technologies must undergo exhaustive testing to ensure safety and efficacy, often requiring multi-phase clinical trials and adherence to stringent international standards. These processes, while essential, are resource-intensive and significantly drive up the final cost to consumers. For instance, the estimated expense of bringing a medical device to market in the United States exceeds $30 million, with a substantial portion dedicated to regulatory compliance. Such expenses exacerbate affordability challenges for regions with already limited financial capacities (Gore et al., 2023; Tenni et al., 2022).
2. **Lack of economies of scale:** Limited adoption of healthcare technologies in LMICs undermines the achievement of economies of scale, a critical mechanism for reducing production and distribution costs. Insufficient demand, driven by financial and infrastructural constraints, deters manufacturers from scaling up production. This, in turn, perpetuates a cycle where high costs restrict adoption, and limited adoption sustains elevated prices (Dutta & Dhar, 2024; Morin et al., 2022). Economies of scale enable reduced unit costs through mass production and streamlined distribution systems. Diagnostic devices such as portable ultrasound machines and glucose monitors exemplify how large-scale manufacturing can significantly lower costs. However, fragmented healthcare systems and constrained market demand in LMICs impede the realization of these efficiencies. The problem is particularly acute in rural and underserved regions, where decentralized procurement systems inflate costs further (E. Kim, 2023; Morin et al., 2022).
3. **Patents and intellectual property rights:** The strict enforcement of patent protections, as outlined by agreements like the Trade-Related Aspects of Intellectual Property Rights (TRIPS), grants exclusive production rights to manufacturers for a minimum of 20 years (E. Kim, 2023; Tenni et al., 2022). While these protections incentivize innovation by securing financial returns for developers, they

also create monopolies that hinder the availability of affordable alternatives. Delayed access to generic or open-source solutions exacerbates the affordability crisis in LMICs (Leevy, 2006; Morin et al., 2022). Flexibilities under TRIPS, such as compulsory licensing and parallel imports, are intended to allow governments to circumvent patents in cases of public health emergencies. However, these measures remain underutilized due to economic constraints and political pressures faced by LMICs (E. Kim, 2023; Tenni et al., 2022). The COVID-19 pandemic starkly illustrated this issue, as initiatives to waive intellectual property rights for vaccines encountered resistance from wealthier nations, delaying equitable distribution. Beyond pharmaceuticals, intellectual property restrictions extend to medical devices and AI-driven diagnostic tools. Proprietary software embedded within advanced technologies often prevents the development of cost-effective alternatives. Although open-source initiatives hold promise for mitigating these barriers, they require sustained financial investment and robust international collaboration to achieve meaningful impact (Dutta & Dhar, 2024; E. Kim, 2023; Leevy, 2006).

4. **Compounding effects on access:** The convergence of high development costs, lack of economies of scale, and restrictive patent frameworks creates systemic barriers to healthcare technology access. These challenges disproportionately affect LMICs, where limited healthcare budgets and significant out-of-pocket expenditures constrain access to life-saving medical tools (Dutta & Dhar, 2024; Morin et al., 2022). For example, the inability to afford advanced diagnostic equipment, such as MRI scanners, limits diagnostic accuracy and delays disease identification and treatment. Similarly, the high costs of essential devices like ventilators have hindered healthcare facilities in LMICs from effectively responding to public health crises, as seen during the COVID-19 pandemic (Gore et al., 2023). Reliance on imported technologies further compounds these issues. Many LMICs lack the infrastructure to manufacture medical devices locally, making them dependent on international suppliers. Import tariffs, currency volatility, and logistical challenges inflate costs, further exacerbating affordability barriers. These systemic obstacles underscore the necessity for targeted policy interventions and collaborative efforts to alleviate financial and structural constraints (Leevy, 2006; Morin et al., 2022).
5. **Broader implications for health equity:** Affordability challenges in healthcare technology are not merely economic issues; they are central to global health equity. The unequal distribution of advanced medical tools reinforces disparities in health outcomes, particularly in LMICs, where populations face greater disease burdens and reduced access to quality care. Addressing these barriers is critical to ensuring that technological advancements benefit all populations equitably, regardless of socioeconomic status or geographic location (Morin et al., 2023). The COVID-19 pandemic exemplified the ramifications of these disparities. While high-income countries swiftly deployed advanced technologies to mitigate the virus's impact, LMICs struggled to procure basic supplies and equipment. This inequity prolonged the pandemic's effects on vulnerable populations and underscored the urgent need for systemic reforms to enhance

affordability and accessibility (Dutta & Dhar, 2024; E. Kim, 2023). Effective strategies to address affordability must consider the broader global health context. Policies that prioritize technology transfer, support local manufacturing, and foster open-source innovation have the potential to reduce costs and expand access. Furthermore, international frameworks must strike a balance between protecting intellectual property rights and prioritizing public health needs, particularly in the face of global emergencies (E. Kim, 2023; Tenni et al., 2022). By addressing these interconnected affordability challenges, stakeholders can foster an inclusive future where healthcare technologies are not only innovative but accessible to all. Ensuring equitable distribution of these advancements will be pivotal in narrowing the health disparities that persist across regions and populations.

13.6 Strategies for Enhancing Accessibility and Affordability

1. **Collaborative innovation:** Public–private partnerships (PPPs) serve as critical mechanisms for addressing the economic and logistical challenges inherent in the development and dissemination of healthcare technologies. By uniting the resources and expertise of governments, private enterprises, and non-governmental organizations, these partnerships foster the creation of scalable, cost-effective solutions tailored to underserved populations. For example, Gavi, the Vaccine Alliance, has successfully utilized collaborative frameworks to negotiate reduced prices for essential vaccines in low- and middle-income countries (LMICs), significantly improving access to life-saving interventions. The efficacy of PPPs demonstrates their potential to enhance both affordability and distribution of advanced healthcare technologies. Strategic collaboration between public and private stakeholders ensures that innovations are not only financially viable but also accessible to those in need (Anawade et al., 2024; Haleem et al., 2021).
2. **Open-source technology:** Open-source platforms offer transformative potential for reducing costs and broadening the adoption of healthcare technologies. By removing proprietary barriers, open-source frameworks enable developers to customize solutions that address specific regional and population needs. For instance, open-source AI models are increasingly utilized in diagnostics, providing cost-effective tools that bridge critical gaps in healthcare delivery within LMICs. These platforms have also proven indispensable in scaling telemedicine and telehealth initiatives. By offering free or low-cost software, open-source technologies empower providers to deliver care in resource-constrained settings. However, the success of such systems necessitates sustained investment in local infrastructure, workforce training, and technical support to ensure optimal implementation (Anawade et al., 2024; *Improving Access to Telehealth | Telehealth.HHS.Gov*, n.d.; *Transforming Lives*, n.d.).

3. **Subsidies and incentives:** Governments and international organizations can employ targeted financial mechanisms, including subsidies, tax breaks, and grants, to mitigate the costs of healthcare technologies in underserved markets. These tools incentivize manufacturers to prioritize affordability while investing in scalable innovations tailored to LMICs. Subsidies can offset production costs, particularly for diagnostic tools and medical devices, enhancing their availability in rural and low-income settings. Additionally, tax incentives have the capacity to attract private sector investment in inclusive and sustainable healthcare solutions. However, such interventions require robust oversight to ensure equitable distribution and long-term impact (Almeman, 2024; Haleem et al., 2021).
4. **Decentralized manufacturing:** Decentralized manufacturing, exemplified by the use of 3D printing, represents a paradigm shift in the production and distribution of healthcare technologies. This model facilitates localized production, minimizing dependence on complex global supply chains and reducing logistical costs. By enabling the on-site production of medical devices and tools, decentralized systems significantly shorten delivery timelines and improve cost-efficiency. During the COVID-19 pandemic, 3D printing emerged as a vital tool for addressing critical shortages of personal protective equipment (PPE) and medical devices. This approach underscored the role of decentralized manufacturing in enhancing healthcare system resilience and ensuring timely access to essential technologies. Localized manufacturing also fosters capacity building within LMICs, empowering these regions to produce essential medical tools independently. However, achieving widespread adoption necessitates investments in workforce development, regulatory infrastructure, and quality assurance systems to maintain safety and efficacy standards (Anawade et al., 2024; *Global Strategy on Digital Health 2020-2025*, 2021; Lee et al., 2023).
5. **Scaling telehealth and mobile health (mHealth):** Telehealth and mobile health (mHealth) platforms provide cost-effective avenues for extending healthcare access to remote and underserved areas. By leveraging existing mobile networks and devices, these technologies deliver care without requiring extensive physical infrastructure. Telehealth consultations, for example, facilitate connections between patients in rural areas and specialists in urban centers, reducing travel burdens and enhancing the quality of care. Similarly, mHealth platforms empower individuals to monitor and manage their health through mobile applications and wearable technologies. These tools generate real-time data and personalized insights, bolstering preventive care and chronic disease management. However, to maximize their potential, telehealth and mHealth programs must address barriers such as digital literacy gaps and inequitable access to technology. Special emphasis should be placed on improving connectivity and device availability for populations in resource-limited settings. The integration of open-source technologies into telehealth initiatives further amplifies their affordability and reach. By reducing licensing costs and enabling localized adaptations, open-source solutions play a crucial role in making telehealth viable in LMICs. Training healthcare providers to effectively utilize these tools is equally impor-

tant for their sustained success (Anawade et al., 2024; Haleem et al., 2021; *Improving Access to Telehealth | Telehealth.HHS.Gov*, n.d.).

13.7 Case Studies of Successful Models

Providing equitable access to healthcare technologies remains a critical challenge, especially in resource-limited settings. However, several innovative and sustainable models have demonstrated that delivering high-quality care at an affordable cost is achievable. By examining and replicating these successful case studies, healthcare providers and policymakers can address systemic barriers and create more equitable healthcare systems.

- **Aravind Eye Care System (India):** The Aravind Eye Care System is an internationally recognized model of accessible and affordable healthcare. Founded in 1976 in Madurai, India, Aravind has revolutionized ophthalmic care by focusing on efficiency and patient-centered innovation. Through its high-volume, low-cost service delivery model, Aravind has performed millions of cataract surgeries at a fraction of the cost charged in high-income countries. Aravind's ability to achieve economies of scale is central to its success. High patient volumes significantly reduce per-procedure costs, allowing the organization to provide affordable care without compromising quality. This efficiency is further enhanced by a well-structured system that reallocates routine tasks from ophthalmologists to mid-level personnel, such as nurses and paramedics. Standardized protocols and rigorous quality assurance processes ensure consistent care delivery ("Case Studies," n.d.; Ravilla & Ramasamy, 2014). Additionally, Aravind emphasizes patient empowerment through counseling and education. These efforts improve compliance with treatment plans and encourage the uptake of services. Aravind's sustainable practices have enabled the replication of its model across more than 300 hospitals in India and internationally, setting a global benchmark for cost-effective healthcare delivery (Liu et al., 2024; Ravilla & Ramasamy, 2014).
- **Portable diagnostic devices in Africa:** In sub-Saharan Africa, the introduction of low-cost, portable diagnostic devices has significantly enhanced access to essential healthcare. These devices rely on simplified designs and local production methods to reduce costs and increase adaptability to regional needs. Portable tools for diagnosing malaria, HIV, and tuberculosis, for example, have enabled healthcare workers to provide timely and accurate care in remote areas. These devices address critical gaps in healthcare infrastructure, overcoming challenges such as unreliable electricity and limited laboratory facilities. By prioritizing affordability and functionality, portable diagnostic technologies have become invaluable in improving health outcomes in underserved communities ("Case Studies," n.d.; DrugPatentWatch, 2017).
- **Telemedicine in rural America:** Telemedicine has proven to be a transformative tool for bridging healthcare gaps in rural and underserved areas of the United

States. By leveraging telecommunication technologies, patients in remote locations can access specialists and advanced medical care without the need for extensive travel. Subsidized programs have further ensured that telemedicine services remain affordable for low-income populations. The success of telemedicine in rural America highlights the potential for integrating technology into healthcare systems to address disparities. Real-time consultations and remote monitoring reduce barriers to access and improve continuity of care. These advancements also underscore the feasibility of scaling similar solutions in LMICs to address widespread inequities ("Case Studies," n.d.; R. Kim et al., 2022).

- **Key considerations for successful implementation:** While these models exemplify the potential for achieving accessibility and affordability, their implementation often requires addressing systemic challenges. For instance, teleophthalmology—the use of telemedicine for eye care—faces barriers such as high initial investment costs, limitations in image quality, personnel shortages, and concerns over data privacy. Overcoming these challenges demands a combination of technical innovation, workforce training, and the establishment of robust regulatory frameworks. Sustainability is also a critical factor in scaling these models. Collaborative innovation, open-source technologies, subsidies and incentives, decentralized manufacturing, and the expansion of telehealth and mobile health (mHealth) platforms are essential components of sustainable healthcare delivery. By integrating these strategies, healthcare systems can ensure cost-effective, equitable, and patient-centered care ("Case Studies," n.d.; R. Kim et al., 2022; Liu et al., 2024).

13.8 Emerging Trends and Future Directions

The intersection of financial technology (fintech) and health care represents a promising frontier for addressing the persistent challenges of accessibility and affordability in medical technologies. Innovations such as AI-driven cost optimization, blockchain integration, and novel financing models are transforming the landscape of healthcare delivery. These advancements, when effectively implemented, can dismantle systemic barriers, foster equity, and create more efficient healthcare systems worldwide.

13.8.1 AI-Powered Cost Optimization

Artificial intelligence (AI) is fundamentally reshaping the financial operations of healthcare systems. AI-driven tools can identify cost-saving opportunities by optimizing resource allocation and predicting maintenance needs for medical equipment, thereby minimizing waste. Predictive analytics powered by AI enable healthcare organizations to analyze expenditure trends, anticipate future costs, and propose personalized savings strategies for individuals and institutions. Additionally, AI enhances operational efficiency by forecasting demand for medical services and technologies. Hospitals and clinics can leverage these insights to allocate resources more effectively, reduce unnecessary expenditures, and improve patient outcomes. By automating routine financial processes, AI reduces administrative burdens, allowing healthcare organizations to reinvest savings into direct patient care (Adewole et al., 2025; Anglen, n.d.; Garza, 2025).

13.8.2 Blockchain for Affordable Healthcare

Blockchain technology is emerging as a transformative tool in health care, offering enhanced transparency, cost reduction, and improved security. By facilitating the secure exchange of medical records and financial information, blockchain mitigates fraud, streamlines administrative tasks, and strengthens patient privacy. One compelling application of blockchain lies in the use of smart contracts to automate claims processing. These contracts execute transactions automatically when predefined conditions are met, expediting reimbursements and reducing human error. Additionally, blockchain streamlines procurement processes for medical supplies, ensuring funds are allocated efficiently and transparently. Blockchain also empowers patients by granting them greater control over their medical data. Through decentralized systems, individuals can securely manage their records and share them with providers or insurers as needed, fostering trust and improving system efficiency (Garza, 2025; Liguori, 2023; Ramezani et al., 2023; Simonoski & Bogatinoska, 2024).

13.8.3 Innovations in Financing Models

Innovative financing models are addressing affordability challenges in health care by expanding access to essential technologies and services for underserved populations. Microinsurance products, for instance, offer cost-effective coverage tailored to low-income settings. These plans, often priced below a dollar per day, provide basic health benefits to individuals who might otherwise lack insurance. Subscription-based healthcare plans have also emerged as a viable solution, allowing individuals to manage predictable, affordable payments for essential medical services. In parallel, Health Savings Accounts (HSAs) enable users to set aside pre-tax income for healthcare expenses, facilitating long-term financial planning. Fintech platforms are further democratizing access to financial tools through mobile applications equipped with automated savings plans, real-time expense tracking, and AI-driven financial guidance. These platforms are particularly impactful in low- and middle-income countries (LMICs), where traditional banking systems may be underdeveloped (*Healthcare Financing Models Innovative Healthcare Financing Models*, n.d.; Kasyapa & Vanmathi, 2024; Omokhoa et al., 2024; Stfalcon.com, 2023).

13.8.4 Integration of Emerging Technologies

The convergence of AI, blockchain, and machine learning is amplifying the capabilities of fintech in health care. Dynamic pricing models for insurance, powered by AI, are becoming increasingly common. These models adjust premiums based on individual risk profiles, ensuring fair and personalized pricing structures. Wearable devices that enable real-time health monitoring are revolutionizing preventive care. These technologies provide continuous data on patients' vital signs and behaviors, allowing healthcare providers to create personalized treatment plans and mitigate long-term costs associated with chronic conditions. Moreover, AI-driven financial advisory tools tailored to healthcare planning are gaining traction. By analyzing users' medical and financial histories, these systems provide customized recommendations for managing expenses and optimizing savings (Anglen, n.d.; Garza, 2025; Omokhoa et al., 2024).

13.8.5 Overcoming Barriers to Implementation

Despite their potential, the widespread adoption of these innovations faces several obstacles. For example, the initial cost of deploying AI and blockchain technologies can deter smaller healthcare organizations. Furthermore, concerns surrounding data security and privacy necessitate the establishment of robust regulatory frameworks to ensure compliance and build trust. Digital literacy and access to technology remain critical challenges in many underserved regions. Bridging these gaps requires targeted investments in infrastructure development, workforce training, and public awareness campaigns. Ensuring that both healthcare providers and patients can effectively utilize these tools is paramount to their success (Adewole et al., 2025; *Healthcare Financing Models Innovative Healthcare Financing Models*, n.d.; Ramezani et al., 2023; Stfalcon.com, 2023).

13.8.6 Case Studies Highlighting Success

Several real-world examples illustrate the transformative potential of these emerging trends. In Mexico, fintech platforms are empowering low-income families to save for medical emergencies through automated micro-savings programs. By combining AI-powered insights with user-friendly interfaces, these tools enhance financial resilience and healthcare accessibility. In India, blockchain technology is being employed to manage the distribution of essential medications in rural areas. Through supply chain tracking and automated payment processes, blockchain ensures that medicines reach their intended recipients while minimizing administrative costs. In the United States, subscription-based telehealth services are gaining traction, offering affordable access to primary and mental health care. These platforms leverage AI and machine learning to match patients with appropriate providers, streamline appointment scheduling, and customize care plans (Anglen, n.d.; Garza, 2025; Simonoski & Bogatinoska, 2024).

13.9 Conclusion

Healthcare technologies hold extraordinary promise, offering solutions to some of the most pressing challenges in global health. Yet, one critical question remains: How can we ensure these advancements are accessible and affordable to everyone, not just a privileged few? Addressing this challenge means breaking down barriers, scaling what works, and embracing innovative solutions that prioritize fairness and inclusion. This chapter has explored how emerging technologies and creative financing are reshaping healthcare systems worldwide. Artificial intelligence, for example, is revolutionizing the way resources are allocated. AI doesn't just predict maintenance needs or streamline equipment usage; it helps hospitals and clinics stretch their budgets further, making care more affordable for patients. These tools enable healthcare providers to focus their energy and resources where they matter most—on the people they serve.

Blockchain technology is equally transformative. By creating secure and transparent systems, blockchain eliminates inefficiencies that often inflate costs, such as redundant paperwork and fraud. Imagine a world where insurance claims process themselves through smart contracts or patients can securely share their medical records with any doctor at the push of a button. Blockchain can help make this vision a reality while empowering patients to take control of their healthcare journey. Financing innovations are proving just as crucial. For many, health care isn't just a matter of access—it's about affordability. Microinsurance plans and subscription-based models are changing the game, especially for individuals in low-resource settings. These approaches make essential health services predictable and manageable, reducing the financial anxiety that prevents so many from seeking care. Fintech platforms that offer automated savings tools and personalized financial advice are putting control back into people's hands, helping them plan for their healthcare needs—even in areas where banks are scarce.

Real-world examples bring these possibilities to life. The Aravind Eye Care System in India shows how a bold idea can change millions of lives. By creating a high-volume, low-cost model for cataract surgeries, Aravind has provided quality care to patients who might otherwise have gone untreated. Their approach—streamlining workflows, empowering mid-level staff, and educating patients—has made eye care accessible in ways few could have imagined. In sub-Saharan Africa, portable diagnostic devices designed for rugged environments are tackling healthcare challenges head-on. These tools are simple, affordable, and life-saving, bringing essential care to communities far from traditional medical infrastructure. In rural America, telemedicine programs connect patients with specialists hundreds of miles away, proving that distance doesn't have to be a barrier to quality care. Despite these successes, challenges remain. Cutting-edge technologies like AI and blockchain can be expensive to implement, putting them out of reach for smaller healthcare organizations. In many parts of the world, digital literacy and infrastructure gaps make it difficult for both patients and providers to fully utilize these tools. Overcoming these hurdles will require focused investments in training, infrastructure, and public education. Collaboration—among governments, businesses, and community groups—will be essential to ensuring these solutions are scaled effectively and sustainably.

As promising as these technologies are, their deployment must be guided by care and responsibility. Data security and ethical considerations aren't optional—they're essential. People need to trust that these tools will protect their privacy and be used to enhance their well-being, not widen existing gaps in access. Developers and policymakers must work together to ensure that these technologies prioritize fairness, transparency, and inclusivity. The ultimate goal is clear: a healthcare system where technology serves as a tool for equity, not a marker of privilege. Achieving this will take a shared commitment to innovation, fairness, and compassion. Governments must lead by creating policies that support affordable healthcare technologies. Companies must prioritize social good alongside profits, designing solutions that meet the needs of those who are often overlooked. Nonprofits and community organizations must advocate for the underserved and step in where the system falls short.

The strategies and models explored in this chapter show us what's possible. Programs like Aravind, portable diagnostics in Africa, and telemedicine in rural America aren't just examples of success—they're blueprints for what can be done when innovation meets a commitment to equity. Emerging technologies like AI and blockchain, paired with creative financing, offer new tools to tackle old problems. But the work isn't done. As we move forward, we must keep people at the center of these innovations. The opportunity to transform health care is within reach, but it requires us to act with intention and focus. By continuing to innovate, collaborate, and prioritize the needs of underserved communities, we can create a healthcare system that doesn't just treat illnesses but also empowers lives. This is the path to a future where health care isn't just advanced—it's accessible to all.

References

Adewole, O. A., Akingbola, A., Adegbesan, A. C., Ndu-Esekea, T. N., & Olalemi, S. B. (2025). Leveraging artificial intelligence to improve health insurance access and address disparities in rural Africa. *Journal of Medicine, Surgery, and Public Health, 5*, 100172. https://doi.org/10.1016/j.glmedi.2024.100172

Ali, G., Khan, I., Amir, M., & Khan, M. N. A. (2021). Prevalence of depression in hospitalized patients of pulmonary tuberculosis. *Pakistan Armed Forces Medical Journal, 71*(1), Article 1. https://doi.org/10.51253/pafmj.v71i1.2862

Ali, H. A., Hartner, A.-M., Echeverria-Londono, S., Roth, J., Li, X., Abbas, K., Portnoy, A., Vynnycky, E., Woodruff, K., Ferguson, N. M., Toor, J., & Gaythorpe, K. A. (2022). Vaccine equity in low and middle income countries: A systematic review and meta-analysis. *International Journal for Equity in Health, 21*(1), 82. https://doi.org/10.1186/s12939-022-01678-5

Almeman, A. (2024). The digital transformation in pharmacy: Embracing online platforms and the cosmeceutical paradigm shift. *Journal of Health, Population and Nutrition, 43*(1), 60. https://doi.org/10.1186/s41043-024-00550-2

Anawade, P. A., Sharma, D., & Gahane, S. (2024). A comprehensive review on exploring the impact of telemedicine on healthcare accessibility. *Cureus, 16*(3), e55996. https://doi.org/10.7759/cureus.55996

Anglen, J. (n.d.). *AI-Driven insurance pricing 2024 | AI agents with machine learning*. RAPID Innovation. Retrieved January 22, 2025, from https://www.rapidinnovation.io/post/ai-powered-insurance-pricing-models

Availability, Accessibility, Acceptability and Quality Framework. (n.d.). Physiopedia. Retrieved January 22, 2025, from https://www.physio-pedia.com/Availability,_Accessibility,_Acceptability_and_Quality_Framework

Boro, E., & Stoll, B. (2022). Barriers to COVID-19 health products in low-and middle-income countries during the covid-19 pandemic: A rapid systematic review and evidence synthesis. *Frontiers in Public Health, 10*. https://doi.org/10.3389/fpubh.2022.928065

Case Studies. (n.d.). *Aravind Eye Care System*. Retrieved January 22, 2025, from https://aravind.org/case-studies/

Chowdhury, A., & Pick, A. (2019). *Current and future technological developments with the potential to improve health outcomes in low- and lower-middle- income countries*.

DrugPatentWatch. (2017, November 27). *Leveraging affordable innovation to tackle India's healthcare challenge*. DrugPatentWatch – Make Better Decisions. https://www.drugpatentwatch.com/blog/leveraging-affordable-innovation-tackle-indias-healthcare-challenge/

Duroseau, B., Kipshidze, N., & Limaye, R. J. (2023). The impact of delayed access to COVID-19 vaccines in low- and lower-middle-income countries. *Frontiers in Public Health, 10*, 1087138. https://doi.org/10.3389/fpubh.2022.1087138

Dutta, N., & Dhar, D. (2024). Understanding medical technology innovation in low- and middle-income countries: Factors, impact, and a model proposal. *She Ji: The Journal of Design, Economics, and Innovation, 10*(2), 192–222. https://doi.org/10.1016/j.sheji.2024.07.002

Garza, G. (2025, January 17). *Financial wellness: How Fintech and Healthcare are joining forces*. Mexico Business. https://mexicobusiness.news/health/news/financial-wellness-how-fintech-and-healthcare-are-joining-forces

Global Strategy on Digital Health 2020-2025 (1st ed). (2021). World Health Organization.

Gore, C., Morin, S., Røttingen, J.-A., & Kieny, M. P. (2023). Negotiating public-health intellectual property licensing agreements to increase access to health technologies: An insider's story. *BMJ Global Health, 8*(9). https://doi.org/10.1136/bmjgh-2023-012964

Haleem, A., Javaid, M., Singh, R. P., & Suman, R. (2021). Telemedicine for healthcare: Capabilities, features, barriers, and applications. *Sensors International, 2*, 100117. https://doi.org/10.1016/j.sintl.2021.100117

Healthcare Financing Models Innovative Healthcare Financing Models: A Guide for Entrepreneurs. (n.d.). FasterCapital. Retrieved January 22, 2025, from https://fastercapital.com/content/

Healthcare-financing-models-Innovative-Healthcare-Financing-Models%2D%2DA-Guide-for-Entrepreneurs.html

Hilabi, B. S., Alghamdi, S. A., & Almanaa, M. (2023). Impact of magnetic resonance imaging on healthcare in low- and middle-income countries. *Cureus, 15*(4), e37698. https://doi.org/10.7759/cureus.37698

Holloway, C., & Dawes, H. (2016). *Disrupting the world of Disability: The next generation of assistive technologies and rehabilitation practices | Request PDF* [Healthcare Technology Letters]. ResearchGate. https://www.researchgate.net/publication/316237130_Disrupting_the_world_of_Disability_The_Next_Generation_of_Assistive_Technologies_and_Rehabilitation_Practices

Iecovich, E., & Carmel, S. (2009). Differences in accessibility, affordability, and availability (AAA) of medical specialists among three age-groups of elderly people in Israel. *Journal of Aging and Health, 21*(5), 776–797. https://doi.org/10.1177/0898264309333322

Improving access to telehealth | Telehealth.HHS.gov. (n.d.). Retrieved January 22, 2025, from https://telehealth.hhs.gov/providers/health-equity-in-telehealth/improving-access-to-telehealth

Jamal, M. H., Aziz, A. F. A., Aizuddin, A. N., & Alijunid, S. M. (2023). Gatekeepers in the health financing scheme: Assessment of knowledge, attitude, practices, and participation of Malaysian private general practitioners in the PeKa B40 scheme. *PLoS One, 18*(10), e0292516. https://doi.org/10.1371/journal.pone.0292516

Kana, B. D., Arbuthnot, P., Botwe, B. K., Choonara, Y. E., Hassan, F., Louzir, H., Matsoso, P., Moore, P. L., Muhairwe, A., Naidoo, K., Ndomondo-Sigonda, M., & Madhi, S. A. (2023). Opportunities and challenges of leveraging COVID-19 vaccine innovation and technologies for developing sustainable vaccine manufacturing capabilities in Africa. *The Lancet Infectious Diseases, 23*(8), e288–e300. https://doi.org/10.1016/S1473-3099(22)00878-7

Kasyapa, M. S. B., & Vanmathi, C. (2024). Blockchain integration in healthcare: A comprehensive investigation of use cases, performance issues, and mitigation strategies. *Frontiers in Digital Health, 6*, 1359858. https://doi.org/10.3389/fdgth.2024.1359858

Khasnabis, C., Holloway, C., & MacLachlan, M. (2020). The digital and assistive technologies for ageing initiative: Learning from the GATE initiative. *The Lancet Healthy Longevity, 1*(3), e94–e95. https://doi.org/10.1016/S2666-7568(20)30049-0

Kim, E. (2023, April 25). *Intellectual property rights in a public health emergency: Direct government support in low and middle income countries as a solution*. HUHPR. http://www.huhpr.org/original-content/2023/10/4/intellectual-property-rights-in-a-public-health-emergency-direct-government-support-in-low-and-middle-income-countries-as-a-solution

Kim, R., Mishra, C., & Sen, S. (2022). The use of teleconsultation and technology by the Aravind Eye Care System, India. *Community Eye Health, 35*(114), 10.

Lee, P., Abernethy, A., Shaywitz, D., Gundlapalli, A. V., Weinstein, J., Doraiswamy, M., Schulman, K., & Madhavan, S. (2023). Digital health COVID-19 impact assessment: Lessons learned and compelling needs. In *Emerging stronger from COVID-19: Priorities for health system transformation* (pp. 177–234). National Academies Press (US). https://doi.org/10.17226/26657

Leevy, T. (2006). Intellectual property and access to medicine for the poor. AMA Journal of Ethics, *8*(12), 834–838. https://doi.org/10.1001/virtualmentor.2006.8.12.hlaw1-0612

Liguori, G. (2023). *Blockchain and generative AI: Transforming the finance and insurance industries | LinkedIn*. The Digital Edge. https://www.linkedin.com/pulse/blockchain-generative-ai-transforming-finance-giuliano-liguori-/

Liu, R., Sule, A. A., Shannon, C. S., Ravilla, T., Taylor, H., Rojas-Carabali, W., Khanna, R. C., Mishra, C., Sen, A., Khatri, A., Tan, A. C. S., Sobrin, L., & Agrawal, R. (2024). A practical model for effective eye care delivery in Southeast Asian rural communities: A proposal built based on experts' recommendations. *Indian Journal of Ophthalmology, 72*(Suppl 3), S473. https://doi.org/10.4103/IJO.IJO_2196_23

Morin, S., Moak, H. B., Bubb-Humfryes, O., Drehle, C. von, Lazarus, J. V., & Burrone, E. (2022). The economic and public health impact of intellectual property licensing of medicines for low-income and middle-income countries: A modelling study. The Lancet Public Health, 7(2), e169–e176. https://doi.org/10.1016/S2468-2667(21)00202-4

Morin, S., Segafredo, G., Piccolis, M., Das, A., Das, M., Loffredi, N., Larbi, A., Mwamelo, K., Villanueva, E., Nobre, S., & Burrone, E. (2023). Expanding access to biotherapeutics in low-income and middle-income countries through public health non-exclusive voluntary intellectual property licensing: Considerations, requirements, and opportunities. *The Lancet Global Health, 11*(1), e145–e154. https://doi.org/10.1016/S2214-109X(22)00460-0

Morris, R., & Becker, J. (2004). *Cognitive neuropsychology of Alzheimer's disease*. OUP Oxford.

Murali, S., Ding, H., Adedeji, F., Qin, C., Obungoloch, J., Asllani, I., Anazodo, U., Ntusi, N. A. B., Mammen, R., Niendorf, T., & Adeleke, S. (2024). Bringing MRI to low- and middle-income countries: Directions, challenges and potential solutions. *NMR in Biomedicine, 37*(7), e4992. https://doi.org/10.1002/nbm.4992

National Academies of Sciences, E., Division, H. and M., Services, B. on H. C., Disabilities, C. on I. H. O. for C. with, Byers, E., Valliere, F. R., & Houtrow, A. J. (2018). Health care programs and services. In *Opportunities for improving programs and services for children with disabilities*. National Academies Press (US). https://www.ncbi.nlm.nih.gov/books/NBK518928/

Omokhoa, H. E., Odionu, C. S., Azubuike, C., & Sule, A. K. (2024). AI-Powered Fintech innovations for credit scoring, debt recovery, and financial access in Microfinance and SMEs. *Gulf Journal of Advance Business Research, 2*(6), Article 6. https://doi.org/10.51594/gjabr.v2i6.55

Paik, K. E., Hicklen, R., Kaggwa, F., Puyat, C. V., Nakayama, L. F., Ong, B. A., Shropshire, J. N. I., & Villanueva, C. (2023). Digital determinants of health: Health data poverty amplifies existing health disparities—A scoping review. PLOS Digital Health, 2 (10), e0000313. https://doi.org/10.1371/journal.pdig.0000313

Pattanshetty, S., Pardesi, M., & Gudi, N. (2023). A comparative analysis on the social determinants of COVID-19 vaccination coverage in fragile and conflict affected settings and non-fragile and conflict affected settings. *International Journal of Health Policy and Management, 12*(Issue 1), 1–9. https://doi.org/10.34172/ijhpm.2022.6830

Ramezani, M., Takian, A., Bakhtiari, A., Rabiee, H. R., Fazaeli, A. A., & Sazgarnejad, S. (2023). The application of artificial intelligence in health financing: A scoping review. *Cost Effectiveness and Resource Allocation, 21*(1), 83. https://doi.org/10.1186/s12962-023-00492-2

Ravilla, T., & Ramasamy, D. (2014). Efficient high-volume cataract services: The Aravind model. *Community Eye Health, 27*(85), 7–8.

Reibling, N., & Wendt, C. (2012). Gatekeeping and provider choice in OECD healthcare systems. *Current Sociology, 60*(4), 489–505. https://doi.org/10.1177/0011392112438333

Roy, D. N., Ferdiousi, N., Mohabbot Hossen, M., Islam, E., & Shah Azam, M. (2024). Global disparities in COVID-19 vaccine booster dose (VBD) acceptance and hesitancy: An updated narrative review. *Vaccine: X, 18*, 100480. https://doi.org/10.1016/j.jvacx.2024.100480

Santangelo, O. E., Provenzano, S., Di Martino, G., & Ferrara, P. (2024). COVID-19 vaccination and public health: Addressing global, regional, and within-country inequalities. *Vaccines, 12*(8), 885. https://doi.org/10.3390/vaccines12080885

Simonoski, O., & Bogatinoska, D. C. (2024). Blockmedcare: Advancing healthcare through blockchain integration with AI and IOT. *International Journal of Network Security & Its Applications, 16*(6), 55–75. https://doi.org/10.5121/ijnsa.2024.16604

Sinuraya, R. K., Nuwarda, R. F., Postma, M. J., & Suwantika, A. A. (2024). Vaccine hesitancy and equity: Lessons learned from the past and how they affect the COVID-19 countermeasure in Indonesia. *Globalization and Health, 20*(1), 11. https://doi.org/10.1186/s12992-023-00987-w

Stfalcon.com. (2023, December 15). Revolutionizing healthcare: How FinTech is transforming the industry. *Medium*. https://stfalconcom.medium.com/revolutionizing-healthcare-how-fintech-is-transforming-the-industry-7bc2f79267ca

Tagoe, E. T., Sheikh, N., Morton, A., Nonvignon, J., Sarker, A. R., Williams, L., & Megiddo, I. (2021). COVID-19 vaccination in lower-middle income countries: National stakeholder views on challenges, barriers, and potential solutions. *Frontiers in Public Health, 9*. https://doi.org/10.3389/fpubh.2021.709127

Tenni, B., Moir, H. V. J., Townsend, B., Kilic, B., Farrell, A.-M., Keegel, T., & Gleeson, D. (2022). What is the impact of intellectual property rules on access to medicines? A systematic review. *Globalization and Health, 18*(1), 40. https://doi.org/10.1186/s12992-022-00826-4

Transforming Lives: The Role of Open Source in Scaling Telemedicine and Telehealth. (n.d.). Nearform. Retrieved January 22, 2025, from https://www.nearform.com/insights/transforming-lives-the-role-of-open-source-in-scaling-telemedicine-and-telehealth

Ukert, B., Lawley, M., & Kum, H.-C. (2024). Geographic disparities in telemedicine mental health use by applying three way ANOVA on Medicaid claims population data. *BMC Health Services Research, 24*(1), 494. https://doi.org/10.1186/s12913-024-10898-0

Valerio, J. E., Ramirez-Velandia, F., Fernandez-Gomez, M. P., Rea, N. S., & Alvarez-Pinzon, A. M. (2024). Bridging the global technology gap in neurosurgery: Disparities in access to advanced tools for brain tumor resection. *Neurosurgery Practice, 5*(2), e00090. https://doi.org/10.1227/neuprac.0000000000000090

Varadarajan, R. (2024). Inclusive innovations in services: Leveraging digital technologies for inclusive innovations in healthcare services. *European Journal of Marketing, 58*(13), 107–129. https://doi.org/10.1108/EJM-02-2023-0080

Chapter 14
Training and Education for Rehabilitation Professionals

14.1 Introduction

In neuropsychological rehabilitation, the interplay between education and training is crucial for cultivating a dual focus on theoretical understanding and practical application. For instance, coursework in "Functional neuroanatomy" might explore the brain's structural organization and its implications for behavior, equipping learners with the knowledge needed to conceptualize neurological disorders. Meanwhile, supervised clinical practice enables trainees to apply this understanding, honing skills in assessment and intervention that directly impact patient outcomes (Smith, 2019). This balance between education and training is exemplified in the process of developing expertise in neuropsychological assessment and intervention techniques. Education provides the theoretical framework for understanding the principles of assessment, while training ensures that practitioners can competently administer and interpret tests in clinical settings. Hands-on practice in administering cognitive tests, interpreting neuroimaging results, and developing treatment plans tailored to individual patient needs (Bahar-Fuchs et al., 2013). These experiences help trainees bridge the gap between theory and practice, fostering a seamless transition into independent professional roles. The critical role of rehabilitation professionals becomes evident in their ability to deliver patient-centered care using advanced technologies, which are rapidly transforming the field. As the landscape of rehabilitation tools and methodologies evolves, traditional education and training paradigms require re-evaluation. Specialized education programs are essential to equip rehabilitation professionals with the knowledge and skills needed to use these advanced technologies effectively. Moreover, integrating technology into training programs is vital for preparing professionals for the demands of the modern healthcare environment (Cunningham et al., 2024).

Education and training in neuropsychological rehabilitation represent two intertwined yet distinct pathways critical to developing competent professionals in this

K. J. Reddy, *Innovations in Neurocognitive Rehabilitation*,
https://doi.org/10.1007/978-3-031-88117-6_14

evolving field. The interplay of these approaches enables practitioners to navigate the complex challenges of assessing and treating individuals with neurological disorders while advancing the broader scientific understanding of brain–behavior relationships. While both education and training serve as essential pillars, they fulfill different roles in shaping expertise, offering a unique blend of theoretical knowledge and practical application.

Education in neuropsychological rehabilitation lays the foundational groundwork, emphasizing a broad spectrum of topics that span the theoretical underpinnings of neuroscience, neuropsychology, and related fields. This comprehensive approach ensures that professionals possess a deep understanding of the brain's structures and functions, the intricacies of behavior, and the interconnections between the two. For instance, foundational courses in "Functional neuroanatomy" and "Neuropsychology of behavior" provide learners with the intellectual tools necessary to conceptualize the mechanisms underlying various neurological conditions. These courses often delve into the "Foundations for the study of brain–behavior relationships," offering theoretical frameworks that guide practitioners in their assessments and interventions (Smith, 2019). Education also plays a crucial role in familiarizing future professionals with research methods, equipping them with the skills to evaluate existing literature critically and contribute to scientific advancements. Through classroom instruction, lectures, and independent study, learners engage with topics such as research design, ethical considerations in neuropsychology, and the integration of evidence-based practices into clinical care. Assessment in educational contexts often takes the form of written examinations, essays, and projects, all designed to evaluate comprehension and foster intellectual growth (Cunningham et al., 2024). This phase of professional preparation typically spans several years, encompassing graduate and doctoral programs that cultivate a deep reservoir of knowledge.

Training in neuropsychological rehabilitation emphasizes the hands-on application of knowledge, focusing on developing clinical skills through supervised practice. Training periods often involve internships, residencies, and practicums, where individuals are immersed in real-world settings to gain experience in patient care. Training programs in neuropsychology may involve direct patient interaction, such as conducting assessments, taking patient histories, and selecting appropriate diagnostic tests. This process helps translate theoretical understanding into practical expertise (Smith, 2019). One of the defining features of training is its reliance on mentorship and guidance. Trainees work closely with experienced practitioners who model best practices and provide feedback on clinical performance. Through this dynamic, trainees refine their skills, build confidence, and develop the nuanced competencies needed to excel in patient care. The evaluation of training outcomes often focuses on clinical performance and the demonstration of specific competencies, ensuring that practitioners are well-equipped to meet the demands of their profession (Stiers & Perry, 2012). The distinctions between education and training become particularly evident when considering their scope, duration, and methods of assessment. Education covers a broader range of topics, fostering a well-rounded understanding of the field, while training hones specialized skills necessary for specific clinical contexts. Education tends to be long-term, spanning years of academic coursework, whereas training often occurs over shorter, focused periods. Finally, assessment in education centers on intellectual mastery through exams and projects, while training assessments prioritize the application of knowledge in practical, patient-centered scenarios (Bahar-Fuchs et al., 2013; Smith, 2019).

Continuous professional development also emerges as a necessary strategy to keep rehabilitation professionals up-to-date with the latest advancements in their field. Strategies such as ongoing education, participation in professional workshops, and earning certifications ensure that practitioners stay current in their expertise. These needs are not limited to high-resource settings; they apply globally, particularly in regions with growing demands for rehabilitation services. For low- and middle-income countries, where access to adequate rehabilitation services and training can be a challenge, addressing these gaps is critical (Cunningham et al., 2024). Another critical aspect of neuropsychological rehabilitation is the focus on interdisciplinary collaboration. Education often introduces learners to the importance of working across disciplines, incorporating insights from psychology, neuroscience, and medicine. Training brings this interdisciplinary approach to life, requiring trainees to engage with colleagues from diverse fields, such as neurologists, occupational therapists, and social workers. This collaborative dynamic is essential for addressing the multifaceted needs of patients with neurological disorders (Molton & Shreman, 2024). The process of acquiring knowledge and skills in neuropsychological rehabilitation also highlights the need for lifelong learning. While education provides a foundational understanding, the rapid advancements in neuroscience and technology necessitate ongoing training to stay abreast of new developments. Continuing education programs, professional workshops, and specialized certifications play a vital role in ensuring that practitioners remain current in their knowledge and skills (Sperling et al., 2017; Van Patten et al., 2022). The integration of education and training into neuropsychological rehabilitation programs underscores their complementary nature. Both are essential for preparing professionals to address the complexities of brain–behavior relationships and deliver effective care. By blending theoretical knowledge with practical expertise, education and training ensure that practitioners are not only equipped to meet the demands of their current roles but are also prepared to adapt to the evolving landscape of the field.

14.2 Core Competencies for Rehabilitation Professionals

14.2.1 Technical Proficiency

Mastering these advanced tools is not optional for neuropsychological rehabilitation professionals—it is essential. Technical proficiency ensures that practitioners can leverage the full potential of these technologies to improve patient outcomes.

Understanding how to calibrate a robotic device for a specific patient's needs or adjust a VR environment to target particular cognitive deficits can make the difference between a good and an exceptional recovery process. Moreover, the use of these tools allows for more naturalistic assessments of cognitive and physical functions. Patients can perform tasks that closely mirror real-life activities, providing therapists with more accurate data to guide treatment. The accessibility of home-based systems also expands the reach of rehabilitation, reducing barriers for patients in remote or underserved areas (Arntz et al., 2023).

- **Robotics**: Robotics has become a cornerstone of rehabilitation, particularly for individuals with motor impairments due to neurological conditions such as spinal cord injuries, strokes, or cerebral palsy. Robotic devices, such as robotic-assisted gait trainers, enable targeted, repetitive movements that are essential for neuroplasticity and recovery. Robotic gait devices have been shown to improve walking abilities in individuals with spinal cord injuries, offering a precise and consistent method for retraining motor skills (Catania et al., 2023). Robotic gloves are another innovation, aiding in the rehabilitation of upper limb function for patients recovering from strokes or traumatic brain injuries. These gloves use sensors to detect movements and provide biofeedback, which guides patients in performing exercises more effectively (Arntz et al., 2023). Exoskeletons, wearable robotic devices designed to support body movements, have further revolutionized care for patients with conditions like cerebral palsy, facilitating improved mobility and independence (Arntz et al., 2023).
- **Virtual Reality (VR)-Immersive Therapy for Cognitive and Functional Skills:** Virtual reality has opened new avenues for addressing cognitive and functional impairments in rehabilitation. VR-based interventions create immersive environments that simulate real-life scenarios, allowing patients to practice skills in a safe and controlled setting. These systems have been particularly effective in rehabilitating executive functions, spatial skills, and language. VR environments designed to mimic daily living activities, such as grocery shopping or navigation through a city, help patients regain independence by improving spatial orientation and decision-making. In language rehabilitation, VR applications are used to engage individuals with aphasia, providing interactive exercises that improve communication skills (Catania et al., 2023). The flexibility of VR systems also allows therapists to adjust scenarios to match the patient's abilities, ensuring a customized and progressive approach to care.
- **Biofeedback:** Enhancing Awareness and Precision: Biofeedback devices are integral to neuropsychological rehabilitation, providing real-time information about physiological functions such as muscle activity, heart rate, or brainwave patterns. These tools empower patients to better understand and control their physical responses, leading to more effective rehabilitation outcomes. Sensor-based biofeedback systems are particularly beneficial for home-based rehabilitation, where patients can perform targeted exercises outside of clinical settings. Biofeedback systems integrated with robotic gloves provide vibrotactile feedback, guiding patients through precise hand movements (Arntz et al., 2023). This

combination of technology not only enhances motor recovery but also fosters greater patient engagement and autonomy.

- **The Integration of Technologies:** The synergy between robotics, VR, and biofeedback has further amplified their impact on neuropsychological rehabilitation. By combining these tools, professionals can create holistic interventions that address multiple aspects of recovery. Gamification has emerged as a powerful strategy in this integration, incorporating gaming elements into rehabilitation programs to increase patient motivation and participation. Exergames, or exercise-based games, are often paired with robotic devices or VR environments, transforming repetitive rehabilitation tasks into engaging challenges. These games encourage patients to push their limits while providing instant feedback on their performance. Sensory feedback is another area where technologies intersect. Sensors embedded in robotic devices or VR systems provide detailed information about a patient's movements, enabling therapists to fine-tune interventions for maximum effectiveness. Brain–computer interfaces (BCIs) also play a role in this integration, allowing individuals with severe impairments to communicate and perform tasks using non-invasive neural control systems (Arntz et al., 2023; Yuan et al., 2021).

14.2.2 Clinical Reasoning

Rehabilitation professionals play a crucial role in promoting recovery and improving the quality of life for patients. To provide the most effective care, it's essential for clinicians to develop robust skills in clinical reasoning and evidence-based practices. These skills ensure interventions are not only scientifically grounded but also tailored to individual patient needs.

- **Mastering Clinical Reasoning for Effective Care:** Clinical reasoning lies at the heart of successful rehabilitation. It involves synthesizing knowledge, patient preferences, and evidence to design interventions that achieve the best possible outcomes. A cornerstone of this competency is the ability to ask the right questions. Formulating clinical questions rooted in patient-specific concerns allows professionals to focus their efforts on identifying relevant solutions (Zhang & He, 2023). Critically evaluating the quality and applicability of evidence is equally vital. Clinicians must navigate the growing body of rehabilitation literature, identifying robust systematic reviews and high-quality randomized controlled trials (McCabe et al., 2009; Zhang & He, 2023). These studies provide a foundation for evidence-based decisions. Furthermore, integrating patient values and professional expertise is essential to creating meaningful, person-centered plans (Cicerone et al., 2019; McCabe et al., 2009).
- **Evidence-Based Practices: Turning Research into Results:** Applying evidence-based practices (EBP) ensures care aligns with the latest research and clinical guidelines. Rehabilitation professionals are encouraged to stay updated

through systematic reviews and meta-analyses, which synthesize findings across studies to highlight the most effective interventions. Resources such as neurorehabilitation labs and clinical practice frameworks provide structured methodologies for evaluating and implementing new approaches. Staying current requires active engagement, including attending conferences, participating in workshops, and reviewing key publications. Such efforts foster continuous learning and adaptability in a rapidly evolving field (Cicerone et al., 2019; Thistlethwaite et al., 2012; Zhang & He, 2023). Professionals can enhance their practice by incorporating outcome assessments, allowing for real-time evaluations of treatment effectiveness (Zhang & He, 2023). The "6A" approach in evidence-based rehabilitation medicine (EBRM)—Ask, Acquire, Appraise, Apply, Assess, and Advance—is a valuable framework for integrating evidence into daily clinical workflow. This cyclical process encourages lifelong learning, where knowledge acquisition directly informs clinical improvement.
- **Patient-Centered Care: Integrating Values and Evidence:** True success in rehabilitation hinges on a collaborative approach. Patient-centered care involves not only addressing clinical symptoms but also considering individual goals, preferences, and values. Engaging patients in setting treatment objectives fosters a sense of ownership and partnership, enhancing motivation and adherence (Thistlethwaite et al., 2012). Clinicians who adopt this approach ensure their interventions are both evidence-based and deeply personalized. Multidisciplinary collaboration further enriches the rehabilitation process. Sharing insights across disciplines broadens perspectives and enhances problem-solving. Standardized outcome measures also play a pivotal role in tracking patient progress and refining interventions (Cicerone et al., 2019; McCabe et al., 2009; Thistlethwaite et al., 2012).
- **Best Practices for Sustained Excellence:** Rehabilitation professionals should prioritize regular reviews of new research, critically appraising findings for clinical relevance. Attending professional gatherings like conferences fosters the exchange of ideas and promotes innovative practices. Using standardized tools to evaluate intervention outcomes ensures consistency and clarity in patient care. By focusing on clinical reasoning, evidence-based practices, and patient-centered care, rehabilitation professionals can provide effective, compassionate, and informed care that continuously evolves to meet the demands of modern health care (McCabe et al., 2009; Zhang & He, 2023).

14.2.3 Interpersonal Skills

The foundation of effective rehabilitation lies in the relationships professionals build with their patients, families, and multidisciplinary teams. Interpersonal skills and the ability to form strong therapeutic alliances are critical for fostering trust, enhancing patient engagement, and achieving successful outcomes. By focusing on culturally competent communication, empathy, and collaboration, rehabilitation professionals can create an environment where patients feel supported and valued.

- **Interpersonal Skills: The Core of Patient-Centered Care:** Interpersonal skills are at the heart of every therapeutic interaction. Rehabilitation professionals must prioritize culturally competent communication, understanding that patients come from diverse backgrounds with unique perspectives and needs (Beamish et al., 2024). Demonstrating empathy, genuine concern, and warmth creates a safe space for patients to express their fears and aspirations, fostering trust and openness. Moreover, tolerance for ambiguity and individual differences allows professionals to navigate the complexities of rehabilitation, where outcomes can be uncertain and progress non-linear. A realistic yet optimistic attitude helps instill hope in patients, motivating them to actively participate in their recovery. Respecting clients' strengths, unique qualities, and personal goals further strengthens the therapeutic relationship (Liberman et al., 2001).
- **Building Collaborative Therapeutic Alliances:** A strong therapeutic alliance is built on collaboration, mutual respect, and shared goals. Engaging patients and their families in goal-setting and treatment planning ensures that care aligns with their values and priorities. This collaborative process empowers patients, giving them a sense of ownership over their rehabilitation journey. Persistence and a collaborative style are essential for addressing challenges that arise during treatment. Professionals must work alongside patients, adapting strategies to meet changing needs while maintaining a focus on continuous healing relationships. This approach not only improves outcomes but also enhances patient satisfaction and adherence to treatment plans (Beamish et al., 2024; Institute of Medicine (US) Committee on the Health Professions Education Summit et al., 2003; Liberman et al., 2001).
- **Collaborating with Multidisciplinary Teams:** Rehabilitation often requires the expertise of diverse professionals working together. Effective interprofessional collaboration ensures comprehensive care, addressing the physical, emotional, and social aspects of recovery. Regularly participating in and leading multidisciplinary team meetings helps streamline communication, align objectives, and optimize patient outcomes. Valuing the unique roles and expertise of team members, from case managers to supported employment specialists, fosters a culture of mutual respect and cooperation. Networking and system coordination skills enable professionals to connect patients with community resources, providing holistic support that extends beyond clinical settings (Beamish et al., 2024; Liberman et al., 2001).
- **Best Practices for Professional Growth:** Continuous learning is key to mastering these core competencies. Communication workshops and team-building activities enhance patient counseling skills and foster effective collaboration. Mentorship programs and leadership trainings provide opportunities for personal and professional growth, preparing professionals to take on greater responsibilities within their teams. Staying informed about evidence-based interventions and psychosocial treatments ensures that care remains aligned with the latest research

and best practices. Demonstrating respect for multidisciplinary team members and supporting their contributions creates a positive work environment, ultimately benefiting both patients and professionals (Anderson & Perlman, 2019; Beamish et al., 2024; Doornebosch et al., 2022; Verma et al., 2006).

14.2.4 Ethical and Cultural Competence

Rehabilitation professionals often encounter complex ethical dilemmas and serve clients from diverse cultural backgrounds. Ethical and cultural competence is crucial to providing equitable, respectful, and effective care. By fostering self-awareness, developing ethical sensitivity, and embracing cultural diversity, professionals can address challenges thoughtfully and inclusively.

- **Navigating Ethical Dilemmas with Sensitivity and Insight:** Ethical dilemmas are an inherent part of rehabilitation work, requiring professionals to balance competing interests while prioritizing patient welfare. Developing strategies to address these dilemmas involves active listening, role-playing, and visualizing different perspectives to foster empathy and creative problem-solving. Ethical sensitivity is key, encompassing skills such as controlling social bias, understanding emotional expression, and interpreting the ethical dimensions of situations. These abilities allow professionals to respond to complex scenarios with compassion and clarity. Prosocial behaviors, such as reflective emotional regulation, further support effective ethical decision-making, ensuring that actions align with both professional standards and patients' best interests (Louw, 2016).
- **Embracing Cultural Competence in a Multicultural World:** In today's diverse society, cultural competence is no longer optional—it's essential. Rehabilitation professionals must recognize that multiculturalism extends beyond race and ethnicity to include factors such as ability, gender, age, socioeconomic status, and religion. This broader understanding ensures that care is inclusive and relevant to each individual's unique context. The journey toward cultural competence begins with self-awareness. Professionals must reflect on their own cultural values and biases to better understand how these influence their interactions with clients. Acquiring knowledge about different cultures and developing skills to communicate effectively in diverse contexts is equally important. This includes mastering both verbal and non-verbal communication strategies to ensure messages are understood and respectful (*Cultural Responsiveness*, n.d.; Louw, 2016).
- **Applying Ethical and Cultural Competence in Practice:** Cultural competence enables rehabilitation professionals to provide services that respect and support clients' values, beliefs, and traditions. Incorporating cultural sensitivity into educational training programs is a vital step in building this competency (Rivera Mindt et al., 2010). Continuous self-reflection and learning, through workshops and cultural immersion activities, can further enhance understanding and empa-

thy. Collaboration is another cornerstone of ethical and culturally competent care. Consulting with colleagues or experts when faced with unfamiliar cultural contexts ensures that decisions are well-informed and appropriate. Involving family members or community representatives in therapeutic processes can also strengthen the relevance and effectiveness of interventions. Professionals should use interpreters and translators when needed, ensuring these individuals are trained in ethical practices to maintain confidentiality and accuracy. Such practices demonstrate respect for clients' communication needs and promote trust (Bray, 2019; *Cultural Responsiveness*, n.d.; Rivera Mindt et al., 2010; Smith, 2019).

- **Best Practices for Continuous Growth:** Ethical and cultural competence requires a commitment to lifelong learning. Engaging in professional development opportunities, such as ethics seminars and cultural diversity workshops, helps professionals stay informed and adaptable. Seeking mentorship and participating in peer discussions can provide valuable perspectives on navigating complex ethical and cultural challenges. Fostering ethical sensitivity and embracing cultural diversity, rehabilitation professionals can create a supportive and inclusive environment where all clients feel valued. These efforts not only improve patient outcomes but also strengthen the integrity and effectiveness of the rehabilitation profession (Bray, 2019; Louw, 2016; Smith, 2019).

14.3 Pedagogical Approaches in Rehabilitation Training

Training programs for rehabilitation professionals have long relied on traditional classroom-based instruction to build foundational knowledge in theories and concepts. However, as the demands of the field grow and technology advances, it is increasingly clear that innovative pedagogical approaches are necessary to prepare practitioners with both theoretical expertise and hands-on skills. By integrating technology-enhanced learning methods and interdisciplinary training, educators can create more interactive and effective learning experiences that align with the evolving needs of the field.

14.3.1 Traditional Methods

Traditional classroom-based learning has been the cornerstone of rehabilitation training for decades. These methods focus on teaching foundational theories, core principles, and essential concepts that form the basis of clinical practice. Lectures, textbooks, and discussions offer students a structured framework for understanding rehabilitation processes and patient care strategies. While traditional approaches are effective for imparting theoretical knowledge, they often fall short in providing hands-on experience with emerging tools and technologies. For instance, practitioners may leave their training programs well-versed in cognitive therapy principles but lack exposure to modern tools like virtual reality (VR) simulations or wearable devices used in neurorehabilitation (Nicolaou et al., 2019).

14.3.2 Technology-Enhanced Learning

Technology-enhanced learning has emerged as a game-changer in rehabilitation education. E-learning platforms provide flexible, on-demand access to training materials, enabling professionals to pursue online certification courses and engage in self-paced learning. These platforms not only enhance accessibility but also allow learners to revisit materials as needed, fostering deeper understanding. Simulation-based training, particularly through virtual and augmented reality (VR/AR), is a standout example of how technology is reshaping the training landscape. VR simulations provide realistic practice scenarios where students can refine their cognitive therapy techniques, problem-solving abilities, and patient interaction skills in a controlled, risk-free environment. For instance, trainees can simulate treating a patient with traumatic brain injury, gaining valuable experience without the potential for harm. Audiovisual media technologies further enrich learning by presenting complex concepts in engaging formats. Videos, animations, and interactive modules make it easier for students to grasp challenging topics, such as neuroanatomy or the mechanics of prosthetic devices (Bremner & Air, 2023; *Immersive Learning, AI and Improvement Science*, n.d.; Nicolaou et al., 2019).

14.3.3 Interdisciplinary Training

Rehabilitation often requires collaboration among professionals from diverse fields, including neuroscience, psychology, and bioengineering. Interdisciplinary training programs are designed to bridge these disciplines, fostering holistic understanding and collaborative problem-solving. Multidisciplinary workshops by physiotherapists and bioengineers on wearable technology provide a platform for sharing expertise and exploring innovative solutions. Such programs not only expand participants' knowledge but also prepare them to work effectively in multidisciplinary teams—a critical skill in modern rehabilitation settings. These collaborative experiences cultivate an appreciation for the unique contributions of each profession, enhancing patient care outcomes (Bremner & Air, 2023; Nicolaou et al., 2019).

14.3.4 Future Modifications in Modern Pedagogical Approaches

While technology-enhanced learning and interdisciplinary training offer significant benefits, it is important to ensure these methods are thoughtfully integrated into existing curricula. Audiovisual technologies, VR/AR simulations, and online platforms should align with training objectives, enhancing rather than replacing traditional methods. Combining classroom instruction with hands-on practice creates a well-rounded educational experience, equipping students with both theoretical knowledge and practical skills. Additionally, educators must consider the accessibility of these tools. Ensuring that students and professionals from all backgrounds can access technology-enhanced learning resources is vital for fostering equity in education (Delungahawatta et al., 2022; Dunn et al., 2013).

14.4 Lifelong Learning and Professional Development

Rehabilitation professionals operate in a field characterized by constant evolution. Advances in technology, new treatment methodologies, and emerging evidence demand a commitment to lifelong learning and professional development. Staying current is not just an aspiration but a necessity to deliver high-quality care and meet the growing complexities of patient needs. By engaging in continuing education, pursuing certifications, participating in mentorship programs, and fostering peer learning, rehabilitation professionals can build a robust framework for career-long growth and excellence.

14.4.1 Continuing Education Programs

Continuing education is the cornerstone of lifelong learning. Workshops, seminars, and conferences are essential for staying abreast of the latest trends, research, and technologies in rehabilitation. These programs provide opportunities to learn about emerging tools, such as robotic-assisted rehabilitation or virtual reality therapies, and how they can be integrated into clinical practice. Annual conferences, like those focused on rehabilitation technologies, serve as platforms for professionals to explore cutting-edge innovations, network with peers, and share insights. Such events are invaluable for reinforcing foundational knowledge while introducing fresh perspectives that can enhance patient care. E-learning platforms also play a vital role by offering flexibility for busy professionals. Online courses and webinars allow learners to update their skills at their own pace, making continuing education more accessible and tailored to individual schedules (*Continuing Education & Professional Development | AOTA*, n.d.; *Professional Education*, n.d.). The rehabilitation landscape is shaped by rapid advancements in technology and research. Staying current is essential not only for maintaining competence but also for delivering the highest standard of care. Lifelong learning helps professionals adapt to innovations like wearable devices, AI-driven diagnostics, and telehealth platforms, which are becoming integral to modern rehabilitation practices. Continuous education ensures that professionals are equipped to implement evidence-based interventions. Understanding the latest findings in neuroplasticity or pain management enables practitioners to refine their approaches and achieve better patient outcomes (Chukwuka et al., 2024; Rollins, 2021).

14.4.2 Certification and Accreditation

In a competitive and ever-changing field, certifications serve as formal recognition of a professional's expertise in specialized areas. They demonstrate a commitment to excellence and ensure that practitioners meet industry standards. Certifications in robotic-assisted rehabilitation or neurorehabilitation underscore a professional's competence in utilizing advanced tools and methodologies. Certification programs also encourage a deeper dive into niche areas, fostering mastery and confidence. Organizations like the Rehabilitation Engineering and Assistive Technology Society of North America (RESNA) and the American Occupational Therapy Association (AOTA) offer numerous certification opportunities tailored to specific interests and career goals. Moreover, certifications often require ongoing maintenance through continuing education, ensuring that professionals remain up-to-date on the latest advancements. This cyclical process not only reinforces knowledge but also cultivates a culture of continuous improvement (*Continuing Education & Professional Development | AOTA*, n.d.; *RESNA > Certification > Continuing Education*, n.d.; Rollins, 2021). To maximize the benefits of lifelong learning, rehabilitation professionals should adopt the following strategies:

- **Participate in Workshops and Conferences:** Regular attendance at industry events, such as conferences on rehabilitation technologies or young investigator seminars, provides exposure to the latest advancements and opportunities to connect with thought leaders in the field
- **Pursue Specialized Certifications:** Certifications in areas like robotic-assisted rehabilitation or cognitive therapy validate expertise and demonstrate a commitment to professional growth.
- **Engage in Mentorship and Peer Learning:** Seek out mentorship opportunities early in your career to gain guidance from experienced professionals. Participate in peer-reviewed case discussions to exchange insights and refine clinical skills.
- **Utilize e-Learning Platforms:** Leverage online resources and webinars to access training materials that fit your schedule. These platforms offer flexible learning options and allow for targeted skill development.

- **Stay Current with Research:** Regularly review academic journals, attend research-focused seminars, and participate in evidence-based practice workshops to keep pace with the latest findings (*Young Cancer Investigator Seminar Series & Career Chats | Montefiore Einstein Comprehensive Cancer Center | Patient Care | Montefiore Einstein*, n.d.).

14.4.3 Mentorship and Peer Learning

Mentorship programs are instrumental in bridging the gap between theory and practice, particularly for early-career professionals. Experienced mentors provide guidance, share clinical insights, and offer invaluable support as mentees navigate the complexities of rehabilitation practice. Peer learning is another powerful tool for professional growth. Participating in peer-reviewed case discussions allows professionals to exchange ideas, evaluate real-world scenarios, and enhance their clinical decision-making skills. Mentorship and peer learning also foster a sense of community within the field. By collaborating and sharing experiences, professionals build networks that extend beyond individual institutions, promoting collective growth and development (*Continuing Education & Professional Development | AOTA*, n.d.; *RESNA > Certification > Continuing Education*, n.d.; Rollins, 2021).

As the field continues to evolve, the importance of lifelong learning will only grow. Rehabilitation professionals must remain proactive in seeking opportunities to expand their knowledge and adapt to emerging challenges. Institutions and organizations also play a critical role by providing accessible and relevant educational programs that cater to diverse needs and career stages. Ultimately, the combination of continuing education, certification, mentorship, and peer learning creates a comprehensive approach to professional development. This model not only enhances individual competence but also strengthens the collective expertise of the rehabilitation community, leading to improved patient care and outcomes. In the words of experts, "Lifelong learning is not just a pathway to professional growth—it is a commitment to excellence, innovation, and the ongoing pursuit of better care for patients" (*Professional Education*, n.d.; Rollins, 2021).

14.5 Challenges in Training and Education

The field of rehabilitation is undergoing a transformative phase, fueled by rapid technological advancements, evolving healthcare needs, and increasing awareness about individualized care. While these developments present exciting opportunities, they also highlight significant challenges in the training and education of rehabilitation professionals. These challenges range from resource constraints to the integration of cutting-edge technologies in curricula. Addressing these issues is crucial to ensuring that professionals are equipped with the necessary skills to meet the growing demands of this dynamic field. Below, we delve into some of the primary challenges and propose actionable solutions to overcome them.

14.5.1 Resource Constraints

Resource limitations remain a persistent challenge, especially in low-resource settings where access to advanced training tools and technologies is restricted. Many rural training institutes lack facilities such as simulation labs, which are vital for providing hands-on experience. Developing countries report that trainees often rely on theoretical knowledge due to the unavailability of practical training environments. These constraints hinder the ability of students to develop the practical skills necessary for effective patient care. To address this issue, innovative approaches such as mobile simulation labs and low-cost virtual training modules can be explored. Partnerships between well-funded urban institutions and rural training centers could bridge the resource gap by sharing equipment and expertise (Mennella et al., 2023).

14.5.2 Balancing Theory and Practice

One of the critical goals in rehabilitation education is achieving an optimal balance between theoretical knowledge and practical application. While a strong theoretical foundation is essential, it must be complemented by opportunities for real-world experience. For instance, integrating clinical internships with academic coursework allows students to apply theoretical concepts in practical settings, enhancing their problem-solving and critical-thinking abilities. In many programs, the lack of structured clinical internships or mentorship opportunities limits students' exposure to real-life rehabilitation scenarios. Establishing robust partnerships with hospitals, clinics, and community-based organizations can provide students with invaluable hands-on experience while fostering collaborative learning environments (Mennella et al., 2023).

14.5.3 Rapid Technological Advancements

The pace of technological advancements in rehabilitation is both a boon and a challenge. Technologies such as artificial intelligence (AI), robotics, and wearable devices have revolutionized the field, offering unprecedented opportunities for individualized care. However, integrating these innovations into training curricula is a daunting task. Many educational institutions struggle to keep their programs updated with the latest developments, leading to a gap between academic training and industry requirements (Akinwalere & Ivanov, 2022). Training professionals to use AI-integrated rehabilitation tools requires not only technical expertise but also a deep understanding of ethical, legal, and regulatory considerations. Incorporating these elements into curricula demands interdisciplinary collaboration and continuous professional development for educators. Regularly updating coursework, organizing workshops, and inviting industry experts as guest lecturers can help bridge this gap. The integration of AI and other emerging technologies in rehabilitation raises complex ethical and regulatory issues. Questions surrounding data privacy, informed consent, and the potential biases in AI algorithms must be addressed to ensure the ethical use of these technologies (Capraro et al., 2024). Training programs need to emphasize the importance of ethical decision-making and equip professionals with the knowledge to navigate these challenges. Case-based learning, which involves analyzing real-world ethical dilemmas, can be an effective pedagogical tool.

14.5.4 Best Practices for Overcoming Challenges

To address the aforementioned challenges, rehabilitation training programs can adopt several best practices:

1. **Implement Hybrid Learning Models**: Combining theoretical instruction with hands-on practice ensures that students develop both foundational knowledge and practical skills. Hybrid models, which blend online learning with in-person training, can be particularly effective in resource-constrained settings (Kushwaha et al., 2024).
2. **Regularly Update Curricula**: Keeping curriculum aligned with the latest technological and methodological advancements is essential. Institutions should establish review committees that include industry experts, alumni, and faculty to ensure that programs remain relevant and forward-looking (Sinha & Lee, 2024).
3. **Expand Access to Training Facilities**: Simulation labs, virtual reality modules, and other training tools should be made accessible to all students, regardless of geographic location. Mobile training units and digital platforms can play a significant role in democratizing access to these resources (Junaid et al., 2022).
4. **Foster Interdisciplinary Training**: Rehabilitation professionals often work in interdisciplinary teams. Training programs should reflect this reality by offering courses that encourage collaboration across disciplines such as engineering, psychology, and occupational therapy (Zhu et al., 2023).
5. **Promote Continuous Professional Development**: The rapid pace of change in the field necessitates lifelong learning. Institutions should provide opportunities for alumni to return for refresher courses and advanced training. Online courses and certifications can be particularly useful for working professionals seeking to update their skills (Coombs et al., 2020).

6. **Emphasize Ethics and Cultural Sensitivity**: Rehabilitation professionals must be equipped to navigate the ethical complexities of their work while remaining culturally sensitive. Training programs should include modules on ethics, cultural competence, and patient-centered care to ensure holistic professional development (Dimitriadou & Lanitis, 2023).

14.6 Global Case Studies

Rehabilitation training is experiencing a transformative shift worldwide, driven by technology, interdisciplinary collaboration, and a commitment to accessibility. From robotic gait training in the United States and India to VR-based simulations in the United Kingdom, global case studies reveal how diverse approaches are enhancing the skills of rehabilitation professionals and improving patient outcomes. These examples highlight innovative strategies that blend traditional practices with cutting-edge technology to address global challenges and opportunities in rehabilitation education.

14.6.1 Training in Robotic Rehabilitation

Robotic technology is revolutionizing rehabilitation by enabling precise, consistent, and repeatable therapeutic interventions. In the United States, hands-on workshops and certification programs are equipping rehabilitation professionals with the skills needed to operate robotic gait trainers effectively. These programs emphasize practical experience, allowing trainees to gain confidence in using advanced equipment. A compelling example comes from Christian Medical College in India, where ten therapists and three doctors received intensive training on robotic equipment. This initiative treated over 150 patients, showcasing the potential of robotic rehabilitation to improve mobility and quality of life. The program underscores the importance of not only introducing new technologies but also ensuring that professionals are adequately trained to use them. Such efforts enhance adoption rates and maximize the efficacy of robotic interventions in clinical settings (*Changing Lives in India with the Latest Robotic Rehabilitation Technology | Basic Page | India*, 2024; Pillai, 2024).

14.6.2 Simulation-Based Learning in Neurorehabilitation

Simulation-based learning, particularly through virtual reality (VR), is gaining traction as an effective training tool in neurorehabilitation. In the United Kingdom, VR-based simulation labs are being used to train professionals in stroke rehabilitation. These simulations provide realistic, immersive scenarios where trainees can practice therapeutic techniques in a controlled environment. The impact of this approach is twofold. First, it enhances the confidence and competence of professionals by allowing them to refine their skills without the risk of causing harm to patients. Second, it bridges the gap between theoretical knowledge and practical application, ensuring that trainees are well-prepared for real-world challenges (Li & Yang, 2024; Pillai, 2024).

14.6.3 E-Learning for Rural Professionals

Access to quality training remains a significant challenge in rural areas, where resources and expertise are often limited. In India, online modules for telerehabilitation practices are addressing this issue by providing flexible, accessible training solutions. These e-learning platforms enable professionals in remote locations to stay updated on best practices, emerging technologies, and treatment methodologies. By bridging the expertise gap between urban and rural practitioners, e-learning empowers professionals to deliver high-quality care regardless of their location. This approach not only enhances the skills of individual practitioners but also improves overall healthcare outcomes in underserved regions. The complexity of rehabilitation often requires input from multiple disciplines, including physiotherapy, engineering, neuroscience, and psychology. Interdisciplinary training programs are emerging as a key strategy for fostering collaboration and holistic understanding among professionals. Joint workshops involving physiotherapists and bioengineers are enabling participants to explore the design, application, and optimization of wearable rehabilitation technologies. Such programs encourage professionals to appreciate the interconnectedness of their roles, facilitating more effective problem-solving and innovation. By breaking down silos between disciplines, interdisciplinary training ensures that rehabilitation interventions are informed by a diverse range of expertise (Bessler et al., 2021; *Case Studies*, n.d.; Pillai, 2024; *Rehabilitation Robots Market Size And Share Report, 2030*, n.d.).

14.7 Future Directions

As the field of rehabilitation evolves, so does the need for innovative and accessible training models. Emerging trends, such as the integration of artificial intelligence (AI), global collaboration, and a focus on sustainability, are reshaping how rehabilitation professionals are educated and equipped to meet the demands of modern health care. These advancements highlight the potential for personalized, technology-driven, and globally connected training programs that emphasize both practical application and ethical considerations.

14.7.1 Integration of AI in Training

Artificial intelligence is transforming the landscape of rehabilitation education by enabling personalized, adaptive learning experiences. AI-driven platforms can analyze student performance data in real-time, identifying knowledge gaps and tailoring content, pace, and difficulty to meet individual needs. Adaptive learning systems provide customized pathways for learners, ensuring they receive targeted instruction in areas where they need the most improvement. AI also facilitates immediate feedback, helping learners refine their skills more efficiently. In rehabilitation training, this could mean personalized assessments in motor skill development, cognitive therapy techniques, or even robotic equipment usage (Halkiopoulos & Gkintoni, 2024; NEA, n.d. ; Strielkowski et al., 2024). These tools not only enhance the learning experience but also allow educators to focus on more complex, hands-on instruction while AI handles routine evaluations. While AI offers immense potential, ethical considerations must remain at the forefront. Ensuring data privacy, security, and transparency in AI-driven education platforms is essential (*The Potential of AI in Personalized Learning | Atlantic International University*, 2024). Developing guidelines for ethical AI use in healthcare education will be critical as these technologies become more integrated into training programs.

As the rehabilitation field becomes more technologically advanced, the need for practical, hands-on training is more critical than ever. Simulation-based learning, particularly through AI and VR technologies, is emerging as a powerful tool for bridging the gap between theoretical knowledge and real-world application. For instance, virtual simulations can replicate complex scenarios, such as treating patients with severe mobility impairments, allowing trainees to practice and refine their skills in a risk-free environment. AI-driven assessment tools are also playing a role in hands-on training by providing detailed analyses of student performance during simulations. These tools help educators identify areas for improvement, enabling more targeted and effective instruction (*Detection of Vitiligo Through Machine Learning and Computer-Aided Techniques: A Systematic Review—Tanvir—2024—BioMed Research International—Wiley Online Library*, n.d.; Omar Jian, 2024; Parvez, 2024).

14.7.2 Global Collaboration

Global collaboration is becoming a cornerstone of rehabilitation training, breaking down geographical barriers and fostering cross-border knowledge sharing. Virtual workshops, online conferences, and international training programs enable professionals from different countries to exchange expertise and learn from diverse perspectives. Cross-border virtual workshops on advanced rehabilitation techniques allow professionals in resource-constrained regions to access world-class training without the need for travel. These programs not only enhance individual competencies but also promote a global standard of care, ensuring that patients worldwide benefit from the latest advancements in rehabilitation. Moreover, collaboration between institutions, governments, and private organizations can drive innovation in training tools and methodologies. By pooling resources and expertise, stakeholders can create more robust and effective training models that address global disparities in healthcare education (Chukwuka et al., 2024; Omar Jian, 2024).

14.7.3 Focus on Sustainability

Sustainability is increasingly shaping the future of rehabilitation education, with a focus on cost-effective and scalable solutions that make advanced training accessible to all. Low-cost simulation tools, for instance, are being developed to provide hands-on training in resource-constrained settings. These tools allow professionals to practice essential skills, such as stroke rehabilitation techniques or robotic-assisted interventions, without the need for expensive equipment or facilities. Online platforms also contribute to sustainability by reducing the logistical and financial barriers associated with traditional in-person training programs. E-learning modules and tele-rehabilitation courses offer flexible, affordable options for professionals in remote areas, ensuring they have access to the same high-quality education as their urban counterparts. By prioritizing sustainable approaches, the rehabilitation field can expand its reach and impact, addressing disparities in training access while minimizing environmental and financial costs. As AI becomes more integrated into rehabilitation training, ethical considerations must guide its development and implementation. Issues such as data privacy, bias in algorithms, and the potential for over-reliance on technology are critical concerns that need to be addressed. AI systems must be designed to respect student confidentiality and ensure that data is used responsibly and transparently. Moreover, educators must strike a balance between leveraging AI and maintaining human-centered learning. While AI can enhance efficiency and personalization, it cannot replace the empathy, intuition, and interpersonal skills that are fundamental to rehabilitation practice (*Detection of Vitiligo Through Machine Learning and Computer-Aided Techniques: A Systematic Review—Tanvir—2024—BioMed Research International—Wiley Online Library*, n.d.; Omar Jian, 2024; Parvez, 2024; Strielkowski et al., 2024).

14.8 Conclusion

The future of rehabilitation training and education lies in its ability to adapt, innovate, and expand access to high-quality, practical learning experiences. Emerging trends, such as the integration of AI-driven personalization, simulation-based learning, and global collaboration, are revolutionizing how rehabilitation professionals acquire knowledge and refine their skills. These advancements are not just technological—they represent a shift toward more inclusive, accessible, and patient-centered approaches to healthcare education. Global case studies have shown the transformative impact of technology, from robotic rehabilitation training in the United States and India to VR-based neurorehabilitation simulations in the United Kingdom. At the same time, e-learning initiatives are bridging the expertise gap between urban and rural professionals, emphasizing the importance of equity in education. By fostering interdisciplinary collaboration, professionals from diverse fields are coming together to address complex challenges, innovate solutions, and deliver holistic care. Sustainability is a guiding principle in the evolution of rehabilitation training, with cost-effective and scalable models ensuring that advancements are accessible to all. This focus on affordability and resource optimization is crucial for extending the benefits of modern rehabilitation techniques to underserved regions and populations. The integration of AI, while promising, also raises important ethical considerations. Addressing concerns around data privacy, algorithmic bias, and over-reliance on technology will require thoughtful policies and a commitment to maintaining the human-centered essence of rehabilitation care. These efforts must balance technological innovation with empathy, interpersonal skills, and cultural competence. The path forward for rehabilitation training and education is one of collaboration, innovation, and inclusivity. By embracing emerging technologies, fostering global partnerships, and prioritizing sustainability, the field can empower professionals to deliver effective, compassionate, and evidence-based care to diverse populations. As the landscape continues to evolve, this commitment to lifelong learning and professional development will ensure that rehabilitation professionals remain equipped to transform lives and advance the frontiers of health care.

References

Akinwalere, S. N., & Ivanov, V. (2022). Artificial intelligence in higher education: Challenges and opportunities. *Border Crossing, 12*(1), article 1. https://doi.org/10.33182/bc.v12i1.2015

Anderson, T., & Perlman, M. R. (2019). C3 Therapeutic interpersonal skills for facilitating the working alliance. In J. N. Fuertes (Ed.), *Working alliance Skills for mental health professionals*. Oxford University Press. https://doi.org/10.1093/med-psych/9780190868529.003.0003

Arntz, A., Weber, F., Handgraaf, M., Lällä, K., Korniloff, K., Murtonen, K.-P., Chichaeva, J., Kidritsch, A., Heller, M., Sakellari, E., Athanasopoulou, C., Lagiou, A., Tzonichaki, I., Salinas-Bueno, I., Martínez-Bueso, P., Velasco-Roldán, O., Schulz, R.-J., & Grüneberg, C. (2023). Technologies in home-based digital rehabilitation: Scoping review. *JMIR Rehabilitation and Assistive Technologies, 10*, e43615. https://doi.org/10.2196/43615

Bahar-Fuchs, A., Clare, L., & Woods, B. (2013). Cognitive training and cognitive rehabilitation for persons with mild to moderate dementia of the Alzheimer's or vascular type: A review. *Alzheimer's Research & Therapy, 5*(4), 35. https://doi.org/10.1186/alzrt189

Beamish, N., Footer, C., Lowe, R., & Cunningham, S. (2024). Rehabilitation professions' core competencies for entry-level professionals: A thematic analysis. *Journal of Interprofessional Care, 38*(1), 32–41. https://doi.org/10.1080/13561820.2023.2241519

Bessler, J., Prange-Lasonder, G. B., Schaake, L., Saenz, J. F., Bidard, C., Fassi, I., Valori, M., Lassen, A. B., & Buurke, J. H. (2021). Safety assessment of rehabilitation robots: A review identifying safety skills and current knowledge gaps. *Frontiers in Robotics and AI, 8*. https://doi.org/10.3389/frobt.2021.602878

Bray, B. (2019). *Multicultural encounters*. https://www.counseling.org/publications/counseling-today-magazine/article-archive/article/legacy/multicultural-encounters

Bremner, P. A. M., & Air, C. (2023). Innovative interdisciplinary pedagogical approaches to enhance students' learning experiences and to benefit them in their future development. *Journal of Applied Research in Higher Education.*, Ahead-of-print. https://doi.org/10.1108/JARHE-07-2023-0290

Capraro, V., Lentsch, A., Acemoglu, D., Akgun, S., Akhmedova, A., Bilancini, E., Bonnefon, J.-F., Brañas-Garza, P., Butera, L., Douglas, K. M., Everett, J. A. C., Gigerenzer, G., Greenhow, C., Hashimoto, D. A., Holt-Lunstad, J., Jetten, J., Johnson, S., Kunz, W. H., Longoni, C., … Viale, R. (2024). The impact of generative artificial intelligence on socioeconomic inequalities and policy making. *PNAS Nexus, 3*(6), 191. https://doi.org/10.1093/pnasnexus/pgae191

Case studies. (n.d.). Imperial College London. Retrieved January 13, 2025, from https://www.imperial.ac.uk/research-and-innovation/about-imperial-research/ref/case-studies/

Catania, V., Rundo, F., Panerai, S., & Ferri, R. (2023). Virtual reality for the rehabilitation of acquired cognitive disorders: A narrative review. *Bioengineering, 11*(1), 35. https://doi.org/10.3390/bioengineering11010035

Changing Lives in India with the Latest Robotic Rehabilitation Technology | Basic Page | India. (2024, October 9). U.S. Agency for International Development. https://www.usaid.gov/work-usaid/business-funding/grant-programs/american-schools-and-hospitals-abroad/stories/changing-lives-in-india-with-robotic-technology

Chukwuka, E., Dependable, A., Okatta, A., Emmanuel, A., Tochi, E., Chiamaka, E., & Ijeoma, E. (2024). The impact of simulation-based training in medical education: A review. *Medicine, 103*(27), e38813. https://doi.org/10.1097/MD.0000000000038813

Cicerone, K. D., Goldin, Y., Ganci, K., Rosenbaum, A., Wethe, J. V., Langenbahn, D. M., Malec, J. F., Bergquist, T. F., Kingsley, K., Nagele, D., Trexler, L., Fraas, M., Bogdanova, Y., & Harley, J. P. (2019). Evidence-based cognitive rehabilitation: Systematic review of the literature from 2009 through 2014. *Archives of Physical Medicine and Rehabilitation, 100*(8), 1515–1533. https://doi.org/10.1016/j.apmr.2019.02.011

Continuing Education & Professional Development | AOTA. (n.d.). Retrieved January 13, 2025, from https://www.aota.org/career/continuing-education

Coombs, C., Hislop, D., Taneva, S. K., & Barnard, S. (2020). The strategic impacts of Intelligent Automation for knowledge and service work: An interdisciplinary review. *The Journal of Strategic Information Systems, 29*(4), 101600. https://doi.org/10.1016/j.jsis.2020.101600

Cultural Responsiveness. (n.d.). American Speech-Language-Hearing Association. Retrieved January 13, 2025, from https://www.asha.org/practice-portal/professional-issues/cultural-responsiveness/?srsltid=AfmBOoqNLriLjbez1JMUf7O0H5AP9s-Y1Qds2_wrpfXyNBPhVii5VJvT

Cunningham, S., Lowe, R., Footer, C., Oo, H. W. M., Omoding, H., Kovalchuk, O., Zehra, N., & Bachani, A. M. (2024). Assessing global access to continuing professional development for rehabilitation professionals: A mixed methods study of current practices and perceived needs. *BMJ Open, 14*(11), e089079. https://doi.org/10.1136/bmjopen-2024-089079

Delungahawatta, T., Dunne, S. S., Hyde, S., Halpenny, L., McGrath, D., O'Regan, A., & Dunne, C. P. (2022). Advances in e-learning in undergraduate clinical medicine: A systematic review. *BMC Medical Education, 22*(1), 711. https://doi.org/10.1186/s12909-022-03773-1

Detection of Vitiligo Through Machine Learning and Computer-Aided Techniques: A Systematic Review—Tanvir—2024—BioMed Research International—Wiley Online Library. (n.d.). Retrieved January 13, 2025, from https://onlinelibrary.wiley.com/doi/full/10.1155/bmri/3277546

Dimitriadou, E., & Lanitis, A. (2023). A critical evaluation, challenges, and future perspectives of using artificial intelligence and emerging technologies in smart classrooms. *Smart Learning Environments, 10*(1), 12. https://doi.org/10.1186/s40561-023-00231-3

Doornebosch, A. J., Smaling, H. J. A., & Achterberg, W. P. (2022). Interprofessional collaboration in long-term care and rehabilitation: A systematic review. *Journal of the American Medical Directors Association, 23*(5), 764–777.e2. https://doi.org/10.1016/j.jamda.2021.12.028

Dunn, D. S., Saville, B. K., Baker, S. C., & Marek, P. (2013). Evidence-based teaching: Tools and techniques that promote learning in the psychology classroom. *Australian Journal of Psychology, 65*(1), 5–13. https://doi.org/10.1111/ajpy.12004

Halkiopoulos, C., & Gkintoni, E. (2024). Leveraging AI in E-learning: Personalized learning and adaptive assessment through cognitive neuropsychology—A systematic analysis. *Electronics, 13*(18), article 18. https://doi.org/10.3390/electronics13183762

Immersive Learning, AI and Improvement Science. (n.d.). Simmons School of Education and Human Development. Retrieved January 13, 2025, from https://www.smu.edu/simmons/research/teil-technology-enhanced-immersive-learning/about-teil-at-smu

Institute of Medicine (US) Committee on the Health Professions Education Summit, Greiner, A. C., & Knebel, E. (2003). The core competencies needed for health care professionals. In *Health professions education: A bridge to quality*. National Academies Press (US). https://www.ncbi.nlm.nih.gov/books/NBK221519/

Junaid, S. B., Imam, A. A., Balogun, A. O., De Silva, L. C., Surakat, Y. A., Kumar, G., Abdulkarim, M., Shuaibu, A. N., Garba, A., Sahalu, Y., Mohammed, A., Mohammed, T. Y., Abdulkadir, B. A., Abba, A. A., Kakumi, N. A. I., & Mahamad, S. (2022). Recent advancements in emerging technologies for healthcare management systems: A survey. *Healthcare, 10*(10), 1940. https://doi.org/10.3390/healthcare10101940

Kushwaha, A., Kushwaha, R. K., & Ahmad, S. (2024). *Transforming learning: The power of educational technology*. Bluerose. https://www.researchgate.net/publication/381408076_Transforming_Learning_The_Power_of_Educational_Technology

Li, R., & Yang, M. (2024). Feasibility study of neurofeedback therapy for Alzheimer's disease. *Advances in Alzheimer's Disease, 13*(3), article 3. https://doi.org/10.4236/aad.2024.133005

Liberman, R. P., Hilty, D. M., Drake, R. E., & Tsang, H. W. H. (2001). Requirements for multidisciplinary teamwork in psychiatric rehabilitation. *Psychiatric Services, 52*(10), 1331–1342. https://doi.org/10.1176/appi.ps.52.10.1331

Louw, B. (2016). Cultural competence and ethical decision making for health care professionals. *Humanities and Social Sciences, 4*(2), article 2. https://doi.org/10.11648/j.hss.s.2016040201.17

McCabe, P., Purcell, A., Baker, E., Madill, C., & Trembath, D. (2009). Case-based learning: One route to evidence-based practice. *Evidence-Based Communication Assessment and Intervention, 3*(4), 208–219. https://doi.org/10.1080/17489530903399145

Mennella, C., Maniscalco, U., De Pietro, G., & Esposito, M. (2023). The role of artificial intelligence in future rehabilitation services: A systematic literature review. *IEEE Access, 11*, 11024–11043. IEEE Access. https://doi.org/10.1109/ACCESS.2023.3236084

Molton, I., & Shreman, J. (2024). *Rehabilitation neuropsychology track—Psychology internship program*. UW Medicine Department of Psychiatry and Behavioral Sciences. https://pip.psychiatry.uw.edu/rehabilitation-neuropsychology-track/

NEA. (n.d.). *IV. Teaching and learning in the age of artificial intelligence | NEA*. Retrieved January 13, 2025, from https://www.nea.org/resource-library/artificial-intelligence-education/iv-teaching-and-learning-age-artificial-intelligence

Nicolaou, C., Matsiola, M., & Kalliris, G. (2019). Technology-enhanced learning and teaching methodologies through audiovisual media. *Education Sciences, 9*(3), article 3. https://doi.org/10.3390/educsci9030196

Omar Jian, M. J. (2024). (PDF) Personalized learning through AI. *Advances in Engineering, 5*(1). https://doi.org/10.54254/2977-3903/5/2023039

Parvez, S. (2024, October 4). How AI-based personalized Learning is transforming education. *SP Cloud Academy*. https://spca.education/how-ai-based-personalized-learning-is-transforming-education-the-future-is-now/

Pillai, I. N. (2024). Robotic rehabilitation in India: A survey of attitudes and perceptions. *Indian Journal of Physical Medicine & Rehabilitation, 34*(2), 103. https://doi.org/10.4103/ijpmr.ijpmr_55_23

Professional Education. (n.d.). UVU Continuing Education. Retrieved January 13, 2025, from https://www.campusce.net/uvu/course/course.aspx?catId=37

Rehabilitation Robots Market Size And Share Report, 2030. (n.d.). Retrieved January 13, 2025, from https://www.grandviewresearch.com/industry-analysis/rehabilitation-robots-market-report

RESNA > Certification > Continuing Education. (n.d.). Retrieved January 13, 2025, from https://www.resna.org/Certification/Continuing-Education

Rivera Mindt, M., Byrd, D., Saez, P., & Manly, J. (2010). Increasing culturally competent neuropsychological services for ethnic minority populations: A call to action. *The Clinical Neuropsychologist, 24*(3), 429–453. https://doi.org/10.1080/13854040903058960

Rollins, J. (2021). *The forces that could shape counseling's future.* https://www.counseling.org/publications/counseling-today-magazine/article-archive/article/legacy/the-forces-that-could-shape-counselings-future

Sinha, S., & Lee, Y. M. (2024). Challenges with developing and deploying AI models and applications in industrial systems. *Discover Artificial Intelligence, 4*(1), 55. https://doi.org/10.1007/s44163-024-00151-2

Smith, G. (2019). Education and training in clinical neuropsychology: Recent developments and documents from the clinical neuropsychology synarchy. *Archives of Clinical Neuropsychology, 34*(3), 418–431. https://doi.org/10.1093/arclin/acy075

Sperling, S. A., Cimino, C. R., Stricker, N. H., Heffelfinger, A. K., Gess, J. L., Osborn, K. E., & Roper, B. L. (2017). Taxonomy for education and training in clinical neuropsychology: Past, present, and future. *The Clinical Neuropsychologist, 31*(5), 817–828. https://doi.org/10.1080/13854046.2017.1314017

Stiers, W., & Perry, K. N. (2012). Education and training in rehabilitation psychology. In Kennedy, P *The Oxford handbook of rehabilitation psychology* (pp. 417–431). Oxford University Press. doi: https://doi.org/10.1093/oxfordhb/9780199733989.013.0023

Strielkowski, W., Grebennikova, V., Lisovskiy, A., Rakhimova, G., & Vasileva, T. (2024). AI-driven adaptive learning for sustainable educational transformation. *Sustainable Development.* https://doi.org/10.1002/sd.3221

The Potential of AI in Personalized Learning | Atlantic International University. (2024, October 29). https://www.aiu.edu/blog/the-potential-of-ai-in-personalized-learning/

Thistlethwaite, J. E., Davies, D., Ekeocha, S., Kidd, J. M., MacDougall, C., Matthews, P., Purkis, J., & Clay, D. (2012). The effectiveness of case-based learning in health professional education. A BEME systematic review: BEME Guide No. 23. *Medical Teacher, 34*(6), e421–e444. https://doi.org/10.3109/0142159X.2012.680939

Van Patten, R., Bellone, J. A., Gaynor, T. R., Gaynor, L., & Block, C. (2022). Digital methods of delivering education and training in neuropsychology. *Archives of Clinical Neuropsychology, 37*(6), 1103–1117. https://doi.org/10.1093/arclin/acac033

Verma, S., Paterson, M., & Medves, J. (2006). Core competencies for health care professionals: What medicine, nursing, occupational therapy, and physiotherapy share. *Journal of Allied Health, 35*(2), 109–115.

Young Cancer Investigator Seminar Series & Career Chats | Montefiore Einstein Comprehensive Cancer Center | Patient Care | Montefiore Einstein. (n.d.). Retrieved January 13, 2025, from https://montefioreeinstein.org/cancer/education/research/young-investigator-seminars

Yuan, F., Klavon, E., Liu, Z., Lopez, R. P., & Zhao, X. (2021). A systematic review of robotic rehabilitation for cognitive training. *Frontiers in Robotics and AI, 8.* https://doi.org/10.3389/frobt.2021.605715

Zhang, J., & He, C. (2023). Evidence-based rehabilitation medicine: Definition, foundation, practice and development. *Medical Review, 4*(1), 42–54. https://doi.org/10.1515/mr-2023-0027

Zhu, S., Yu, T., Xu, T., Chen, H., Dustdar, S., Gigan, S., Gunduz, D., Hossain, E., Jin, Y., Lin, F., Liu, B., Wan, Z., Zhang, J., Zhao, Z., Zhu, W., Chen, Z., Durrani, T. S., Wang, H., Wu, J., … Pan, Y. (2023). Intelligent computing: The latest advances, challenges, and future. *Intelligent Computing*, *2*, 0006. https://doi.org/10.34133/icomputing.0006

Chapter 15
Integrating Technology into Clinical Practice

15.1 Introduction

Advancements in technology are fundamentally changing how health care is delivered, making it more precise, accessible, and tailored to individual needs. Tools such as telemedicine, artificial intelligence (AI), virtual reality (VR), and robotic systems are not only enhancing diagnostic accuracy but also allowing patients to receive care in ways that were once unimaginable. For example, telemedicine now enables remote consultations, making healthcare accessible to those in underserved areas or with mobility challenges. These innovations are part of a larger shift toward evidence-based medicine, where clinical decisions are informed by rigorous research to achieve better outcomes, improve efficiency, and enhance patient satisfaction (Junaid et al., 2022). In the realm of neuropsychological rehabilitation, technology is paving the way for more effective interventions and recovery strategies, particularly through the use of robotic-assisted therapy (RAT). This innovative approach is proving to be a game-changer for patients recovering from conditions such as strokes, brain injuries, and spinal cord injuries. RAT devices are designed with advanced features, including highly repetitive movement patterns, intelligent control systems, and real-time sensorimotor feedback. These features allow therapists to customize therapy sessions according to the specific needs and abilities of each patient. The result is a more targeted approach to rehabilitation, which can lead to improved motor function, reduced fatigue for both patients and therapists, and greater focus on individualized recovery goals (Chua & Kuah, 2017).

What sets RAT apart from traditional therapies is its ability to deliver consistent and precise therapy over extended periods. For example, patients recovering from a stroke often require intensive, repetitive movements to relearn motor skills. A therapist working manually may become fatigued over time, potentially compromising the quality of care. In contrast, RAT devices can maintain the same level of precision and intensity throughout the session, ensuring consistent therapy. Additionally,

K. J. Reddy, *Innovations in Neurocognitive Rehabilitation*,
https://doi.org/10.1007/978-3-031-88117-6_15

these devices allow therapists to redirect their focus to higher-order tasks, such as assessing functional improvements or addressing specific cognitive and emotional needs of patients (Chua & Kuah, 2017). Despite the undeniable benefits, integrating technologies like RAT into clinical practice presents unique challenges. One of the primary obstacles is the variability in outcomes when comparing RAT to conventional therapy. While some patients experience significant functional gains, others show little to no improvement. This inconsistency can lead to skepticism among clinicians and patients, creating barriers to adoption. Another significant challenge is the cost of these advanced systems. Many RAT devices are expensive, and the associated costs—such as maintenance, training, and upgrades—are often not covered by insurance. This can make it difficult for hospitals, clinics, and rehabilitation centers to justify their investment, especially in resource-limited settings (Chua & Kuah, 2017; Javaid et al., 2024).

Managing expectations is another crucial aspect of successfully implementing RAT. Patients and their families may have high hopes for rapid or dramatic improvements, which may not always align with clinical realities. Similarly, healthcare staff may feel overwhelmed or unsure about using new technologies, particularly if they have not received adequate training. To address these concerns, clear communication, ongoing education, and hands-on training programs are essential. These efforts not only help build confidence among users but also ensure that the technology is being utilized to its full potential (Chua & Kuah, 2017; Javaid et al., 2024; Junaid et al., 2022). The integration of technology into neuropsychological rehabilitation also requires careful consideration of the broader healthcare ecosystem. For example, the implementation of RAT must be aligned with existing workflows and resources. This includes ensuring compatibility with electronic health record (EHR) systems, providing adequate space and infrastructure for the devices, and developing protocols to guide their use. Without a well-thought-out strategy, there is a risk of these technologies becoming underutilized or causing disruptions to care delivery (Arshi & Chaudhary, 2024).

Beyond these logistical challenges, there are ethical considerations that must be addressed. For instance, data security is a significant concern when using devices that collect and store patient information. Robust measures must be in place to protect this data from breaches and unauthorized access. Additionally, questions around equity and access remain central to the discussion. Advanced technologies like RAT are often more readily available in urban, well-funded healthcare settings, leaving rural and underserved populations at a disadvantage. Bridging this gap requires policies and initiatives aimed at making these tools more affordable and accessible to all patients, regardless of their socioeconomic background (Capecci et al., 2024; Sharma et al., 2018). The transformative potential of RAT and similar technologies is further supported by a growing body of research. For example, studies have highlighted how RAT's ability to provide consistent, repetitive therapy can lead to better motor recovery in stroke patients compared to traditional methods. However, there is also a need for more comprehensive research on its cost-effectiveness to build a stronger case for adoption among healthcare institutions and third-party payers. Demonstrating the long-term benefits of these technologies, both in terms of patient outcomes and cost savings, will be crucial in driving wider adoption (Adeghe et al., 2024; Sutton et al., 2020).

Another critical factor in the successful integration of RAT is the need for ongoing training and upskilling of healthcare professionals. As these technologies become more sophisticated, therapists will require not only technical knowledge but also an understanding of how to interpret data and adjust treatment plans accordingly. Developing comprehensive training programs and integrating them into clinical education will ensure that therapists are well-prepared to use these tools effectively. Moreover, fostering a culture of openness to innovation within healthcare teams can help overcome resistance to change and encourage the active adoption of new technologies (Bonnechère et al., 2023). While challenges remain, the ongoing refinement of these technologies, coupled with strategic implementation and a patient-centered approach, has the potential to redefine how rehabilitation is delivered and experienced. The future of neuropsychological rehabilitation lies at the intersection of technology and compassion, where innovative tools empower both clinicians and patients to achieve better outcomes together.

15.2 Technologies Transforming Clinical Practice

15.2.1 Telemedicine

Telemedicine has revolutionized healthcare delivery by enabling patients and clinicians to connect remotely, making care more accessible and efficient. During the COVID-19 pandemic, telemedicine played a critical role in maintaining continuity of care despite physical distancing restrictions, allowing patients to access consultations, follow-ups, and chronic disease management from the safety of their homes. This remote care model minimized disruptions to essential services and reduced the risk of virus transmission (Al Kuwaiti et al., 2023; Wu & Luo, 2019). One of telemedicine's most significant benefits is its ability to overcome geographical barriers. Patients in rural or underserved areas can consult specialists without the need for extensive travel, improving access to high-quality care. For instance, a patient in a remote area can receive expert cardiology advice through a simple video consultation. Similarly, telemedicine increases convenience for urban patients by eliminating long wait times and travel expenses, enhancing patient adherence to medical advice. Telemedicine also optimizes resource utilization by allowing healthcare providers to focus on critical in-person cases. Routine consultations and follow-ups, which previously occupied valuable clinical hours, can now be conducted virtually, freeing up capacity in hospitals and clinics. Remote patient monitoring devices further extend the reach of telemedicine by enabling continuous tracking of chronic conditions, facilitating timely interventions (Bajwa et al., 2021; LaBoone & Marques, 2024). Despite these advantages, challenges persist. Ensuring equitable access to telemedicine services requires addressing disparities in digital literacy and internet connectivity. Data security and privacy concerns also need to be managed to protect sensitive patient information. Nevertheless, with ongoing advancements in technology and policy frameworks, telemedicine is poised to become a cornerstone of patient-centered health care (Dinh et al., 2023; Olawade et al., 2025).

15.2.2 Artificial Intelligence (AI)

Artificial intelligence (AI) is transforming clinical practice by improving decision-making, streamlining workflows, and enhancing diagnostic precision. At its core, AI leverages algorithms to analyze large data sets, detect patterns, and generate insights. One notable example is the use of AI algorithms for detecting diabetic retinopathy through retinal imaging. Studies have demonstrated that AI systems can match or exceed the diagnostic accuracy of human specialists, making early detection more accessible and cost-effective (Daich Varela et al., 2024; Iqbal et al., 2024). AI's applications extend beyond diagnostics to predictive analytics, where it forecasts disease progression and tailors treatment plans. This approach enables personalized medicine, improving patient outcomes while minimizing trial-and-error treatments. Additionally, AI reduces human error, especially in high-stakes scenarios such as surgery or emergency care, where real-time decision support can be critical (Iqbal et al., 2024). Operationally, AI enhances efficiency by automating administrative tasks like scheduling, billing, and record management, allowing clinicians to dedicate more time to patient care. Recent developments, such as cloud-based AI platforms, have increased the scalability and accessibility of these tools, enabling their integration into diverse healthcare settings (Daich Varela et al., 2024; Iqbal et al., 2024). However, AI's integration is not without challenges. Ethical concerns, including biases in algorithms and the need for transparent decision-making, remain significant hurdles. Data privacy and compliance with regulations like HIPAA and GDPR are also critical considerations. Nonetheless, as research advances and implementation strategies mature, AI's potential to revolutionize health care continues to grow (Christopoulou, 2024; Li, 2021).

15.2.3 Virtual Reality (VR) and Augmented Reality (AR)

Virtual reality (VR) and augmented reality (AR) are transforming health care by offering immersive, interactive environments for training and therapy. These technologies allow clinicians to simulate real-world scenarios, providing a risk-free space to develop and refine skills. For example, surgical trainees can practice complex procedures using AR-guided simulations, enhancing their technical proficiency and confidence without endangering patients (Li, 2021; Tuborgh et al., 2021). For patients, VR and AR provide novel therapeutic interventions. VR-based therapy has proven particularly effective for managing conditions like anxiety, phobias, and post-traumatic stress disorder (PTSD). By immersing patients in controlled virtual environments, VR helps them confront and manage their triggers in a safe and gradual manner. In rehabilitation, VR games are being used to enhance motor recovery in stroke patients, turning repetitive exercises into engaging activities. These technologies also enhance patient engagement. For instance, AR can provide real-time visualizations during physical therapy sessions, helping patients understand their progress and motivating them to stay committed to their recovery goals. Similarly, VR-based pain management techniques have shown promise in reducing discomfort by distracting patients during medical procedures. Challenges such as high costs, the need for specialized equipment, and the integration of VR/AR into existing healthcare systems remain barriers to widespread adoption. However, as these technologies become more affordable and accessible, they are likely to play an increasingly central role in both clinical practice and patient care (Bonnechère et al., 2023).

15.2.4 Wearable Devices

Wearable devices have become essential tools in health care, providing real-time monitoring of physiological metrics like heart rate, activity levels, and sleep patterns. These devices, ranging from smartwatches to advanced medical sensors, empower individuals to take charge of their health while offering clinicians valuable data for decision-making. For example, wearable devices with electrocardiogram (ECG) functionality can detect arrhythmias, enabling early intervention and potentially saving lives. One of the most significant advantages of wearable devices is their role in promoting proactive health management. By continuously monitoring health metrics, these devices alert users to potential issues before they become serious. For instance, sudden changes in heart rate variability might prompt users to seek medical attention, potentially catching conditions like atrial fibrillation early. Wearables have also proven invaluable for chronic disease management, with glucose monitors helping diabetes patients maintain better control over their condition (Tuborgh et al., 2021). Recent advancements have expanded the applications of wearable technology. Smart sensors designed for elderly care can detect falls and monitor daily activity patterns, providing peace of mind to families and caregivers. During the COVID-19 pandemic, wearables were used to monitor symptoms and predict disease onset, showcasing their potential in public health crises. Furthermore, the integration of AI and machine learning into wearables has enhanced their ability to analyze data and deliver personalized health insights. While challenges like data privacy, cost, and integration with healthcare systems remain, wearable devices continue to evolve. With improved designs and functionalities, they are increasingly becoming indispensable tools for enhancing patient care and empowering individuals to manage their health effectively (Bonnechère et al., 2023).

15.3 Implementation Strategies

The integration of emerging technologies into health care presents transformative opportunities, but successful adoption requires careful planning and implementation. From assessing clinical needs to ongoing evaluation, a systematic approach ensures that new technologies align with healthcare goals, meet patient needs, and fit seamlessly into existing workflows. Below, key strategies for implementing these technologies are discussed in detail, supported by evidence from recent literature.

15.3.1 Needs Assessment

The foundation of successful technology integration lies in a comprehensive needs assessment. This involves evaluating clinical requirements, patient demographics, and technological capabilities to ensure that the selected tools address the organization's specific needs (Gemert-Pijnen, 2022). For example, when implementing telemedicine platforms, understanding patient access to digital devices and internet connectivity is essential. Similarly, assessing the clinical workflows and identifying potential bottlenecks ensures that the technology complements rather than disrupts care delivery (Junaid et al., 2022). In addition to clinical considerations, involving key stakeholders early in the process is critical. Stakeholders—including clinicians, administrative staff, IT professionals, and patients—bring diverse perspectives that can uncover potential challenges and opportunities. For instance, healthcare providers may highlight operational issues, while patients can provide insights into usability and accessibility (Gemert-Pijnen, 2022). Identifying influential stakeholders also facilitates strategic planning and promotes buy-in across the organization. To ensure a tailored approach, organizations should evaluate their readiness for change by examining resources, infrastructure, and workforce capacity. For instance, a small rural clinic implementing AI-driven diagnostic tools might require additional IT support and training compared to a large urban hospital with existing digital infrastructure (Stefano, 2023). Conducting such assessments early in the process helps mitigate risks and sets the stage for successful implementation.

15.3.2 Training and Support

Once a technology is selected, providing robust training and support is essential for ensuring its effective use. Healthcare providers often face challenges such as unfamiliarity with new tools, concerns about job security, and doubts about the technology's benefits. Addressing these issues through targeted training programs can significantly ease the transition (Stefano, 2023). Hands-on workshops, e-learning modules, and simulation-based training sessions are proven strategies to build confidence among staff. For instance, when introducing AI-driven diagnostic software, clinicians may benefit from interactive sessions that demonstrate the tool's functionality and offer opportunities for practical application. Additionally, continuous learning opportunities, such as webinars and refresher courses, help staff stay updated on system enhancements (Gemert-Pijnen, 2022). Beyond technical training, it's crucial to address emotional and cultural barriers. Concerns about loss of control or increased workloads are common when adopting new technologies. Clear communication about how the technology will enhance—not replace—the clinician's role can alleviate these fears. For example, AI systems designed to analyze imaging data should be presented as tools to augment a radiologist's expertise rather than a replacement for their judgment (Stefano, 2023). Providing ongoing support, such as access to help desks or dedicated IT teams, ensures that users have resources to troubleshoot issues as they arise. This continuous support fosters confidence and encourages adoption, making technology a trusted part of clinical workflows (Holmgren et al., 2023).

15.3.3 Workflow Integration

Seamlessly integrating new technologies into existing workflows is one of the most challenging yet critical aspects of implementation. Disruptions caused by poorly aligned tools can lead to inefficiencies, frustration, and even reduced quality of care (Gemert-Pijnen, 2022). Successful integration begins with a thorough mapping of current workflows to identify where the new technology can add value. For example, incorporating telemedicine into a clinic's operations requires more than just deploying video conferencing software. Scheduling systems, patient intake processes, and follow-up protocols must be adapted to accommodate virtual appointments. Integrating telemedicine platforms with electronic health records (EHRs) ensures that patient information is easily accessible during consultations, streamlining the provider's workflow (Gemert-Pijnen, 2022; Subbaiah, 2023). Aligning new technologies with the organization's culture and values is equally important. Resistance to change often stems from a perception that the technology conflicts with established practices. By involving staff in the design and implementation process, organizations can foster a sense of ownership and reduce resistance. For instance, engaging nurses and physicians in customizing an AI-based decision-support tool can ensure that it aligns with their clinical needs and preferences (Stefano, 2023). Another key consideration is scalability. As organizations grow or adapt to changing healthcare demands, the technology should be able to scale without significant disruptions. Cloud-based platforms and modular systems offer flexibility, enabling healthcare providers to expand their capabilities over time (Subbaiah, 2023).

15.3.4 Monitoring and Evaluation

Effective implementation does not end with deployment; it requires continuous monitoring and evaluation to ensure long-term success. Regular assessments of effectiveness, efficiency, and user satisfaction provide valuable insights into how well the technology is performing and where improvements are needed (Gemert-Pijnen, 2022). Using a combination of quantitative and qualitative methods can offer a comprehensive understanding of the technology's impact. For example, tracking clinical outcomes, patient satisfaction scores, and cost-effectiveness metrics can reveal the tangible benefits of the technology. Simultaneously, gathering feedback through surveys, focus groups, and interviews can highlight user experiences and areas for refinement (Junaid et al., 2022). Real-time data collection is particularly useful for identifying and addressing issues early. For instance, monitoring usage patterns of a telemedicine platform can reveal whether certain features are underutilized or if users encounter technical difficulties. These insights enable organizations to adjust their implementation strategies proactively, improving adoption and satisfaction. Importantly, evaluation should be an ongoing process rather than a one-time activity. As technologies evolve, so do their capabilities and user expectations. Regular updates and refinements ensure that the tools remain relevant and effective in meeting healthcare needs. Moreover, iterative improvements based on real-world data demonstrate a commitment to excellence, fostering trust among users (Gemert-Pijnen & JEWC (Lisette), 2022; Junaid et al., 2022; Subbaiah, 2023).

15.4 Challenges in Technology Integration

The integration of emerging technologies in health care has the potential to revolutionize patient care, enhance efficiency, and improve outcomes. However, the process is fraught with challenges that need to be carefully addressed for successful implementation. From overcoming resistance to change to navigating technical limitations, healthcare providers must tackle a range of barriers to realize the full potential of digital solutions. Below, the most pressing challenges in technology integration are explored, along with strategies to address them.

15.4.1 Resistance to Change

One of the most common barriers to adopting new technologies in health care is resistance to change. Clinicians, staff, and even patients may hesitate to embrace digital solutions due to skepticism, fear of the unknown, or a lack of familiarity. For example, some healthcare professionals may worry that AI tools will replace their expertise, while patients may be concerned about the reliability of telemedicine platforms compared to in-person consultations (Abernethy et al., 2022; Villar, 2024). Overcoming this resistance requires clear communication and education. Healthcare providers need to demonstrate the reliability and benefits of new technologies, using real-world examples and evidence-based outcomes. For instance, showcasing how AI-powered diagnostic tools can improve accuracy and save time may help ease concerns among clinicians. Additionally, involving staff early in the implementation process fosters a sense of ownership and collaboration, which can reduce apprehension (Shojaei et al., 2024). Addressing emotional concerns is equally important. Many employees worry about job security, loss of control, or increased workloads with the introduction of new tools. Providing reassurance that these technologies are designed to augment rather than replace human expertise can alleviate these fears. Furthermore, hands-on training and ongoing support help build confidence and competence, making the transition smoother for everyone involved (Shojaei et al., 2024).

15.4.2 Data Security and Privacy

The integration of digital technologies into healthcare systems introduces significant challenges related to data security and privacy. With patient information increasingly stored and transmitted electronically, safeguarding sensitive data is a top priority. Cybersecurity breaches can have devastating consequences, including loss of trust, legal repercussions, and compromised patient safety. To address these concerns, healthcare organizations must implement robust security measures. Encryption protocols, multi-factor authentication, and regular security audits are essential for protecting patient data. Adherence to regulations such as the Health Insurance Portability and Accountability Act (HIPAA) in the United States and the General Data Protection Regulation (GDPR) in Europe ensures that healthcare providers comply with legal standards for data protection (Abernethy et al., 2022; *Digital Transformation in Healthcare*, 2024; Yigzaw et al., 2022). Building a culture of cybersecurity awareness is also critical. Staff should be trained to recognize potential threats, such as phishing emails or weak passwords, and encouraged to report suspicious activities. Regular updates to software and hardware systems further mitigate vulnerabilities, ensuring that healthcare organizations stay ahead of emerging threats. Despite these efforts, data security remains a moving target due to the rapidly evolving nature of cyber threats. Collaborative efforts between technology providers, regulators, and healthcare organizations are essential to develop innovative solutions that address both current and future challenges in digital health security (*Digital Transformation in Healthcare*, 2024; Yigzaw et al., 2022).

15.4.3 Cost and Accessibility

Financial constraints are a significant hurdle in adopting emerging healthcare technologies. Advanced tools, such as robotic surgery systems or AI-driven diagnostics, often require substantial upfront investments in equipment, infrastructure, and training. For smaller healthcare providers or those serving underserved communities, these costs can be prohibitive. One solution is to prioritize technologies that offer a clear return on investment (ROI). For example, telemedicine platforms can reduce operational costs by minimizing patient no-shows and enabling more efficient use of clinical resources. Similarly, wearable devices that facilitate remote monitoring can help prevent costly hospital readmissions, ultimately saving money for both providers and patients (Anawade et al., 2024; Villar, 2024). Subsidies and public funding programs can also play a vital role in increasing accessibility. Governments and non-profit organizations can provide financial support to healthcare providers in low-resource settings, ensuring that digital innovations reach underserved populations. For example, subsidizing telehealth services for rural areas can help bridge the gap in access to specialist care. Equitable access remains a critical consideration in addressing healthcare disparities. Tailoring technologies to meet the needs of diverse patient populations—such as creating user-friendly interfaces for older adults or ensuring compatibility with low-cost devices—ensures that digital health solutions benefit everyone, regardless of socioeconomic status (*Digital Transformation in Healthcare*, 2024; Villar, 2024).

15.4.4 Technical Limitations

The technical complexities of integrating new technologies into healthcare systems can pose significant challenges. Issues such as interoperability, the integration of legacy systems, and the lack of universal standards for data sharing often hinder seamless adoption. These challenges are particularly pronounced in fragmented healthcare environments where multiple vendors and platforms coexist. Interoperability—the ability of different systems to communicate and exchange information effectively—is a key area of concern. For example, integrating electronic health records (EHRs) with telemedicine platforms or wearable devices often requires custom solutions, which can be time-consuming and expensive. Developing universal data standards and fostering collaboration between technology providers and healthcare organizations can help address these challenges (*Digital Transformation in Healthcare*, 2024; Villar, 2024). Legacy systems, which are often outdated but still critical to healthcare operations, present another barrier. Upgrading or replacing these systems to accommodate new technologies requires careful planning and significant investment. A phased approach to integration, where legacy systems are gradually replaced while maintaining essential functionality, can help minimize disruptions. In addition to technical challenges, usability is an important consideration. Technologies that are difficult to use or poorly designed can lead to frustration among clinicians and patients, reducing adoption rates. User-centered design, which involves direct input from end-users during the development process, ensures that digital tools are intuitive and meet the practical needs of their intended audience (*Digital Transformation in Healthcare*, 2024).

15.5 Case Studies

The adoption of emerging technologies in health care is reshaping how care is delivered, improving patient outcomes, and streamlining operations. Case studies from diverse settings provide concrete examples of how these innovations are being implemented, the benefits they offer, and the challenges they present. This section focuses on real-world applications of artificial intelligence (AI), telemedicine, and virtual reality (VR), demonstrating their transformative potential in health care.

15.5.1 AI in Radiology

Case Study: AI-Powered Diagnostics in the United States

In a large hospital network in the United States, AI-powered radiology tools were implemented to assist in diagnosing breast cancer through mammography. Using deep-learning algorithms, the system analyzed mammograms to detect subtle signs of malignancy that could be missed by the human eye. The results were striking: AI not only improved diagnostic accuracy by 15% but also reduced the time required for image analysis by 30% (Najjar, 2023). The AI tool was particularly effective in triaging high-risk cases, allowing radiologists to prioritize their workload. This streamlined workflow led to faster diagnosis and earlier treatment for patients with suspicious findings. Moreover, the system provided explainable outputs, including annotated images highlighting areas of concern, which helped radiologists validate the AI's recommendations (Stoumpos et al., 2023). Despite these successes, challenges arose. The hospital faced initial skepticism from radiologists, who were concerned about the system's reliability and the "black box" nature of its decision-making. To address these concerns, extensive training sessions were conducted, and the AI tool was positioned as a decision-support system rather than a replacement for human expertise. This approach fostered trust and acceptance among clinicians, ultimately leading to widespread adoption (Strika et al., 2025).

15.5.2 *Telemedicine in Rural Health Care*

Case Study: Government Telemedicine Program in India
In rural India, a government-supported telemedicine initiative was launched to address the severe shortage of healthcare providers in remote areas. The program established telemedicine hubs in district hospitals, connecting them with specialists in urban centers. Using video consultations and electronic health records, patients received expert diagnoses and treatment recommendations without needing to travel long distances (Strika et al., 2025). One notable success story involved a rural clinic in the state of Tamil Nadu, where telemedicine significantly improved access to care for diabetic patients. Specialists provided guidance on managing the disease, while local healthcare workers monitored patients' vitals and relayed the data through digital platforms. Over a 2-year period, the program reported a 40% reduction in diabetes-related complications and hospitalizations (Palozzi et al., 2020). The initiative also proved cost-effective. By reducing travel expenses for patients and optimizing the use of specialist time, the program saved millions of rupees annually. However, challenges such as poor internet connectivity and limited digital literacy among patients initially hindered adoption. To address these issues, the government invested in infrastructure upgrades and conducted community education campaigns, leading to improved outcomes and higher patient satisfaction (Gomez-Gonzalez & Gomez-gutierrez, 2020).

15.5.3 *VR in Pain Management*

Case Study: VR Therapy for Chronic Pain in the United Kingdom
A healthcare trust in the United Kingdom implemented virtual reality (VR) therapy as a non-invasive approach to managing chronic pain. The program targeted patients suffering from conditions such as fibromyalgia and complex regional pain syndrome. Participants used VR headsets to immerse themselves in calming, interactive environments, such as serene beaches or tranquil forests, during therapy sessions (Secinaro et al., 2021). The impact of this intervention was profound. Patients reported a significant reduction in pain perception, with many describing the experience as a welcome distraction from their discomfort. Anxiety levels also decreased, as the immersive environments helped patients feel more relaxed and engaged. Clinicians noted that patients who used VR therapy were more willing to participate in physical therapy sessions, further supporting their recovery. The program faced logistical challenges, including the high cost of VR equipment and the need for staff training. To mitigate these issues, the healthcare trust partnered with technology providers to secure discounted devices and conducted hands-on workshops for clinicians. These efforts ensured the program's sustainability and scalability, making VR therapy an integral part of the trust's pain management strategy (Gomez-Gonzalez & Gomez-gutierrez, 2020; Secinaro et al., 2021).

15.6 Ethical and Regulatory Considerations

The integration of artificial intelligence (AI) and digital health tools into health care has brought tremendous opportunities, including enhanced diagnostic accuracy, personalized treatment plans, and optimized workflows. However, these advancements also raise significant ethical and regulatory challenges. Addressing these issues is critical to ensuring that these technologies are implemented responsibly, transparently, and equitably.

1. **Informed Consent:** One of the foundational principles in health care is ensuring that patients fully understand and agree to the use of AI and digital tools in their care. Many patients may lack familiarity with AI-driven systems, making it essential to provide clear and accessible information about how these tools work and their implications. For example, explaining how an AI model analyzes patient data and the role it plays in decision-making fosters trust and empowers patients to make informed choices (Gerke et al., 2020; Hanna et al., 2024).
2. **Algorithmic Bias:** AI systems are only as unbiased as the data they are trained on. Biases in data sets can lead to unfair outcomes, such as misdiagnoses or disparities in care for certain populations. For instance, underrepresentation of minority groups in training data sets has resulted in lower diagnostic accuracy for these populations. Addressing this issue requires the use of diverse data sets and continuous monitoring to identify and mitigate biases (Chukwuka et al., 2024; Gerke et al., 2020; Mennella et al., 2023).
3. **Data Ownership and Usage:** The question of who owns patient data is a significant ethical concern. Clear guidelines on how data is collected, stored, shared, and used are essential. Patients must retain control over their data, with explicit consent mechanisms in place to ensure transparency. Ethical considerations also extend to the secondary use of data, such as in research or for improving AI algorithms (*Data Privacy and AI*, 2024; Gerke et al., 2020).
4. **Equity and Access:** Digital health technologies must be accessible to all, regardless of socioeconomic status or geographic location. Unequal access can exacerbate existing healthcare disparities, leaving underserved populations behind. Ethical deployment of these tools involves ensuring that cost, infrastructure, and literacy barriers are addressed to promote equitable access (Gerke et al., 2020; Mennella et al., 2023).

1. **Frameworks for Ethical Regulation**

   (a) **Legal Challenges:**

   AI and digital health tools face legal hurdles, including ensuring safety, effectiveness, and accountability. Questions of liability—whether it lies with the developer, healthcare provider, or institution—must be clarified. Regulations also need to address data protection, cybersecurity, and intellectual property rights to safeguard patients and providers (Gerke et al., 2020; Mennella et al., 2023).

(b) **Regulatory Adaptation:**

Existing healthcare regulations often lag behind technological advancements. Adapting these frameworks to accommodate the unique challenges posed by AI, such as explainability and continuous learning algorithms, is critical. For instance, regulatory bodies like the U.S. Food and Drug Administration (FDA) and the European Medicines Agency (EMA) are developing guidelines for AI in health care to ensure both safety and innovation (Gerke et al., 2020).

(c) **International Cooperation:**

Health care is a global concern, and many challenges, such as cross-border data sharing and algorithmic biases, require international collaboration. Developing adaptable, harmonized regulations can help address these global challenges while fostering innovation. For example, the World Health Organization (WHO) is working on establishing international standards for digital health technologies (*Data Privacy and AI*, 2024; Gerke et al., 2020).

2. **Best Practices for Ethical AI Integration**

(a) **Transparency and Explainability:**

AI systems must be designed to provide clear explanations for their decisions. This is especially important in health care, where clinicians and patients need to understand how an AI system arrived at a particular recommendation. Transparency builds trust and allows clinicians to integrate AI insights into their expertise effectively (Chukwuka et al., 2024; *Ethical Issues of AI in Healthcare*, 2024).

(b) **Accountability Frameworks:**

Clear accountability structures are necessary to define responsibilities among manufacturers, institutions, and professionals. For example, if an AI system makes a diagnostic error, it must be clear who is accountable—whether it's the software developer, the hospital, or the overseeing physician. These frameworks ensure that ethical and legal responsibilities are upheld (Chukwuka et al., 2024; *Ethical Issues of AI in Healthcare*, 2024).

(c) **Privacy and Security Measures:**

Implementing robust privacy measures, such as encryption and anonymization, is critical to protect sensitive patient data. Regular audits and compliance with regulations like the Health Insurance Portability and Accountability Act (HIPAA) and the General Data Protection Regulation (GDPR) ensure that digital health tools operate securely (Chukwuka et al., 2024; *Ethical Issues of AI in Healthcare*, 2024).

(d) **Bias Mitigation:**

Using diverse and representative data sets is key to minimizing biases in AI models. Ongoing monitoring and validation ensure that systems remain equitable as they evolve. Additionally, involving diverse stakeholders in the development process can help identify potential biases early (Chukwuka et al., 2024; Marketing, 2023).

(e) **Ethical Guidelines:**

Developing accessible and regularly updated ethical guidelines helps align stakeholders around best practices. These guidelines should address key areas such as consent, equity, bias, and accountability, ensuring that ethical principles are consistently applied (Chukwuka et al., 2024; *Ethical Issues of AI in Healthcare*, 2024).

3. **Societal Implications**

(a) **Impact on the Workforce:**

The adoption of AI in health care raises concerns about job displacement and changes in professional roles. While AI can handle routine tasks, such as analyzing medical images, it also creates opportunities for clinicians to focus on more complex and human-centered aspects of care. Preparing the workforce through training and education ensures that AI complements rather than replaces human expertise (Gerke et al., 2020).

(b) **Public Discourse:**

Engaging the public in discussions about AI ethics in health care is crucial for building trust and understanding. Initiatives that involve patients, advocacy groups, and other stakeholders in shaping the ethical use of AI can help address societal concerns and foster acceptance.

(c) **Economic Benefits:**

AI and digital health tools offer significant economic benefits, such as reducing healthcare costs and improving efficiency. However, these gains must be balanced with ethical considerations to ensure that the focus remains on improving patient care rather than solely maximizing profits.

15.7 Future Directions

The future of healthcare technology integration is defined by a shift toward personalized, accessible, and sustainable care. Innovations such as AI-driven personalized medicine, expanded remote care, collaborative ecosystems, and sustainability initiatives are transforming the way health care is delivered. These advancements promise to improve outcomes, enhance efficiency, and make health care more equitable. Below, we explore these emerging directions and their potential impact on the healthcare industry.

15.7.1 AI-Driven Personalized Medicine

Personalized medicine, enabled by AI and machine learning, is revolutionizing health care by tailoring treatments to individual genetic profiles and health data. By analyzing large data sets, AI systems can predict disease risk, recommend customized treatment plans, and optimize medication regimens. For example, AI tools can analyze a patient's genetic makeup to determine the most effective cancer treatment, reducing trial-and-error approaches. One prominent application is in pharmacogenomics, where AI identifies how genetic variations influence drug efficacy and side effects. This capability not only enhances treatment outcomes but also minimizes adverse reactions. Additionally, AI-powered diagnostic tools are helping clinicians detect diseases at earlier stages, enabling timely interventions and improving survival rates. However, challenges remain. The "black box" nature of AI decision-making makes it difficult for clinicians to understand how algorithms reach their conclusions, potentially hindering trust and adoption. Furthermore, data quality issues, such as incomplete or biased data sets, can compromise the accuracy of AI predictions. Addressing these concerns through transparency, robust data collection practices, and continuous validation will be crucial for realizing the full potential of personalized medicine.

15.7.2 Expansion of Remote Care

The integration of 5G and the Internet of Things (IoT) is driving the expansion of remote care, making telemedicine more efficient, reliable, and accessible. With faster data transmission and reduced latency, 5G enables seamless video consultations, real-time monitoring, and remote surgeries, bridging gaps in healthcare access. Remote care solutions are particularly beneficial for managing chronic diseases, where continuous monitoring and timely interventions are critical. For instance, wearable devices connected to IoT platforms can track patients' vitals and alert healthcare providers to potential issues. These innovations reduce hospital readmissions and improve patient outcomes, especially in underserved or rural areas. Economic and sustainability benefits further highlight the value of remote care. Studies have shown that telemedicine reduces healthcare costs by minimizing travel and optimizing resource utilization. As the global population ages and healthcare demands increase, expanding remote care capabilities will be essential for creating a scalable and sustainable healthcare model. Challenges include ensuring widespread 5G coverage, addressing bandwidth requirements, and maintaining device compatibility across locations. Healthcare organizations must also invest in training programs to equip clinicians with the skills needed to leverage these technologies effectively.

15.7.3 Collaborative Ecosystems

Collaboration between technology companies, healthcare providers, and regulatory bodies is vital for accelerating innovation and fostering the adoption of new technologies. These partnerships enable the development of integrated solutions that address complex healthcare challenges. For example, collaborations between AI developers and hospitals have resulted in decision-support tools that enhance clinical workflows and improve patient outcomes. International cooperation is also playing a key role in advancing healthcare technology. Shared knowledge and resources across borders are accelerating the adoption of 5G and AI-driven solutions. For instance, global initiatives to standardize interoperability protocols are facilitating seamless data exchange between healthcare systems, enabling more coordinated and effective care. Strategic planning and investment in interoperability infrastructure are crucial for promoting integration. Open standards, modular systems, and data normalization tools can help healthcare organizations overcome silos and create cohesive ecosystems. By fostering collaboration, the industry can accelerate innovation while ensuring that technologies are accessible and user-friendly.

15.7.4 Sustainability Initiatives

As healthcare systems face increasing pressure to reduce their environmental impact, sustainable technologies are gaining prominence. Eco-friendly innovations, such as energy-efficient medical devices and digital tools, are helping healthcare organizations minimize their carbon footprint. For example, telemedicine reduces the need for patient travel, lowering emissions and resource consumption. Scalable infrastructure and modular solutions are also being developed to maximize efficiency and reduce costs. By adopting open standards and interoperability initiatives, healthcare organizations can build flexible systems that adapt to evolving needs without requiring significant overhauls. These efforts not only enhance sustainability but also improve the affordability and accessibility of healthcare technologies. Healthcare organizations are also exploring ways to integrate sustainability into daily operations. Initiatives such as reducing waste, optimizing supply chains, and implementing green building practices are aligning healthcare delivery with environmental goals. These efforts demonstrate that technological innovation and sustainability can go hand in hand, creating a more resilient and responsible healthcare system.

15.8 Conclusion

The integration of emerging technologies into healthcare represents a transformative shift in how care is delivered, making it more personalized, accessible, efficient, and sustainable. From artificial intelligence (AI) and telemedicine to virtual reality (VR), wearable devices, and the Internet of Things (IoT), these innovations are reshaping the landscape of clinical practice. As we embrace these advancements, it is imperative to consider not only their immense potential but also the challenges they present, such as ethical concerns, regulatory barriers, and disparities in access.

AI-driven personalized medicine is one of the most promising developments in health care. By leveraging machine learning and predictive analytics, AI enables clinicians to tailor treatments to individual genetic profiles and health conditions. This approach has demonstrated remarkable success in areas like pharmacogenomics, where it enhances the efficacy of treatments and minimizes adverse effects. AI is also revolutionizing diagnostics, with tools that can detect diseases at early stages, leading to better outcomes and more effective interventions. However, challenges such as the "black box" nature of AI decision-making and data quality issues need to be addressed. Ensuring transparency, explainability, and continuous monitoring of AI systems is essential for building trust among clinicians and patients.

Telemedicine has emerged as a powerful tool for bridging gaps in healthcare access, particularly in underserved and rural areas. It allows patients to consult with healthcare providers remotely, reducing the need for travel and making specialist care more accessible. During the COVID-19 pandemic, telemedicine proved invaluable in maintaining continuity of care. Beyond its role in immediate healthcare delivery, telemedicine also offers significant economic and sustainability benefits by optimizing resource utilization and reducing the environmental impact of healthcare operations. However, barriers such as limited internet connectivity, digital literacy challenges, and data privacy concerns remain. Addressing these issues through investments in infrastructure, training programs, and robust cybersecurity measures will be critical to fully realizing the potential of remote care. Virtual reality and augmented reality are also making strides in health care, offering innovative solutions for both patient care and clinician training. VR-based therapies have been particularly effective in pain management and rehabilitation, providing immersive environments that distract patients from discomfort or enhance their engagement in therapy sessions. These technologies also play a crucial role in medical education, allowing clinicians to practice complex procedures in a risk-free setting. Despite their promise, the high cost of VR and AR systems and the need for specialized training pose challenges to widespread adoption. Partnerships between healthcare providers and technology developers, along with efforts to reduce costs, are essential for integrating these tools into mainstream practice.

Wearable devices and IoT-enabled health care are empowering patients to take a more active role in managing their health. These devices monitor physiological data in real-time, providing insights that enable early intervention and proactive disease management. For example, wearable ECG monitors can detect arrhythmias, and glucose monitoring systems help diabetes patients manage their condition more effectively. During the COVID-19 pandemic, wearable technologies were also instrumental in tracking symptoms and predicting disease progression. However, challenges such as data interoperability, privacy concerns, and equitable access need to be addressed to ensure that these innovations benefit all segments of the population. Collaborative ecosystems are driving much of the progress in healthcare technology integration. Partnerships between tech companies, healthcare providers, regulatory bodies, and researchers are fostering innovation and enabling the development of solutions that address complex healthcare challenges. International collaboration is particularly crucial in establishing global standards for data sharing and interoperability, which are essential for scaling technologies across borders. These efforts also underscore the importance of a patient-centered approach, ensuring that technologies are not only effective but also accessible and aligned with the needs of diverse populations.

Sustainability has become an increasingly important consideration in healthcare technology development. As the industry seeks to minimize its environmental impact, efforts are being made to develop eco-friendly and cost-effective solutions. Telemedicine reduces the need for patient travel, while energy-efficient devices and digital tools contribute to greener healthcare operations. Scalable infrastructure and modular systems further support sustainability by allowing healthcare providers to adapt to changing needs without significant overhauls. By aligning technological advancements with environmental goals, the healthcare industry can create a more resilient and responsible system. While the potential of these technologies is vast, their integration is not without challenges. Ethical concerns such as informed consent, algorithmic bias, data ownership, and equitable access must be carefully managed. Regulatory frameworks need to evolve to address the unique demands of AI, IoT, and other digital tools, ensuring safety, effectiveness, and accountability. Furthermore, the healthcare workforce must be prepared for this technological shift through training and education, ensuring that clinicians can effectively leverage these tools to enhance patient care.

The path forward requires a holistic approach that balances innovation with ethical considerations, regulatory compliance, and sustainability. Healthcare providers must engage in continuous dialogue with patients, policymakers, and technology developers to address emerging challenges and align solutions with societal values. Transparency, collaboration, and adaptability will be key to navigating this complex landscape. The integration of emerging technologies into health care is not just about adopting new tools; it is about reimagining the way care is delivered. By focusing on patient-centered solutions, fostering collaboration, and addressing barriers to adoption, the healthcare industry can create a system that is more equitable, efficient, and responsive to the needs of individuals and communities. As technology continues to evolve, the potential to improve health outcomes and enhance the quality of life for people around the world is limitless.

References

Abernethy, A., Adams, L., Barrett, M., Bechtel, C., Brennan, P., Butte, A., Faulkner, J., Fontaine, E., Friedhoff, S., Halamka, J., Howell, M., Johnson, K., Long, P., McGraw, D., Miller, R., Lee, P., Perlin, J., Rucker, D., Sandy, L., et al. (2022). The promise of digital health: Then, now, and the future. *NAM Perspectives, 202206e*. https://doi.org/10.31478/202206e

Adeghe, E. P., Okolo, C. A., & Ojeyinka, O. T. (2024). A review of wearable technology in healthcare: Monitoring patient health and enhancing outcomes. *Open Access Research Journal of Multidisciplinary Studies, 7*(1), 142–148. https://doi.org/10.53022/oarjms.2024.7.1.0019

Al Kuwaiti, A., Nazer, K., Al-Reedy, A., Al-Shehri, S., Al-Muhanna, A., Subbarayalu, A. V., Al Muhanna, D., & Al-Muhanna, F. A. (2023). A review of the role of artificial intelligence in healthcare. *Journal of Personalized Medicine, 13*(6), Article 6. https://doi.org/10.3390/jpm13060951

Anawade, P. A., Sharma, D., & Gahane, S. (2024). A comprehensive review on exploring the impact of telemedicine on healthcare accessibility. *Cureus, 16*(3), e55996. https://doi.org/10.7759/cureus.55996

Arshi, O., & Chaudhary, A. (2024). Smart healthcare: Integration of AI and brain cells for advanced healthcare applications. *Open Health, 5*(1). https://doi.org/10.1515/ohe-2023-0029

Bajwa, J., Munir, U., Nori, A., & Williams, B. (2021). Artificial intelligence in healthcare: Transforming the practice of medicine. *Future Healthcare Journal, 8*(2), e188–e194. https://doi.org/10.7861/fhj.2021-0095

Bonnechère, B., Timmermans, A., & Michiels, S. (2023). Current technology developments can improve the quality of research and level of evidence for rehabilitation interventions: A narrative review. *Sensors, 23*(2), Article 2. https://doi.org/10.3390/s23020875

Capecci, M., Gandolfi, M., Straudi, S., Calabrò, R. S., Baldini, N., Pepa, L., Andrenelli, E., Smania, N., Ceravolo, M. G., Morone, G., Bonaiuti, D., & on behalf of HTA section of SIMFER (Italian Society of Physical and Rehabilitation Medicine). (2024). Advancing public health through technological rehabilitation: Insights from a national clinician survey. *BMC Health Services Research, 24*(1), 1626. https://doi.org/10.1186/s12913-024-11991-0

Christopoulou, S. C. (2024). Machine learning models and technologies for evidence-based telehealth and smart care: A review. *BioMedInformatics, 4*(1), Article 1. https://doi.org/10.3390/biomedinformatics4010042

Chua, K. S. G., & Kuah, C. W. K. (2017). Innovating with rehabilitation technology in the real world: Promises, potentials, and perspectives. *American Journal of Physical Medicine & Rehabilitation, 96*(10), S150. https://doi.org/10.1097/PHM.0000000000000799

Chukwuka, E., Dependable, A., Okatta, A., Emmanuel, A., Tochi, E., Chiamaka, E., & Ijeoma, E. (2024). The impact of simulation-based training in medical education: A review. *Medicine, 103*(27), e38813. https://doi.org/10.1097/MD.0000000000038813

Daich Varela, M., Sanders Villa, A., Pontikos, N., Crossland, M. D., & Michaelides, M. (2024). Digital health and wearable devices for retinal disease monitoring. *Graefe's Archive for Clinical and Experimental Ophthalmology*. https://doi.org/10.1007/s00417-024-06634-3

Data privacy and AI: Ethical considerations and 8 best practices. (2024, November 7). https://community.trustcloud.ai/docs/grc-launchpad/grc-101/governance/data-privacy-and-ai-ethical-considerations-and-best-practices/

Digital Transformation in Healthcare. (2024, June 3). https://www.newmetrics.net/insights/digital-transformation-in-healthcare/

Dinh, H. Q., Walker, N., Hodges, L. F., Song, C., & Kobayashi, A. (2023). Evaluating the effectiveness of virtual reality exposure therapy for treating anxiety disorders: A meta-analysis. *Journal of Anxiety Disorders, 92*, 102634. https://doi.org/10.1016/j.janxdis.2022.102634

Ethical Issues of AI in Healthcare. (2024, June 1). https://redresscompliance.com/ethical-issues-ai-healthcare/

Junaid, S. B., Imam, A. A., Balogun, A. O., De Silva, L. C., Surakat, Y. A., Kumar, G., Abdulkarim, M., Shuaibu, A. N., Garba, A., Sahalu, Y., Mohammed, A., Mohammed, T. Y., Abdulkadir, B. A., Abba, A. A., Kakumi, N. A. I., & Mahamad, S. (2022). Recent advancements in emerging technologies for healthcare management systems: A survey. *Healthcare, 10*(10), 1940. https://doi.org/10.3390/healthcare10101940

Gemert-Pijnen, V., & JEWC (Lisette). (2022). Implementation of health technology: Directions for research and practice. *Frontiers in Digital Health, 4*. https://doi.org/10.3389/fdgth.2022.1030194

Gerke, S., Minssen, T., & Cohen, G. (2020). Chapter 12—Ethical and legal challenges of artificial intelligence-driven healthcare. In A. Bohr & K. Memarzadeh (Eds.), *Artificial intelligence in healthcare* (pp. 295–336). Academic Press. https://doi.org/10.1016/B978-0-12-818438-7.00012-5

Gomez-Gonzalez, E., & Gomez-gutierrez, E. (2020). *Artificial intelligence in medicine and healthcare: Applications, availability and societal impact* (EUR 3019 EN). https://doi.org/10.2760/047666

Hanna, M., Pantanowitz, L., Jackson, B., Palmer, O., Visweswaran, S., Pantanowitz, J., Deebajah, M., & Rashidi, H. (2024). Ethical and bias considerations in artificial intelligence (AI)/machine learning. *Modern Pathology*, 100686. https://doi.org/10.1016/j.modpat.2024.100686

Holmgren, A. J., McBride, S., Gale, B., & Mossburg, S. (2023). Technology as a tool for improving patient safety. *Technology as a Tool for Improving Patient Safety*. https://psnet.ahrq.gov/perspective/technology-tool-improving-patient-safety

Iqbal, A. I., Aamir, A., Hammad, A., Hafsa, H., Basit, A., Oduoye, M. O., Anis, M. W., Ahmed, S., Younus, M. I., & Jabeen, S. (2024). Immersive technologies in healthcare: An in-depth exploration of virtual reality and augmented reality in enhancing patient care, medical education, and training paradigms. *Journal of Primary Care & Community Health, 15*, 21501319241293311. https://doi.org/10.1177/21501319241293311

Javaid, M., Haleem, A., & Singh, R. P. (2024). Health informatics to enhance the healthcare industry's culture: An extensive analysis of its features, contributions, applications and limitations. *Informatics and Health, 1*(2), 123–148. https://doi.org/10.1016/j.infoh.2024.05.001

LaBoone, P. A., & Marques, O. (2024). Overview of the future impact of wearables and artificial intelligence in healthcare workflows and technology. *International Journal of Information Management Data Insights, 4*(2), 100294. https://doi.org/10.1016/j.jjimei.2024.100294

Li, F. (2021). The beneficial role of vitamin B12 in injury induced by ischemia/reperfusion: Beyond scavenging superoxide? *Journal of Experimental Nephrology, 2*(1), 3–6. https://doi.org/10.46439/nephrology.2.007

Marketing. (2023, October 19). *Exploring AI Ethics: Decoding the Digital Dilemma*. https://www.datadynamicsinc.com/blog-decoding-the-digital-dilemma-navigating-the-ethical-landscape-of-ai/

Mennella, C., Maniscalco, U., De Pietro, G., & Esposito, M. (2023). The role of artificial intelligence in future rehabilitation services: A systematic literature review. *IEEE Access, 11*, 11024–11043. https://doi.org/10.1109/ACCESS.2023.3236084

Najjar, R. (2023). Redefining radiology: A review of artificial intelligence integration in medical imaging. *Diagnostics, 13*(17), 2760. https://doi.org/10.3390/diagnostics13172760

Olawade, D. B., David-Olawade, A. C., Adereni, T., Egbon, E., Teke, J., & Boussios, S. (2025). Integrating AI into cancer immunotherapy—A narrative review of current applications and future directions. *Diseases, 13*(1), 24. https://doi.org/10.3390/diseases13010024

Palozzi, G., Schettini, I., & Chirico, A. (2020). Enhancing the sustainable goal of access to healthcare: Findings from a literature review on telemedicine employment in rural areas. *Sustainability, 12*(8), Article 8. https://doi.org/10.3390/su12083318

Secinaro, S., Calandra, D., Secinaro, A., Muthurangu, V., & Biancone, P. (2021). The role of artificial intelligence in healthcare: A structured literature review. *BMC Medical Informatics and Decision Making, 21*(1), 125. https://doi.org/10.1186/s12911-021-01488-9

Sharma, A., Harrington, R. A., McClellan, M. B., Turakhia, M. P., Eapen, Z. J., Steinhubl, S., Mault, J. R., Majmudar, M. D., Roessig, L., Chandross, K. J., Green, E. M., Patel, B., Hamer, A., Olgin, J., Rumsfeld, J. S., Roe, M. T., & Peterson, E. D. (2018). Using digital health technology to better generate evidence and deliver evidence-based care. *Journal of the American College of Cardiology, 71*(23), 2680–2690. https://doi.org/10.1016/j.jacc.2018.03.523

Shojaei, P., Vlahu-Gjorgievska, E., & Chow, Y.-W. (2024). Security and privacy of technologies in health information systems: A systematic literature review. *Computers, 13*(2), Article 2. https://doi.org/10.3390/computers13020041

Stefano, G. B. (2023). Artificial intelligence as a tool for the diagnosis and treatment of neurodegenerative diseases. *Brain Sciences, 13*(6), Article 6. https://doi.org/10.3390/brainsci13060938

Stoumpos, A. I., Kitsios, F., & Talias, M. A. (2023). Digital transformation in healthcare: Technology acceptance and its applications. *International Journal of Environmental Research and Public Health, 20*(4), Article 4. https://doi.org/10.3390/ijerph20043407

Strika, Z., Petkovic, K., Likic, R., & Batenburg, R. (2025). Bridging healthcare gaps: A scoping review on the role of artificial intelligence, deep learning, and large language models in alleviating problems in medical deserts. *Postgraduate Medical Journal, 101*(1191), 4–16. https://doi.org/10.1093/postmj/qgae122

Subbaiah, V. (2023, September 1). *5 Best practices for successful health IT implementation*. Asahi Technologies - Custom Software Development Company in New York. https://www.asahitechnologies.com/blog/best-practices-for-successful-health-it-implementation/

Sutton, R. T., Pincock, D., Baumgart, D. C., Sadowski, D. C., Fedorak, R. N., & Kroeker, K. I. (2020). An overview of clinical decision support systems: Benefits, risks, and strategies for success. *Npj Digital Medicine, 3*(1), 1–10. https://doi.org/10.1038/s41746-020-0221-y

Tuborgh, A., Svendsen, S. W., Elklit, A., Hunter, J., Jensen, J. S., Schröder, A., Nielsen, J. F., Thastum, M. M., Næss-Schmidt, E. T., & Rask, C. U. (2021). Attachment and symptom reporting in adolescents and young adults after a concussion. *Journal of Psychosomatic Research, 150*, 110603. https://doi.org/10.1016/j.jpsychores.2021.110603

Villar, R. (2024, April 28). *Overcoming the 6 big barriers to digital health technology adoption—calcium*. Calciumhealth. https://calciumhealth.com/overcoming-the-6-big-barriers-to-digital-health-technology-adoption/

Wu, M., & Luo, J. (2019, November 25). *Wearable technology applications in healthcare: A literature review | HIMSS*. HIMSS. https://gkc.himss.org/resources/wearable-technology-applications-healthcare-literature-review

Yigzaw, K. Y., Olabarriaga, S. D., Michalas, A., Marco-Ruiz, L., Hillen, C., Verginadis, Y., de Oliveira, M. T., Krefting, D., Penzel, T., Bowden, J., Bellika, J. G., & Chomutare, T. (2022). Chapter 14 - Health data security and privacy: Challenges and solutions for the future. In E. Hovenga & H. Grain (Eds.), *Roadmap to successful digital health ecosystems* (pp. 335–362). Academic Press. https://doi.org/10.1016/B978-0-12-823413-6.00014-8

Chapter 16
Future Directions and Innovations

16.1 Introduction

The field of rehabilitation is undergoing a remarkable transformation, fueled by rapid advancements in technology and innovative solutions. Tools such as artificial intelligence (AI), virtual reality (VR), augmented reality (AR), brain–computer interfaces (BCIs), and machine- learning (ML) algorithms are revolutionizing the way rehabilitation services are designed, delivered, and experienced. These developments hold immense potential to enhance patient outcomes, optimize clinical workflows, and expand access to high-quality care, especially for individuals with neurological and physical disorders. As these technologies evolve, they are reshaping traditional rehabilitation paradigms and unlocking new possibilities for both clinicians and patients (Catania et al., 2023; Georgiev et al., 2021; Ying et al., 2024).

Among these technological advancements, AI has emerged as a cornerstone, driving innovation in rehabilitation by enabling a shift toward highly personalized approaches. Through the analysis of real-time data, AI systems can dynamically adapt interventions to meet the unique needs of individual patients. This capacity for customization marks a significant departure from the one-size-fits-all approach that has historically characterized rehabilitation. AI-driven tools not only enhance the precision of care but also enable continuous monitoring and feedback, ensuring that therapies remain effective and engaging (Catania et al., 2023; Wankhede et al., 2025). For instance, AI integration into VR-based cognitive rehabilitation allows patients to receive real-time guidance, making therapy sessions more interactive and tailored to their progress. Similarly, AI-powered natural language processing and speech recognition technologies are facilitating advancements in language therapy, enabling more effective communication exercises for patients with speech impairments (Catania et al., 2023; Georgiev et al., 2021).

In addition to personalization, AI's ability to analyze complex data sets and identify patterns has made it a powerful tool for clinicians. Machine-learning algorithms,

K. J. Reddy, *Innovations in Neurocognitive Rehabilitation*,
https://doi.org/10.1007/978-3-031-88117-6_16

for example, can analyze patient data to provide actionable insights, helping clinicians develop more effective treatment plans and predict patient outcomes. These algorithms are also being used to detect subtle changes in a patient's condition, enabling early interventions that could significantly improve recovery trajectories. Such advancements align closely with the principles of personalized medicine, tailoring treatments to the biological, psychological, and environmental attributes of each patient (Calderone et al., 2024; Ying et al., 2024). This level of customization not only enhances the efficacy of rehabilitation interventions but also reduces the trial-and-error approach often associated with traditional methods.

VR and AR technologies are also playing pivotal roles in shaping the future of rehabilitation. These tools offer immersive and interactive therapeutic environments that engage patients and promote active participation in their recovery. By simulating real-world scenarios, AR and VR help bridge the gap between therapy and everyday life, enabling patients to practice skills in a safe and controlled setting before applying them in the real world (Calderone et al., 2024; Ying et al., 2024). For example, AR can be used to overlay real-time guidance and feedback onto the user's physical environment, creating opportunities for cognitive and motor skill training. These technologies are not only effective but are also becoming more accessible and cost-effective, making them viable options for widespread adoption in clinical and home-based settings (Georgiev et al., 2021; Prats-Bisbe et al., 2024).

The integration of AI and VR into rehabilitation has also paved the way for tele-rehabilitation, a rapidly growing field that leverages technology to deliver care remotely. Tele-rehabilitation platforms powered by AI and supported by advancements in 5G connectivity allow patients to access high-quality care regardless of their geographic location. This is particularly valuable for individuals living in underserved or remote areas, where access to specialized rehabilitation services may be limited. By providing real-time, adaptive interventions, tele-rehabilitation systems ensure that patients receive consistent and effective care while reducing the need for travel and in-person visits (Catania et al., 2023; Georgiev et al., 2021; Prats-Bisbe et al., 2024).

Despite the promise of these technologies, their adoption comes with significant ethical and practical considerations. One of the primary challenges is ensuring data privacy and security, particularly when dealing with sensitive patient information. AI systems rely on large data sets to function effectively, and safeguarding this data from breaches or misuse is critical to maintaining patient trust (Calderone et al., 2024; Catania et al., 2023). Algorithmic transparency is another pressing issue, as the decision-making processes of AI systems must be understandable and free from biases that could lead to inequities in care delivery. Addressing these challenges requires a collaborative approach involving healthcare professionals, technologists, and policymakers to establish guidelines and regulations that prioritize patient safety and equity (Abedi et al., 2024; Calderone et al., 2024).

Another key consideration is the need for large-scale validation studies to assess the efficacy and safety of AI-driven rehabilitation tools. While early results are promising, more extensive clinical trials are necessary to ensure that these technologies are effective across diverse patient populations and conditions. Such studies will also help identify potential limitations and areas for improvement, enabling the development of more robust and reliable systems (Abedi et al., 2024; Calderone et al., 2024). Furthermore, efforts must be made to design user-friendly interfaces that integrate seamlessly into clinical workflows. For these technologies to be widely adopted, they must be intuitive for both clinicians and patients, minimizing the learning curve and maximizing their impact (Calderone et al., 2024; Georgiev et al., 2021).

The potential of these innovations to improve diagnostic accuracy, expand access to care, and enhance patient outcomes is immense. For example, AI and ML can be used to identify early signs of neurological disorders, enabling timely interventions that could significantly improve recovery prospects. Similarly, home-based rehabilitation systems powered by AI and VR can democratize access to advanced therapies, making them available to individuals who might otherwise be unable to afford or access specialized care (Catania et al., 2023; Georgiev et al., 2021; Prats-Bisbe et al., 2024). By integrating these technologies into existing healthcare systems, clinicians can provide more efficient, effective, and equitable care, ultimately improving the quality of life for patients worldwide (Medenica et al., 2024; Rizzo et al., 2004; Schiza et al., 2019).

Looking ahead, the future of rehabilitation lies in the seamless integration of cutting-edge technologies with traditional care practices. This will require a careful balance between innovation and ethical responsibility, ensuring that these advancements are accessible, equitable, and centered around the needs of patients. Interdisciplinary collaboration will play a critical role in achieving this vision, bringing together healthcare professionals, researchers, technologists, and policymakers to develop solutions that are both innovative and compassionate. By addressing these challenges and embracing the opportunities presented by technology, the field of rehabilitation can achieve transformative breakthroughs that will redefine the standards of care for years to come (Medenica et al., 2024; Rizzo et al., 2004; Schiza et al., 2019). The integration of AI, VR, AR, and other advanced technologies into rehabilitation represents a paradigm shift that has the potential to revolutionize patient care. These tools offer new possibilities for personalization, engagement, and accessibility, empowering patients to achieve better outcomes and a higher quality of life. However, realizing this potential will require a thoughtful and collaborative approach, one that prioritizes ethical considerations, rigorous validation, and the development of user-friendly systems. By doing so, the future of rehabilitation can be both innovative and deeply human-centered, creating a more inclusive and effective healthcare landscape.

16.2 Emerging Technologies in Rehabilitation

The field of rehabilitation is at the forefront of a transformative technological revolution. Emerging innovations such as neuroadaptive systems, precision rehabilitation, augmented and mixed reality (AR/MR), and advanced wearable devices are reshaping the landscape of therapy. These tools not only enhance clinical precision but also address long-standing challenges by providing personalized, effective, and engaging solutions tailored to the unique needs of each patient. By leveraging real-time feedback and advanced analytics, these technologies are opening new doors for recovery and improving patient outcomes.

16.2.1 Neuroadaptive Systems

Neuroadaptive systems represent a cutting-edge breakthrough in rehabilitation by employing real-time neurofeedback to create therapies uniquely tailored to individual patients. Central to these systems are brain–computer interfaces (BCIs), which decode brain signals and translate them into actionable commands. For patients with motor impairments, BCIs enable control of assistive devices such as prosthetic limbs or communication through text-to-speech systems. This has revolutionized opportunities for individuals with conditions like spinal cord injuries or strokes (Georgiev et al., 2021; Hariharan et al., 2024). For example, BCIs have facilitated motor recovery in stroke survivors by providing precise neural feedback, allowing targeted rehabilitation efforts that adapt dynamically to individual progress. Beyond motor recovery, neuroadaptive systems can enhance cognitive training, as they leverage a patient's brain activity to guide exercises that strengthen neural pathways. The ability of these systems to adjust in real-time ensures optimal effectiveness, making them invaluable for addressing complex conditions such as neurodegenerative diseases (Catania et al., 2023; *VR in Rehab*, n.d.). The adaptability of neuroadaptive systems empowers clinicians to deliver interventions that respond to the unique neural responses of their patients. By reinforcing neural activity patterns linked to desired outcomes, such as movement or speech, these systems accelerate recovery while improving patient engagement. This intersection of neuroscience and technology is poised to become a cornerstone of personalized rehabilitation.

16.2.2 Precision Rehabilitation

Precision rehabilitation takes the concept of individualized care to a new level by integrating genomics, biometrics, and advanced analytics. By understanding the genetic and biological profiles of patients, clinicians can tailor rehabilitation plans to align with each individual's recovery potential. For instance, genetic testing can predict how a patient may respond to specific interventions, particularly in cases such as stroke recovery. This eliminates much of the trial-and-error associated with traditional approaches, streamlining the path to effective outcomes (Singh et al., 2024). Biomarkers and genetic insights form the foundation of precision rehabilitation. These tools allow clinicians to develop treatment strategies informed by the patient's unique biological composition. For example, genetic markers associated with neuroplasticity can help predict recovery outcomes, while biometrics like muscle activation patterns and heart rate variability offer real-time data to fine-tune interventions. This integration of biological and physiological data ensures therapies are both targeted and adaptive (Worlikar et al., 2023). The benefits of precision rehabilitation extend beyond efficacy. By reducing unnecessary interventions and focusing on evidence-based approaches, this model minimizes the risk of adverse effects while fostering patient satisfaction. Moreover, precision rehabilitation represents a shift toward a proactive, patient-centered paradigm where therapies are designed not only to address symptoms but to optimize long-term functional outcomes (Xie et al., 2024).

16.2.3 Augmented Reality (AR) and Mixed Reality (MR)

AR and MR technologies are redefining rehabilitation by creating immersive, interactive environments that bridge the gap between clinical settings and real-world applications. By overlaying virtual elements onto physical environments, these tools enable patients to engage in scenarios that mimic everyday tasks. Such environments enhance both cognitive and motor training, fostering skill development and seamless transition to daily activities (Singh et al., 2024; *VR in Rehab*, n.d.). One compelling application of AR is in gait training for patients with Parkinson's disease. By projecting visual cues onto the ground, AR assists patients in improving their walking patterns, reducing the risk of falls, and enhancing mobility. Similarly, MR platforms allow stroke survivors or individuals recovering from traumatic brain injuries to practice tasks such as preparing meals or navigating complex environments within safe, controlled settings (Digital Health Center of Excellence, 2025; Sokołowska, 2024). The interactive and engaging nature of AR and MR amplifies their effectiveness. During therapy sessions, patients receive real-time feedback, enabling them to make precise adjustments and gain confidence in their progress. For example, a patient practicing fine motor skills can instantly see how slight changes in their movements impact their outcomes, which accelerates learning and enhances motivation. This combination of technology and human-centered design ensures that therapy is not only effective but also empowering (Worlikar et al., 2023).

16.2.4 Advanced Wearable Devices

Advanced wearable devices are revolutionizing rehabilitation by providing continuous monitoring of physiological and biomechanical data. These devices, such as smart garments and wearable sensors, capture metrics like muscle activity, joint angles, and posture during therapy. The data they generate is invaluable for clinicians, enabling them to refine rehabilitation protocols and measure progress with unprecedented accuracy (Catania et al., 2023; *VR in Rehab*, n.d.). For example, smart garments equipped with electromyography (EMG) sensors allow real-time monitoring of muscle activity, helping therapists identify underutilized muscle groups and adjust exercises accordingly. This ensures that therapy targets the area most in need of intervention, optimizing outcomes. Wearables also empower patients by offering immediate feedback, encouraging them to stay engaged and adhere to their treatment plans (Xie et al., 2024). Beyond in-person therapy, wearable devices play a critical role in expanding access to care through telehealth applications. Patients in remote or underserved areas can benefit from wearable technologies that transmit data to clinicians in real time. This enables therapists to provide feedback, modify interventions, and monitor progress without the need for physical visits, reducing barriers to specialized care and fostering equity in rehabilitation services (Catania et al., 2023; *VR in Rehab*, n.d.).

Enhancing Rehabilitation Through Emerging Technologies
The integration of neuroadaptive systems, precision rehabilitation, AR/MR, and wearable devices is transforming the field of rehabilitation into a data-driven, patient-centered discipline. These technologies not only address the limitations of traditional approaches but also create opportunities to personalize care, improve patient outcomes, and enhance the overall therapy experience. By combining real-time feedback, advanced analytics, and user-friendly designs, they are paving the way for a new era of accessible and effective rehabilitation interventions. However, the adoption of these technologies is not without challenges. Ethical concerns, such as data privacy and the potential for algorithmic biases, must be addressed to build trust in AI-driven systems. Furthermore, rigorous clinical trials are needed to validate the efficacy and safety of these innovations across diverse patient populations. Lastly, ensuring that these tools are intuitive and seamlessly integrated into clinical workflows is critical for their widespread acceptance and success (Georgiev et al., 2021; Hariharan et al., 2024; Sokołowska, 2024). By embracing advancements such as neuroadaptive systems, precision rehabilitation, AR/MR, and advanced wearable devices, the field is transitioning toward a future where therapy is not only more effective but also deeply personalized and empowering. As these technologies continue to mature, they promise to redefine the possibilities of recovery and enhance the quality of life for countless individuals.

16.3 Expanding Frontiers in Artificial Intelligence

Artificial intelligence (AI) is revolutionizing the field of rehabilitation, providing tools that make therapies more personalized, accessible, and effective. With the ability to analyze complex data, offer real-time feedback, and adapt interventions dynamically, AI is reshaping how rehabilitation services are delivered and experienced. By integrating predictive capabilities and virtual coaching, AI bridges gaps in accessibility, precision, and efficiency, fundamentally transforming the patient and clinician experience. This section explores AI's potential through predictive analytics and autonomous virtual coaches, two innovations that stand at the forefront of this transformation.

16.3.1 Predictive Analytics

Predictive analytics powered by AI has become a game-changer in rehabilitation, enabling clinicians to anticipate patient needs and tailor interventions with unprecedented accuracy. By analyzing patient data, AI systems can forecast potential risks, recommend targeted treatments, and support personalized recovery plans. For example, machine-learning algorithms are increasingly being used to predict fall risks in elderly populations. These algorithms identify subtle patterns in movement data that may indicate an elevated likelihood of falling, allowing clinicians to intervene proactively with preventive strategies (Abedi et al., 2024; Khalid et al., 2024). Within virtual rehabilitation (VRehab) environments, predictive analytics integrates data from wearable devices, imaging systems, and patient-reported outcomes to create a comprehensive understanding of a patient's condition. This multi-modal approach allows for more precise and adaptive interventions tailored to individual progress and needs (Mennella et al., 2023; Yang et al., 2024). For instance, wearable devices might monitor a patient's gait or posture, generating real-time data that AI systems analyze to recommend corrective exercises. Imaging data can further assist in tracking recovery progress, enabling clinicians to adjust treatment plans dynamically based on measurable outcomes. Beyond prevention, predictive analytics enhances the overall patient experience by personalizing care and optimizing resource allocation. For instance, patients recovering from strokes can benefit from AI-driven models that evaluate motor function improvement rates and recommend appropriate therapy intensities. This level of precision ensures that interventions align closely with a patient's specific recovery goals, reducing trial-and-error approaches and improving outcomes. Predictive analytics is fundamentally transforming rehabilitation into a dynamic, data-driven field where interventions are not only proactive but also highly responsive to individual needs.

16.3.2 Autonomous Virtual Coaches

Autonomous virtual coaches, powered by AI, are redefining how patients engage with rehabilitation therapies. These virtual assistants guide patients through exercises, provide real-time feedback, and offer personalized support to ensure adherence to treatment plans. For example, chatbots embedded in home-based rehabilitation platforms can monitor a patient's movements, evaluate the quality of their exercises, and offer corrective feedback when necessary (Abedi et al., 2024). Designed to replicate the guidance of in-person therapists, these systems provide consistent care even when patients are outside clinical settings. One of the most significant advantages of virtual coaches is their ability to enhance accessibility. Patients facing challenges such as mobility limitations, geographic barriers, or scheduling conflicts can still receive high-quality, consistent care through remote rehabilitation programs. Virtual coaches bridge these gaps by delivering therapy directly to patients' homes, ensuring that logistical challenges do not hinder progress (Abedi et al., 2024; Khalid et al., 2024). For instance, a patient recovering from knee surgery can rely on a virtual coach to guide them through prescribed exercises, while the AI tracks their progress and provides detailed feedback to optimize recovery. In addition to accessibility, virtual coaches create an engaging and interactive therapy experience. AI algorithms within these systems adapt to a patient's progress, adjusting the difficulty of exercises and providing tailored encouragement to sustain motivation. For example, when a patient successfully completes a set of exercises, the system may introduce more challenging tasks to maintain engagement and foster continued improvement.

Conversely, if the patient struggles with a task, the AI can simplify it or provide step-by-step guidance to build confidence. This dynamic adaptability ensures that therapy is both challenging and rewarding, helping patients stay committed to their recovery journey (Abedi et al., 2024; Mennella et al., 2023). Virtual coaches also address the emotional and psychological dimensions of rehabilitation. By providing positive reinforcement and real-time feedback, these systems can boost a patient's morale, helping to alleviate frustration or feelings of isolation often associated with prolonged recovery periods. This holistic approach fosters a sense of empowerment, enabling patients to take an active role in their healing process. The potential of virtual coaches extends beyond physical rehabilitation. As AI technologies continue to advance, these systems are expected to expand their scope, incorporating more sophisticated feedback mechanisms and supporting a wider range of therapeutic applications. For example, virtual coaches could integrate cognitive rehabilitation exercises for patients recovering from brain injuries or mental health support for those dealing with psychological challenges during recovery.

16.4 Global Trends and Innovations

Advancements in rehabilitation technology are reshaping how care is delivered, making it more accessible, personalized, and effective. Innovations such as telerehabilitation, bioprinting, and robotics offer new opportunities to address long-standing challenges while improving patient outcomes. This section explores these key developments and their transformative potential.

16.4.1 Tele-Rehabilitation Expansion

Tele-rehabilitation has emerged as a vital solution for delivering care remotely, leveraging advancements in connectivity such as 5G-enabled platforms. These platforms support high-quality video streaming and sensor integration, enabling real-time monitoring and feedback for patients at a distance. By bridging gaps in access, tele-rehabilitation significantly reduces care disparities, particularly in underserved and rural areas (Elendu et al., 2023; Ettefagh & Roshan Fekr, 2024; *Lifesaving Telesurgery in the Age of 5G*, n.d.). For example, wearable devices integrated into tele-rehabilitation systems can monitor physiological parameters during therapy sessions, ensuring proper technique and tracking progress. Clinicians can adjust treatment plans remotely, offering personalized interventions that enhance adherence and outcomes. This approach not only expands access to specialized services but also reduces the need for frequent travel, making therapy more convenient for patients (*Empower EMR*, n.d.; *Lifesaving Telesurgery in the Age of 5G*, n.d.).

16.4.2 Bioprinting and Regenerative Medicine

Bioprinting is revolutionizing regenerative medicine by enabling the creation of customized tissue scaffolds for rehabilitation. Through 3D bioprinting technology, personalized implants can be designed to match a patient's unique anatomy, promoting faster healing and functional restoration. This is particularly valuable for orthopedic recovery, where tailored implants enhance integration and recovery outcomes (Reilly, 2023; *Tech Innovations and Physical Therapy*, n.d.). For instance, bioprinted scaffolds can support bone regeneration for patients recovering from fractures or joint replacements, significantly reducing recovery times compared to traditional methods. Beyond orthopedic applications, bioprinting offers innovative solutions for soft tissue repair, addressing chronic wounds or traumatic injuries with precision-engineered tissues that mimic natural structures. These advancements open new possibilities for personalized rehabilitation strategies that prioritize both function and aesthetics (Ettefagh & Roshan Fekr, 2024; *Tech Innovations and Physical Therapy*, n.d.).

16.4.3 Integration of Robotics

The integration of robotics into rehabilitation practices is offering transformative tools for physical therapy and mobility training. Robotic systems, including exoskeletons and robotic-assisted devices, are designed to assist patients with severe motor impairments, enabling them to regain mobility and independence. These technologies provide intensive and effective therapy sessions, addressing limitations of traditional approaches (*Empower EMR*, n.d.; Reilly, 2023; *Tech Innovations and Physical Therapy*, n.d.). Exoskeletons, for example, deliver mechanical support for patients with neuromuscular conditions or spinal cord injuries, facilitating gait training in a controlled and supportive environment. Studies have shown that exoskeletons not only improve mobility but also enhance muscle strength and cardiovascular health. Robotic-assisted devices ensure precise, repeatable movements during therapy, reducing the risk of injury and optimizing outcomes. Additionally, real-time feedback mechanisms keep patients motivated by tracking progress and demonstrating tangible improvements (*Empower EMR*, n.d.; Haseltine, n.d.).

16.5 Ethical and Societal Considerations in Rehabilitation Technology

As advanced technologies revolutionize rehabilitation practices, they bring with them critical ethical and societal challenges that must be addressed to ensure equitable, effective, and responsible integration. The use of artificial intelligence (AI), robotics, wearable devices, and tele-rehabilitation platforms offers immense potential for improving patient outcomes and expanding access to care. However, these advancements also raise significant concerns around equity, privacy, human-centered care, and long-term societal impacts. This section explores the key ethical and societal considerations related to the integration of technology into rehabilitation and offers insights into how these challenges can be addressed.

16.5.1 Addressing Equity and Accessibility

One of the most pressing issues in technology-driven rehabilitation is ensuring that innovations are accessible to all patients, regardless of socioeconomic status. Advanced rehabilitation tools such as AI-driven systems, wearable devices, and tele-rehabilitation platforms are often expensive, creating barriers for low-income communities. To address this, healthcare systems must implement subsidized programs that provide access to these technologies for underserved populations (Dhirani et al., 2023; Elendu et al., 2023; Mondal & Mondal, 2024).

For example, wearable devices that monitor physiological parameters during rehabilitation can significantly enhance outcomes but remain unaffordable for many patients. Governments and healthcare institutions should explore partnerships with technology manufacturers to subsidize costs and ensure equitable access. Additionally, digital infrastructure must be improved in underserved regions to facilitate broader access to tele-rehabilitation platforms. Bridging the digital divide is essential to ensure that technology does not exacerbate existing disparities in healthcare delivery (ForumCosmos, 2023; Todorovsa, n.d.). Efforts to improve accessibility also include designing simpler, more affordable versions of advanced rehabilitation tools. Simplifying technology without sacrificing functionality can make it accessible to a wider range of patients, ensuring that innovations benefit all segments of society.

16.5.2 Privacy and Security

The integration of advanced technologies in rehabilitation involves the collection and processing of sensitive patient data, raising significant privacy and security concerns. AI-driven systems and tele-rehabilitation platforms often require extensive data sharing, making them vulnerable to breaches and misuse. To safeguard patient information, robust data protection measures must be implemented, including encryption protocols and secure data storage solutions. Blockchain technology offers a promising solution for secure data sharing. By creating decentralized and tamper-proof records, blockchain can ensure transparency and accountability while protecting sensitive information. Additionally, healthcare institutions must develop clear data retention policies to determine how long patient data should be stored after the completion of rehabilitation. These policies should strike a balance between ensuring future accessibility to important data and mitigating risks of misuse or breaches. Education also plays a role in ensuring security. Patients and clinicians alike must be trained to recognize potential data vulnerabilities, empowering them to take proactive steps to protect sensitive information (Capitol Technology University, 2023; ForumCosmos, 2023; Wangmo et al., 2019).

16.5.3 Balancing Technology and Human-Centered Care

While technology can enhance the efficiency and effectiveness of rehabilitation, over-reliance on these tools risks compromising the human-centric nature of care. Rehabilitation is inherently personal, often requiring empathy, trust, and interpersonal connection between clinicians and patients. As AI and robotics play an increasing role in rehabilitation, it is crucial to maintain a balance between technological solutions and human interaction (Elendu et al., 2023; Todorovsa, n.d.). Training programs for clinicians should focus on integrating technology into their practices without sacrificing empathy and personalized care. For example, clinicians using AI-driven virtual assistants must ensure that the technology complements their work rather than replacing the human touch. Maintaining this balance is essential for building patient trust and fostering positive therapeutic relationships. Similarly, healthcare providers must actively seek feedback from patients to ensure their experiences remain at the forefront of care delivery. Furthermore, incorporating patient stories and qualitative feedback into technology design processes can ensure that these tools remain empathetic and user-friendly (Pazzanese, 2020).

16.5.4 Ethical Considerations

The ethical implications of technology in rehabilitation extend beyond privacy and accessibility. AI-driven systems, for instance, are susceptible to algorithmic bias, which can result in unequal treatment outcomes for different patient groups. Ensuring transparency and explainability in AI decision-making processes is critical to addressing these biases and building trust in the technology. Clear frameworks of accountability must also be established to define the responsibilities of manufacturers, healthcare institutions, and professionals. For example, when an AI system provides an inaccurate recommendation, it must be clear who bears responsibility for the error. Ethical guidelines should also address the broader societal implications of automation, including the potential job displacement of rehabilitation professionals. A proactive approach to addressing these issues will help mitigate negative consequences and promote a fair and equitable integration of technology (Capitol Technology University, 2023; ForumCosmos, 2023; Wangmo et al., 2019).

16.5.5 Societal Implications

Advanced rehabilitation technologies have the potential to reshape societal perceptions of health care and recovery. However, they also raise concerns about the digital divide, as underserved communities may lack the infrastructure or digital literacy to access these tools. Bridging this gap requires investments in digital literacy programs and equitable infrastructure development. Policymakers must ensure that resources are allocated to train both patients and healthcare providers in the effective use of technology. Additionally, the automation of certain rehabilitation processes raises questions about the future role of human professionals. While technologies like robotics and AI can enhance efficiency, they should be viewed as tools to support, rather than replace, clinicians. Policymakers and healthcare leaders must engage in proactive discussions to address the societal impacts of automation and ensure that technology complements the human workforce. By doing so, rehabilitation professionals can focus on delivering personalized care while leveraging technology to enhance outcomes (Dhirani et al., 2023; ForumCosmos, 2023; Pazzanese, 2020; Wangmo et al., 2019).

16.5.6 Regulatory Challenges

The rapid pace of technological advancement in rehabilitation poses significant regulatory challenges. Legal frameworks must adapt to address issues such as liability, intellectual property, and cross-border data sharing. For example, when using AI-driven rehabilitation tools, clear guidelines are needed to determine who is liable in cases of malfunction or data breaches. International cooperation is also essential for developing consistent ethical standards and policies. Collaborative efforts among governments, regulatory bodies, and industry leaders can help establish global benchmarks for the ethical integration of technology in rehabilitation. These efforts will ensure that patients worldwide benefit from advancements while adhering to uniform standards of safety and fairness (Mennella et al., 2023; Mondal & Mondal, 2024; Todorovsa, n.d.).

16.5.7 Long-Term Impacts

The long-term reliance on technology in rehabilitation has implications for patient autonomy, healthcare disparities, and overall well-being. While advanced tools can enhance recovery, there is a risk that heavy reliance on technology could diminish the role of human judgment and creativity in rehabilitation practices. To mitigate this, healthcare systems must emphasize patient autonomy by ensuring that individuals retain control over their rehabilitation processes and data. Patients should be provided with the necessary skills and knowledge to effectively use technology-enhanced tools, empowering them to take an active role in their recovery. Emphasizing human-centered care alongside technological innovations will help maintain a balance between efficiency and empathy (Capitol Technology University, 2023; ForumCosmos, 2023; Lee, 2016; Pazzanese, 2020).

16.5.8 Patient Autonomy

Maintaining patient autonomy is a cornerstone of ethical rehabilitation practices. Technology should enhance, rather than diminish, a patient's ability to make decisions about their care. For example, patients should have control over how their data is used and shared, as well as the ability to opt out of certain technological interventions if they prefer traditional methods. Education plays a critical role in preserving autonomy. Patients must be provided with clear, accessible information about how technology works, its benefits, and potential risks. This ensures informed decision-making and fosters trust in technology-enhanced rehabilitation processes. Providing patients with the tools to make informed choices will ultimately lead to more successful and satisfactory outcomes. Another ethical concern is how long personal and medical data should be retained after a patient's rehabilitation is complete. Prolonged data retention increases the risk of misuse or breaches, making it essential to establish clear guidelines for data deletion. Policies should balance the need for data to inform future care with the importance of protecting patient privacy. Transparency about data retention practices will help build patient trust and address ethical concerns (Dhirani et al., 2023; Elendu et al., 2023; ForumCosmos, 2023; Pazzanese, 2020).

Addressing the ethical and societal challenges of rehabilitation technology requires global collaboration. Consistent ethical standards and guidelines must be developed to ensure that patients receive equitable and safe care worldwide. International cooperation can facilitate knowledge sharing, support cross-border regulatory frameworks, and promote ethical innovation in rehabilitation practices. By working together, global stakeholders can ensure that technological advancements benefit all patients equitably. The integration of advanced technology into rehabilitation practices holds tremendous promise for improving patient outcomes and expanding access to care. However, these innovations also bring ethical and societal challenges that must be addressed to ensure responsible and equitable implementation. By prioritizing equity, privacy, human-centered care, and patient autonomy, stakeholders can create a future where technology enhances rehabilitation without compromising ethical principles. Ongoing dialogue among policymakers, healthcare professionals, technologists, and patients is essential to navigate these challenges. Through collaboration, transparency, and proactive regulation, the field of rehabilitation can embrace technological advancements while upholding the values of inclusivity, empathy, and fairness (Mennella et al., 2023; Mondal & Mondal, 2024).

16.6 Future Research Directions

Rehabilitation technologies are advancing at an unprecedented pace, creating exciting opportunities to improve patient outcomes, expand accessibility, and tailor therapies to individual needs. To maximize the impact of these innovations, targeted research is crucial to address existing gaps and guide future developments. This chapter highlights key areas for future exploration, focusing on long-term efficacy studies, interdisciplinary collaborations, and advancements in human–technology interfaces.

16.6.1 Long-Term Efficacy Studies

Understanding the long-term effectiveness of rehabilitation technologies is essential to ensure their sustained impact on patient outcomes. While innovations like VR-based cognitive rehabilitation programs have shown promising short-term results, their durability over extended periods, such as a decade or more, remains unclear. Conducting longitudinal studies is vital to evaluate these technologies' sustainability, adaptability, and real-world effectiveness. For instance, a study assessing the 10-year impact of VR-based therapies for stroke survivors could provide valuable insights into whether patients maintain their functional and cognitive gains. Such studies can also identify how these technologies adapt to patients' evolving needs and whether adherence diminishes over time. Similarly, research on wearable devices and AI-driven platforms could uncover patterns of long-term use, barriers to sustained engagement, and areas for technological refinement. By building a robust evidence base through longitudinal studies, researchers can establish best practices and ensure these innovations deliver consistent, meaningful outcomes across diverse populations (Aderinto et al., 2023; Catania et al., 2023; Prats-Bisbe et al., 2024).

16.6.2 Interdisciplinary Collaborations

The development and successful implementation of rehabilitation technologies require the combined expertise of technologists, clinicians, and policymakers. Interdisciplinary collaborations bridge the gap between cutting-edge technological innovation and practical clinical application, ensuring that advancements align with real-world patient needs. Joint research initiatives, such as those focusing on AI-driven rehabilitation platforms, exemplify the transformative potential of these partnerships. Engineers and physiotherapists, for example, can work together to design robotic systems tailored to specific therapeutic goals, such as improving motor function in stroke patients. Policymakers, in turn, can establish regulatory frameworks that facilitate the integration of these technologies into healthcare systems while safeguarding patient safety and privacy. These collaborations also encourage diverse perspectives, fostering innovation and enabling the creation of solutions that are both technically robust and clinically relevant. Strengthening these partnerships ensures that rehabilitation technologies address practical challenges while maintaining ethical and safety standards (Aderinto et al., 2023; Khan et al., 2024).

16.6.3 Advancements in Human–Technology Interfaces

Improving the usability and intuitiveness of rehabilitation devices is another critical area for research. Complex or poorly designed systems can hinder patient engagement, particularly for individuals with physical or cognitive impairments. To overcome these challenges, research should prioritize the development of user-friendly, adaptable, and accessible interfaces that accommodate the needs of all patients, including those with limited mobility. Voice-controlled systems offer a promising avenue for innovation, particularly for patients with motor impairments or limited dexterity. These systems enable users to navigate rehabilitation exercises and interact with technology through simple voice commands, reducing frustration and increasing accessibility. Similarly, advancements in haptic feedback and augmented reality (AR) interfaces can create immersive and interactive therapy experiences. For example, AR environments can simulate everyday tasks, such as cooking or grocery shopping, allowing patients to practice essential skills in a controlled and supportive setting. User-centered design is fundamental to these advancements. Actively involving patients in the development process ensures that technologies align with their preferences and needs. Wearable devices, for instance, can be optimized for comfort, durability, and ease of use, encouraging long-term adherence. By creating intuitive and engaging interfaces, researchers can enhance patient satisfaction and improve therapy outcomes, fostering sustained engagement with rehabilitation programs (Akbar et al., 2024; Prats-Bisbe et al., 2024).

16.7 Conclusion

The integration of advanced technologies into rehabilitation represents a transformative shift in how health care is delivered and experienced. From artificial intelligence and virtual reality to robotics and wearable devices, these innovations are redefining what is possible in terms of personalization, accessibility, and effectiveness. They provide tools that not only address the limitations of traditional rehabilitation practices but also create entirely new possibilities for patient care. Yet, while the potential of these technologies is immense, their successful implementation demands careful consideration of ethical, societal, and practical challenges, as well as a commitment to continuous research and improvement. One of the most compelling aspects of technology-driven rehabilitation is its ability to provide highly personalized care. AI-driven systems, for example, can analyze complex data sets to tailor therapies to individual needs, optimizing outcomes and improving patient engagement. Virtual reality and augmented reality platforms create immersive environments where patients can practice real-world tasks in safe and controlled settings, making therapy more engaging and relevant to their daily lives. Robotics and neuroadaptive systems offer precision and adaptability, responding in real-time to patients' movements and neural activity to deliver targeted interventions. These advancements are not just enhancing the effectiveness of rehabilitation but are also empowering patients to take a more active role in their recovery journeys.

However, with great innovation comes great responsibility. The adoption of these technologies raises critical ethical and societal considerations that must be addressed to ensure equitable and responsible implementation. Issues such as data privacy, algorithmic bias, and the digital divide highlight the need for robust frameworks that prioritize patient rights and accessibility. For example, AI systems must be transparent and explainable, ensuring that their recommendations can be understood and trusted by both clinicians and patients. Similarly, efforts must be made to bridge gaps in access to technology, particularly in underserved communities where the cost of advanced tools may create barriers to care. Subsidized programs and partnerships between healthcare providers and technology developers are essential for making these innovations accessible to all. Another challenge lies in maintaining the human element of care in an increasingly technology-driven field. Rehabilitation is fundamentally a human experience, requiring empathy, trust, and personal connection between clinicians and patients. While technologies like AI and robotics can enhance the efficiency and precision of care, they should complement rather than replace the clinician–patient relationship. Training programs must equip healthcare professionals with the skills to integrate these tools into their practices without compromising the interpersonal aspects of care. By striking this balance, clinicians can harness the benefits of technology while preserving the emotional and psychological support that is so crucial to recovery.

The importance of ongoing research cannot be overstated in this context. Longitudinal studies are needed to evaluate the long-term efficacy and sustainability of these technologies. For instance, while VR-based cognitive rehabilitation programs may show promising short-term results, their impact over a decade or more remains unclear. Similarly, research into the synergistic effects of combining different technologies, such as robotic therapy and neural interfaces, could uncover new therapeutic pathways and enhance outcomes for complex conditions. Interdisciplinary collaborations between technologists, clinicians, and policymakers will be key to driving innovation and ensuring that new developments align with clinical realities and patient needs. Advancements in human–technology interfaces represent another critical area for exploration. Usability and intuitiveness are paramount for ensuring that patients can engage effectively with these tools. Complex or poorly designed systems can create barriers to adoption, particularly for individuals with physical or cognitive impairments. Voice-controlled systems, wearable devices optimized for comfort, and augmented reality interfaces are examples of how user-centered design can make rehabilitation technologies more accessible and engaging. Incorporating patient feedback during the development process ensures that these innovations are not only functional but also aligned with the lived experiences of those they are designed to help.

Looking ahead, the integration of advanced technologies into rehabilitation must be guided by a shared vision of inclusivity, innovation, and ethical responsibility. Policymakers have a critical role to play in establishing regulatory frameworks that support the safe and effective deployment of these tools. International collaboration is also essential for developing consistent standards and guidelines, ensuring that the benefits of technological advancements are distributed equitably across the globe. By addressing these challenges collectively, stakeholders can create a future where rehabilitation is not only more effective but also more equitable and compassionate. Ultimately, the promise of technology-driven rehabilitation lies in its potential to transform lives. For patients, these innovations mean faster recovery, greater independence, and improved quality of life. For clinicians, they offer tools to deliver more precise, efficient, and impactful care. And for society as a whole, they represent a step toward a healthcare system that is more adaptive, inclusive, and forward-thinking. The journey ahead will require collaboration, creativity, and a steadfast commitment to placing patients at the center of every innovation. But with these principles guiding the way, the future of rehabilitation holds extraordinary promise.

References

Abedi, A., Colella, T. J. F., Pakosh, M., & Khan, S. S. (2024). Artificial intelligence-driven virtual rehabilitation for people living in the community: A scoping review. *Npj Digital Medicine, 7*(1), 1–13. https://doi.org/10.1038/s41746-024-00998-w

Aderinto, N., Olatunji, G., Abdulbasit, M. O., Edun, M., Aboderin, G., & Egbunu, E. (2023). Exploring the efficacy of virtual reality-based rehabilitation in stroke: A narrative review of current evidence. *Annals of Medicine, 55*(2), 2285907. https://doi.org/10.1080/07853890.2023.2285907

Akbar, K., Passaro, A., Di Gioia, M., Martini, E., Dragone, M., Zullo, A., & Stasolla, F. (2024). Reviewing the horizon: The future of extended reality and artificial intelligence in neurorehabilitation for brain injury recovery. *Information, 15*(8), Article 8. https://doi.org/10.3390/info15080501

Calderone, A., Latella, D., Bonanno, M., Quartarone, A., Mojdehdehbaher, S., Celesti, A., & Calabrò, R. S. (2024). Towards transforming neurorehabilitation: The impact of artificial intelligence on diagnosis and treatment of neurological disorders. *Biomedicines, 12*(10), Article 10. https://doi.org/10.3390/biomedicines12102415

Capitol Technology University. (2023). *The ethical considerations of artificial intelligence | Capitol Technology University*. Retrieved January 25, 2025, from https://www.captechu.edu/blog/ethical-considerations-of-artificial-intelligence

Catania, V., Rundo, F., Panerai, S., & Ferri, R. (2023). Virtual reality for the rehabilitation of acquired cognitive disorders: A narrative review. *Bioengineering, 11*(1), 35. https://doi.org/10.3390/bioengineering11010035

Dhirani, L. L., Mukhtiar, N., Chowdhry, B. S., & Newe, T. (2023). Ethical dilemmas and privacy issues in emerging technologies: A review. *Sensors (Basel, Switzerland), 23*(3), 1151. https://doi.org/10.3390/s23031151

Digital Health Center of Excellence. (2025). Augmented reality and virtual reality in medical devices. *FDA*. https://www.fda.gov/medical-devices/digital-health-center-excellence/augmented-reality-and-virtual-reality-medical-devices

Elendu, C., Amaechi, D. C., Elendu, T. C., Jingwa, K. A., Okoye, O. K., John Okah, M., Ladele, J. A., Farah, A. H., & Alimi, H. A. (2023). Ethical implications of AI and robotics in healthcare: A review. *Medicine, 102*(50), e36671. https://doi.org/10.1097/MD.0000000000036671

Empower EMR. (n.d.). Retrieved January 25, 2025, from https://www.empoweremr.com/blog/exoskeleton-rehabilitation

Ettefagh, A., & Roshan Fekr, A. (2024). Technological advances in lower-limb tele-rehabilitation: A review of literature. *Journal of Rehabilitation and Assistive Technologies Engineering, 11*. https://doi.org/10.1177/20556683241259256

ForumCosmos. (2023, July 13). Ethical considerations in data privacy and security. *Medium*. https://medium.com/@armaanakhan91/ethical-considerations-in-data-privacy-and-security-1874a10061f0

Georgiev, D. D., Georgieva, I., Gong, Z., Nanjappan, V., & Georgiev, G. V. (2021). Virtual reality for neurorehabilitation and cognitive enhancement. *Brain Sciences, 11*(2), 221. https://doi.org/10.3390/brainsci11020221

Hariharan, V., Malini Prithiva Kumari, P. K., & Rajanandh, M. G. (2024). Extended reality in revolutionizing neurological disease: A new era for chronic condition treatment. *Cureus, 16*(8), e67633. https://doi.org/10.7759/cureus.67633

Haseltine, W. A. (n.d.). Revolutionizing rehabilitation: How exoskeletons are restoring mobility for neuromuscular patients. *Forbes*. Retrieved January 25, 2025, from https://www.forbes.com/sites/williamhaseltine/2023/07/31/revolutionizing-rehabilitation-how-exoskeletons-are-restoring-mobility-for-neuromuscular-patients/

Khalid, U. B., Naeem, M., Stasolla, F., Syed, M. H., Abbas, M., & Coronato, A. (2024). Impact of AI-powered solutions in rehabilitation process: Recent improvements and future trends. *International Journal of General Medicine, 17*, 943–969. https://doi.org/10.2147/IJGM.S453903

Khan, H. U., Ali, Y., Khan, F., & Al-antari, M. A. (2024). A comprehensive study on unraveling the advances of immersive technologies (VR/AR/MR/XR) in the healthcare sector during the COVID-19: Challenges and solutions. *Heliyon, 10*(15), e35037. https://doi.org/10.1016/j.heliyon.2024.e35037

Lee, W. W. (2016). *Volume 6 An ethical approach to data privacy protection*. ISACA. Retrieved February 6, 2025, from https://www.isaca.org/resources/isaca-journal/issues/2016/volume-6/an-ethical-approach-to-data-privacy-protection

Lifesaving telesurgery in the age of 5G. (n.d.). Ericsson.Com. Retrieved January 25, 2025, from https://www.ericsson.com/en/blog/2024/3/cutting-the-cord-lifesaving-telesurgery-in-the-age-of-5g

Medenica, V., Ivanovic, L., & Milosevic, N. (2024). Applicability of artificial intelligence in neuropsychological rehabilitation of patients with brain injury. *Applied Neuropsychology: Adult*, 1–28. https://doi.org/10.1080/23279095.2024.2364229

Mennella, C., Maniscalco, U., De Pietro, G., & Esposito, M. (2023). The role of artificial intelligence in future rehabilitation services: A systematic literature review. *IEEE Access, 11*, 11024–11043. https://doi.org/10.1109/ACCESS.2023.3236084. IEEE Access.

Mondal, H., & Mondal, S. (2024). Chapter thirteen—Ethical and social issues related to AI in healthcare. In A. Srivastava & V. Mishra (Eds.), *Methods in microbiology* (Vol. 55, pp. 247–281). Academic Press. https://doi.org/10.1016/bs.mim.2024.05.009

Pazzanese, C. (2020, October 26). Ethical concerns mount as AI takes bigger decision-making role. *Harvard Gazette*. https://news.harvard.edu/gazette/story/2020/10/ethical-concerns-mount-as-ai-takes-bigger-decision-making-role/

Prats-Bisbe, A., López-Carballo, J., García-Molina, A., Leno-Colorado, D., García-Rudolph, A., Opisso, E., & Jané, R. (2024). Virtual reality-based neurorehabilitation support tool for people with cognitive impairments resulting from an acquired brain injury: Usability and feasibility study. *JMIR Neurotechnology, 3*(1), e50538. https://doi.org/10.2196/50538

Reilly, P. A. (2023). Optimizing readiness…the power of military medical research. *Military Medicine, 188*(Suppl_6), 2–2. https://doi.org/10.1093/milmed/usad185

Rizzo, A. A., Schultheis, M., Kerns, K. A., & Mateer, C. (2004). Analysis of assets for virtual reality applications in neuropsychology. *Neuropsychological Rehabilitation, 14*(1–2), 207–239. https://doi.org/10.1080/09602010343000183

Schiza, E., Matsangidou, M., Neokleous, K., & Pattichis, C. S. (2019). Virtual reality applications for neurological disease: A review. *Frontiers in Robotics and AI, 6*. https://doi.org/10.3389/frobt.2019.00100

Singh, M. K., Shin, Y., Ju, S., Han, S., Kim, S. S., & Kang, I. (2024). Comprehensive overview of Alzheimer's disease: Etiological insights and degradation strategies. *International Journal of Molecular Sciences, 25*(13), Article 13. https://doi.org/10.3390/ijms25136901

Sokołowska, B. (2024). Being in virtual reality and its influence on brain health—An overview of benefits, limitations and prospects. *Brain Sciences, 14*(1), Article 1. https://doi.org/10.3390/brainsci14010072

Tech Innovations and Physical Therapy. (n.d.). Retrieved January 25, 2025, from https://eksobionics.com/tech-innovations-and-physical-therapy/

Todorovsa, I. (n.d.). Ethical issues in information technology—Balancing innovation and responsibility. *Barrage*. Retrieved January 25, 2025, from https://www.barrage.net/blog/strategy/the-importance-of-ethics-in-technology

VR in Rehab: Exploring the Latest Technological Advances. (n.d.). Neuro Rehab VR. Retrieved January 25, 2025, from https://neurorehabvr.com/blog/tech-advancements

Wangmo, T., Lipps, M., Kressig, R. W., & Ienca, M. (2019). Ethical concerns with the use of intelligent assistive technology: Findings from a qualitative study with professional stakeholders. *BMC Medical Ethics, 20*(1), 98. https://doi.org/10.1186/s12910-019-0437-z

Wankhede, N. L., Koppula, S., Ballal, S., Doshi, H., Kumawat, R., Raju, S. S., Arora, I., Sammeta, S. S., Khalid, M., Zafar, A., Taksande, B. G., Upaganlawar, A. B., Gulati, M., Umekar, M. J., Kopalli, S. R., & Kale, M. B. (2025). Virtual reality modulating dynamics of neuroplasticity: Innovations in neuro-motor rehabilitation. *Neuroscience, 566*, 97–111. https://doi.org/10.1016/j.neuroscience.2024.12.040

Worlikar, H., Coleman, S., Kelly, J., O'Connor, S., Murray, A., McVeigh, T., Doran, J., McCabe, I., & O'Keeffe, D. (2023). Mixed reality platforms in telehealth delivery: Scoping review. *JMIR Biomedical Engineering, 8*(1), e42709. https://doi.org/10.2196/42709

Xie, X., Shi, R., Yu, H., Wan, X., Liu, T., Duan, D., Li, D., & Wen, D. (2024). Executive function rehabilitation and evaluation based on brain-computer interface and virtual reality: Our opinion. *Frontiers in Neuroscience, 18*. https://doi.org/10.3389/fnins.2024.1377097

Yang, R., Yuan, Q., Zhang, W., Cai, H., & Wu, Y. (2024). Application of artificial intelligence in rehabilitation science: A scientometric investigation utilizing citespace. *SLAS Technology, 29*(4). https://doi.org/10.1016/j.slast.2024.100162

Ying, H., Pranolo, A., Nuryana, Z., & Syafitri, A. I. (2024). Emerging trends in the evolution of neuropsychology and artificial intelligence: A comprehensive analysis. *Telematics and Informatics Reports, 16*, 100171. https://doi.org/10.1016/j.teler.2024.100171

Index

K. J. Reddy, *Innovations in Neurocognitive Rehabilitation*,
https://doi.org/10.1007/978-3-031-88117-6

E

G

H

I

L

M

N

O

P

Q

R

S

T

MIX
Papier aus verantwortungsvollen Quellen
Paper from responsible sources
FSC® C105338

If you have any concerns about our products,
you can contact us on
ProductSafety@springernature.com

In case Publisher is established outside the EU,
the EU authorized representative is:
Springer Nature Customer Service Center GmbH
Europaplatz 3, 69115 Heidelberg, Germany

Printed by Libri Plureos GmbH
in Hamburg, Germany